HIGH VOLTAG

Second Edition

About the Authors

M S NAIDU is Professor in the Department of High Voltage Engineering, Indian Institute of Science, Bangalore. A Ph D from the University of Liverpool, he served as a visiting scientist at the High Voltage Laboratory of the Eindhoven University of Technology, Netherlands. He has also lectured at many high voltage laboratories in West Germany, Switzerland and France.

Prof. Naidu is a Chartered Engineer and a Fellow of the Institution of Engineers (India) and also a Fellow of the National Academy of Engineering. His research interests include gaseous insulation, circuit breaker arcs, pollution under HVDC etc. He has published many research papers and has authored *Advances in High Voltage Breakdown and Arc Interruption in SF_6 and Vacuum* (Pergamon Press, 1981).

V KAMARAJU obtained his Ph D in High Voltage Engineering from the Indian Institute of Science, Bangalore and is currently a Professor of Electrical Engineering at the Engineering College, Kakinada, Andhra Pradesh. He has done extensive research in the area of liquid and solid dielectrics, composite insulation and partial discharge. He is a Chartered Engineer and a Fellow of the Institution of Engineers (India). He has published many research papers and has been a consultant to various industries and to the Andhra Pradesh State Electricity Board.

HIGH VOLTAGE ENGINEERING

Second Edition

M S Naidu

Department of High Voltage Engineering
Indian Institute of Science
Bangalore

V Kamaraju

Department of Electrical Engineering
College of Engineering
Jawaharlal Nehru Technological University
Kakinada

McGraw-Hill

New York   San Francisco   Washington, D.C.   Auckland   Bogotá
Caracas   Lisbon   London   Madrid   Mexico City   Milan
Montreal   New Delhi   San Juan   Singapore
Sydney   Tokyo   Toronto

McGraw-Hill

A Division of The McGraw·Hill Companies

First published © 1995, Tata McGraw-Hill Publishing Company Limited

Copyright © 1996 by The McGraw-Hill Companies, Inc. All rights reserved. Printed in the United States of America. Except as permitted under the United States Copyright Act of 1976, no part of this publication may be reproduced or distributed in any form or by any means, or stored in a data base or retrieval system, without the prior written permission of the publisher.

11 12 13 DOD 16 15 14

ISBN 0-07-136108-1

Printed and bound by RR Donnelley

To

Our Children

*May their world be filled
with understanding, love and peace*

Preface

The demand for the generation and transmission of large amounts of electric power today, necessitates its transmission at extra-high voltages. In the developed countries like USA, power transmission voltages have reached 765 kV or 1100 kV, and 1500 kV systems are also being built. In our country, 400 kV a.c. power systems have already come into operation, and in another 10 years time every state is expected to be linked by a National Power Grid operating at 400 kV or at 800 kV. At this juncture, a practising electrical engineer or a student of electrical engineering is expected to possess a knowledge of high voltage techniques and should have sufficient background in high voltage engineering. Unfortunately, at present only very few textbooks in high voltage engineering are available, compared to those in other areas of electrical engineering; even among these, no single book has covered broadly the entire range of topics in high voltage engineering and presented the material in a lucid manner. Therefore, an attempt has been made in this book, to bring together different topics in high voltage engineering to serve as a single semester course for final year undergraduate students or postgraduate students studying this subject. This book is also intended to serve power engineers in industry who are involved in the design and development of electrical equipment and also engineers in the electricity supply and utility establishments. It provides all the latest information on insulating materials, breakdown phenomena, overvoltages, and testing techniques.

The material in this book has been organized into five sections, namely, (i) insulating materials and their applications in electrical and electronic engineering, (ii) breakdown phenomena in insulating materials—solids, liquids, and gases, (iii) generation and measurement of high d.c., a.c., and impulse voltages and currents, (iv) overvoltage phenomena in electrical power transmission systems and insulation coordination, and (v) high voltage testing techniques, testing of apparatus and equipment, and planning of high voltage laboratories. Much of the information on these topics has been drawn from standard textbooks and reference books, which is simplified and reorganized to suit the needs of the students and graduate engineers. Many research publications have also been referred to, and relevant standard specifications have been quoted to help the reader to gain an easy access to the original references.

We have been associated with the subject of High Voltage Engineering for the last 30 years, both as teachers and researchers. This book is useful for undergraduate students of Electrical Engineering, and postgraduate students of Electrical Engineering, Electronics and Applied Physics. It is also useful for self study by engineers in the field of electricity utilities and in the design, development and testing of electrical apparatus, transmission line hardware, particle accelerators, etc.

Major changes incorporated in the second edition are:

* Chapter 2 has been expanded to include vacuum insulation, including vacuum breakdown and practical applications of vacuum insulation.
* Chapter 4 includes various aspects of breakdown of composite insulation/ insulation systems.
* Chapter 8 incorporates many new aspects of high voltage and extra high voltage AC power transmission.
* In Chapters 6 and 7, certain aspects of production and measurement of high voltages have been deleted; instead, the recent developments have been incorporated.

Many smaller changes have been made throughout the book to update the material and improve the clarity of presentation.

The authors acknowledge with thanks the permission given by the Bureau of Indian Standards, New Delhi for permitting them to refer to their various specifications and to include the following figures and table in this book.

(i) Fig. 6.14: *Impulse waveform and its definitions*, from IS: 2071 Part II–1973.
(ii) Fig. 10.1: *Computation of absolute humidity*, and Fig. 10.2: *Humidity correction factor* from IS: 731–1971.
(iii) Table 7.6: *Relationship between correction factor K and air density factor d*, from IS: 2071 Part I–1973.

We also wish to express our thanks to the persons who helped us during the preparation of this second edition. Mr. Mohamed Saleem and Mrs Meena helped with the typing work, while Mr. Dinesh Bhat and Mr. S.T. Paramesh helped with the technical preparation of the manuscript. Technical information derived from various research publications is gratefully acknowledged. We owe our special gratitude to the Director, Indian Institute of Science, Bangalore and to the Vice-Chancellor, Jawaharlal Nehru Technological University, Hyderabad for their encouragement.

M S NAIDU
V KAMARAJU

Contents

1
Introduction

In modern times, high voltages are used for a wide variety of applications covering the power systems, industry, and research laboratories. Such applications have become essential to sustain modern civilization. High voltages are applied in laboratories in nuclear research, in particle accelerators, and Van de Graaff generators. For transmission of large bulks of power over long distances, high voltages are indispensable. Also, voltages up to 100 kV are used in electrostatic precipitators, in automobile ignition coils, etc. X-ray equipment for medical and industrial applications also uses high voltages. Modern high voltage test laboratories employ voltages up to 6 MV or more. The diverse conditions under which a high voltage apparatus is used necessitate careful design of its insulation and the electrostatic field profiles. The principal media of insulation used are gases, vacuum, solid, and liquid, or a combination of these. For achieving reliability and economy, a knowledge of the causes of deterioration is essential, and the tendency to increase the voltage stress for optimum design calls for judicious selection of insulation in relation to the dielectric strength, corona discharges, and other relevant factors. In this chapter some of the general principles used in high voltage technology are discussed.

1.1 ELECTRIC FIELD STRESSES

Like in mechanical designs where the criterion for design depends on the mechanical strength of the materials and the stresses that are generated during their operation, in high voltage applications, the dielectric strength of insulating materials and the electric field stresses developed in them when subjected to high voltages are the important factors in high voltage systems. In a high voltage apparatus the important materials used are conductors and insulators. While the conductors carry the current, the insulators prevent the flow of currents in undesired paths. The electric stress to which an insulating material is subjected to is numerically equal to the voltage gradient, and is equal to the electric field intensity,

$$E = - \nabla \varphi \qquad (1.1)$$

where E is the electric field intensity, φ is the applied voltage, and ∇ (read del) operator is defined as

$$\nabla \equiv a_x \frac{\partial}{\partial x} + a_y \frac{\partial}{\partial y} + a_z \frac{\partial}{\partial z}$$

where a_x, a_y, and a_z are components of position vector $r = a_x x + a_y y + a_z z$.

As already mentioned, the most important material used in a high voltage apparatus is the insulation. The dielectric strength of an insulating material can be defined as the maximum dielectric stress which the material can withstand. It can also be defined as the voltage at which the current starts increasing to very high values unless controlled by the external impedance of the circuit. The electric breakdown strength of insulating materials depends on a variety of parameters, such as pressure, temperature, humidity, field configurations, nature of applied voltage, imperfections in dielectric materials, material of electrodes, and surface conditions of electrodes, etc. An understanding of the failure of the insulation will be possible by the study of the possible mechanisms by which the failure can occur.

The most common cause of insulation failure is the presence of discharges either within the voids in the insulation or over the surface of the insulation. The probability of failure will be greatly reduced if such discharges could be eliminated at the normal working voltage. Then, failure can occur as a result of thermal or electrochemical deterioration of the insulation.

1.2 GAS/VACUUM AS INSULATOR

Air at atmospheric pressure is the most common gaseous insulation. The breakdown of air is of considerable practical importance to the design engineers of power transmission lines and power apparatus. Breakdown occurs in gases due to the process of collisional ionization. Electrons get multiplied in an exponential manner, and if the applied voltage is sufficiently large, breakdown occurs. In some gases, free electrons are removed by attachment to neutral gas molecules; the breakdown strength of such gases is substantially large. An example of such a gas with larger dielectric strength is sulphur hexaflouride (SF_6).

The breakdown strength of gases increases steadily with the gap distance between the electrodes; but the breakdown voltage gradient reduces from 3 MV/m for uniform fields and small distances to about 0.6 MV/m for large gaps of several metres. For very large gaps as in lightning, the average gradient reduces to 0.1 to 0.3 MV/m.

High pressure gas provides a flexible and reliable medium for high voltage insulation. Using gases at high pressures, field gradients up to 25 MV/m have been realized. Nitrogen (N_2) was the gas first used at high pressures because of its inertness and chemical stability, but its dielectric strength is the same as that of air. Other important practical insulating gases are carbon-dioxide (CO_2), dichlorodifluoro-methane (CCl_2F_2) (popularly known as freon), and sulphur hexafluoride (SF_6). Investigations are continuing with more complex and heavier gases to be adopted as possible insulators. SF_6 has been found to maintain its insulation superiority, about 2.5 times over N_2 and CO_2 at atmospheric pressure, the ratio increasing at higher pressures. SF_6 gas was also observed to have superior arc quenching properties over any other gas. The breakdown voltage at higher pressures in gases shows an increasing dependence on the nature and smoothness of the electrode material. It is relevant to point out that, of the gases examined to-date, SF_6 has probably the most attractive overall dielectric and arc quenching properties for gas insulated high voltage systems.

Ideally, vacuum is the best insulator with field strengths up to 10^7 V/cm, limited only by emissions from the electrode surfaces. This decreases to less than 10^5

V/cm for gaps of several centimetres. Under high vacuum conditions, where the pressures are below 10^{-4} torr*, the breakdown cannot occur due to collisional processes like in gases, and hence the breakdown strength is quite high. Vacuum insulation is used in particle accelerators, x-ray and field emission tubes, electron microscopes, capacitors, and circuit breakers.

1.3 LIQUID BREAKDOWN

Liquids are used in high voltage equipment to serve the dual purpose of insulation and heat conduction. They have the advantage that a puncture path is self-healing. Temporary failures due to overvoltages are reinsulated quickly by liquid flow to the attacked area. However, the products of the discharges may deposit on solid insulation supports and may lead to surface breakdown over these solid supports.

Highly purified liquids have dielectric strengths as high as 1 MV/cm. Under actual service conditions, the breakdown strength reduces considerably due to the presence of impurities. The breakdown mechanism in the case of very pure liquids is the same as the gas breakdown, but in commercial liquids, the breakdown mechanisms are significantly altered by the presence of the solid impurities and dissolved gases.

Petroleum oils are the commonest insulating liquids. However, askarels, fluorocarbons, silicones, and organic esters including castor oil are used in significant quantities. A number of considerations enter into the selection of any dielectric liquid. The important electricial properties of the liquid include the dielectric strength, conductivity, flash point, gas content, viscosity, dielectric constant, dissipation factor, stability, etc. Because of their low dissipation factor and other excellent characteristics, polybutanes are being increasingly used in the electrical industry. Askarels and silicones are particularly useful in transformers and capacitors and can be used at temperatures of 200°C and higher. Castor oil is a good dielectric for high voltage energy storage capacitors because of its high corona resistance, high dielectric constant, non-toxicity, and high flash point.

In practical applications liquids are normally used at voltage stresses of about 50-60 kV/cm when the equipment is continuously operated. On the other hand, in applications like high voltage bushings, where the liquid only fills up the voids in the solid dielectric, it can be used at stresses as high as 100-200 kV/cm.

1.4 SOLID BREAKDOWN

If the solid insulating material is truly homogeneous and is free from imperfections, its breakdown stress will be as high as 10 MV/cm. This is the 'intrinsic breakdown strength', and can be obtained only under carefully controlled laboratory conditions. However, in practice, the breakdown fields obtained are very much lower than this value. The breakdown occurs due to many mechanisms. In general, the breakdown occurs over the surface than in the solid itself, and the surface insulation failure is the most frequent cause of trouble in practice.

---

*1 torr = 1 mm of Hg.

The breakdown of insulation can occur due to mechanical failure caused by the mechanical stresses produced by the electrical fields. This is called "electromechanical" breakdown.

On the other hand, breakdown can also occur due to chemical degradation caused by the heat generated due to dielectric losses in the insulating material. This process is cumulative and is more severe in the presence of air and moisture.

When breakdown occurs on the surface of an insulator, it can be a simple flashover or formation of a conducting path on the surface. When the conducting path is formed, it is called "tracking", and results in the degradation of the material. Surface flashover normally occurs when the solid insulator is immersed in a liquid dielectric. Surface flashover, as already mentioned, is the most frequent cause of trouble in practice. Porcelain insulators for use on transmission lines must therefore be designed to have a long path over the surface. Surface contamination of electrical insulation exists almost everywhere to some degree. In porcelain high voltage insulators of the suspension type, the length of the path over the surface will be 20 to 30 times greater than that through the solid. Even there, surface breakdown is the commonest form of failure.

The failure of solid insulation by discharges which may occur in the internal voids and cavities of the dielectric, called partial discharges, is receiving much attention today, mostly because it determines the life versus stress characteristics of the material. The energy dissipated in the partial discharges causes further deterioration of the cavity walls and gives rise to further evolution of gas. This is a cumulative process eventually leading to breakdown. In practice, it is not possible to completely eliminate partial discharges, but a level of partial discharges is fixed depending on the expected operating life of the equipment. Also, the insulation engineer should attempt to raise the discharge inception level, by carefully choosing electric field distributions and eliminating voids, particularly from high field systems. This requires a very high quality control during manufacture and assembly. In some applications, the effect of the partial discharges can be minimized by vacuum impregnation of the insulation. For high voltage applications, cast epoxy resin is solving many problems, but great care should be exercised during casting. High voltage switchgear, bushings, cables, and transformers are typical devices for which partial discharge effects should be considered in design.

So far, the various mechanisms that cause breakdown in dielectrics have been discussed. It is the intensity of the electric field that determines the onset of breakdown and the rate of increase of current before breakdown. Therefore, it is very essential that the electric stress should be properly estimated and its distribution known in a high voltage apparatus. Special care should be exercised in eliminating the stress in the regions where it is expected to be maximum, such as in the presence of sharp points.

1.5 ESTIMATION AND CONTROL OF ELECTRIC STRESS

The electric field distribution is governed by the Poisson's equation:

$$\nabla^2 \varphi = -\frac{\rho}{\varepsilon_0} \tag{1.2}$$

where φ is the the potential at a given point, ρ is the space charge density in the region, and ε_0 is the electric permittivity of free space (vacuum). However, in most of the high voltage apparatus, space charges are not normally present, and hence the potential distribution is governed by the Laplace's equation:

$$\nabla^2 \varphi = 0 \tag{1.3}$$

In Eqs. (1.2) and (1.3) the operator ∇^2 is called the Laplacian and is a scalar with properties

$$\nabla \cdot \nabla = \nabla^2 = \frac{\partial^2}{\partial x^2} + \frac{\partial^2}{\partial y^2} + \frac{\partial^2}{\partial z^2}$$

There are many methods available for determining the potential distribution, the most commonly used methods being,
  (i) the electrolytic tank method, and
  (ii) the method using digital computers.
The potential distribution can also be calculated directly. Howevei, this is very difficult except for simple geometries. In many practical cases, a good understanding of the problem is possible by using some simple rules to sketch the field lines and equipotentials. The important rules are
  (i)   the equipotentials cut the field lines at right angles,
  (ii) when the equipotentials and field lines are drawn to form curvilinear squares, the density of the field lines is an indication of the electric stress in a given region, and
  (iii) in any region, the maximum electric field is given by dv/dx, where dv is the voltage difference between two successive equipotentials dx apart.
Considerable amount of labour and time can be saved by properly choosing the planes of symmetry and shaping the electrodes accordingly. Once the voltage distribution of a given geometry is established, it is easy to refashion or redesign the electrodes to minimize the stresses so that the onset of corona is prevented. This is a case normally encountered in high voltage electrodes of the bushings, standard capacitors, etc. When two dielectrics of widely different permittivities are in a series, the electric stress is very much higher in the medium of lower permittivity. Considering a solid insulation in a gas medium, the stress in the gas becomes ε_r times that in the solid dielectric, where ε_r is the relative permittivity of the solid dielectric. This enhanced stress occurs

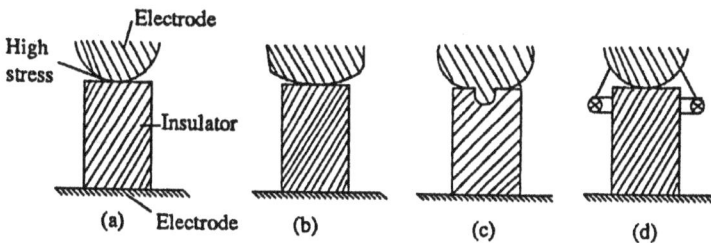

Fig. 1.1   Control of stress at an electrode edge

at the electrode edges and one method of overcoming this is to increase the electrode diameter. Other methods of stress control are shown in Fig. 1.1.

1.5.1 Electric Field

A brief review of the concepts of electric fields is presented, since it is essential for high voltage engineers to have a knowledge of the field intensities in various media under electric stresses. It also helps in choosing proper electrode configurations and economical dimensioning of the insulation, such that highly stressed regions are not formed and reliable operation of the equipment results in its anticipated life.

The field intensity E at any location in an electrostatic field is the ratio of the force on an infinitely small charge at that location to the charge itself as the charge decreases to zero. The force F on any charge q at that point in the field is given by

$$F = q E \tag{1.4}$$

The electric flux density D associated with the field intensity E is

$$D = \varepsilon E \tag{1.5}$$

where ε is the permittivity of the medium in which the electric field exists. The work done on a charge when moved in an electric field is defined as the potential. The potential φ is equal to

$$\varphi = - \int_l E \, dl \tag{1.6}$$

where l is the path through which the charge is moved.

Several relationships between the various quantities in the electric field are summarized as follows:

$$D = \varepsilon E \tag{1.5}$$

$$\varphi = - \int_l E \cdot dl \ (\text{or } E = - \nabla\varphi) \tag{1.6}$$

$$E = \frac{F}{q} \tag{1.7}$$

$$\iint_S E \cdot dS = \frac{q}{\varepsilon_0} \ (\text{Gauss theorem}) \tag{1.8}$$

$$\nabla \cdot D = \rho \ (\text{Charge density}) \tag{1.9}$$

$$\nabla^2 \varphi = -\frac{\rho}{\varepsilon_0} \ (\text{Poisson's equation}) \tag{1.10}$$

$$\nabla^2 \varphi = 0 \ (\text{Laplace's equation}) \tag{1.11}$$

where F is the force exerted on a charge q in the electric field E, and S is the closed surface contianing charge q.

1.5.2 Electric Field in a Single Dielectric Medium

When several conductors are situated in an electric field with the conductors charged, a definite relationship exists among the potentials of the conductors, the charges on them, and the physical location of the conductors with respect to each other.

In a conductor, electrons can move freely under the influence of an electric field. This means that the charges are distributed inside the substance and over the surface such that, $E = 0$ everywhere inside the conductor. Since $E = -\nabla\varphi = 0$, it is necessary that φ is constant inside and on the surface of the conductor. Thus, the conductor is an equipotential surface.

A dielectric material contains an array of charges which remain in equilibrium when an electric field is not zero within the substance. Therefore, a non-conductor or dielectric material is one that does contain free electrons or charges in appreciable number.

A simple capacitor consists of two conductors which are separated by a dielectric. If the two conductors contain a charge $+Q$ and $-Q$ and the potential difference between them is φ_{12}, the capacitance of such a capacitor is defined as the ratio of charge Q to the potential difference φ_{12}. Thus $C = Q/\varphi_{12}$. If the charge is not distributed uniformly over the two conductor surfaces, and if the charge density is ρ and the electric field in the dielectric is E,

then,
$$C = \iint\limits_S \rho dS \Big/ \int_1^2 E \cdot dl \tag{1.12}$$

When several conductors are present with charges $Q_1, Q_2, \ldots Q_n$ on them and their respective potentials are $\varphi_1, \varphi_2, \ldots \varphi_n$ the relationships between the charges and the potentials are given by

$$
\begin{bmatrix} Q_1 \\ Q_2 \\ \cdot \\ \cdot \\ \cdot \\ Q_n \end{bmatrix}
=
\begin{bmatrix} C_{11} & C_{12} & \cdots & C_{1n} \\ C_{21} & C_{22} & \cdots & C_{2n} \\ \cdot & \cdot & & \cdot \\ \cdot & \cdot & & \cdot \\ \cdot & \cdot & & \cdot \\ C_{n1} & C_{n2} & & C_{nn} \end{bmatrix}
\begin{bmatrix} \varphi_1 \\ \varphi_2 \\ \cdot \\ \cdot \\ \cdot \\ \varphi_n \end{bmatrix}
\tag{1.13}
$$

where $C_{11}, C_{22}, \ldots, C_{ii}, \ldots, C_{nn}$ are called capacitance coefficients, and $C_{12}, C_{21}, \ldots, C_{ij}, C_{ji}$ are called induction coefficients. Here C_{ij} is the quantity of charge on the ith conductor, which will charge the jth conductor to unity potential when all other conductors are kept at zero potential.

These coefficients are geometric factors, and can be estimated from the configuration of the conductors. The reciprocity property holds good for coefficients of induction and $C_{ij} = C_{ji}$. The self-capacitance of a conductor i is

$$C_{ii} = \sum_{j=1}^{n} C_{ij} \tag{1.14}$$

The mutual capacitance between two conductors i and j is

$$C_{ij} = C_{ji} \qquad (1.15)$$

This concept is very useful in the calculation of either potentials or charges in an electric field with known potential or charge distributions. In simple cases the electric field problems are solved, using Laplace or Poisson equation for the potential φ with the given boundary conditions. The electric field is estimated from the potential φ, and hence the charge distribution is obtained.

1.5.3 Electric Field in Mixed Dielectrics

When more than one dielectric material is present in any region of an electric field, the boundary conditions satisfied by the electric field intensity E at the dielectric boundary are

$$E_{t1} = E_{t2}; \ \varepsilon_1 \, E_{n1} = \varepsilon_2 E_{n2}; \ \text{and} \ \frac{\tan\alpha_1}{\tan\alpha_2} = \frac{\varepsilon_1}{\varepsilon_2} \qquad (1.16)$$

where E_{t1} and E_{t2} are the tangential components of the electric field, E_{n1} and E_{n2} are the normal components of the electric field, α_1 and α_2 are the angle of incidence and angle of refraction with the normal direction at the boundary, and ε_1 and ε_2 are the permittivities of the two dielectrics at the boundary.

Normally, all dielectrics are good insulators at lower magnitudes of field intensities. But as the electric field increases, the electrons bound to the molecules of the dielectric will be subjected to higher forces, and some of them are freed from their molecular bonding. The electrons move in the opposite direction to the electric field and thus create conduction current. This dissociation is temporary in gases in which a combination occurs when the field is removed, whereas it is a partial or permanent feature in liquids and solids. Also, this phenomenon depends on a number of factors like impurities present in the substance, temperature, humidity, length of time for which an electric field is present, etc. The phenomenon is called dielectric breakdown, and the magnitude of an electric field that gives rise to the dielectric breakdown and destroys the property of insulation in dielectric materials is called the dielectric breakdown strength. Breakdown strength is usually expressed in kV/cm or MV/metre. Detailed study of the breakdown phenomena in various dielectric media is presented in the following chapters.

1.5.4 Estimation of Electric Field in Some Geometric Boundaries

It has been shown that the maximum electric field E_m in a given electric field configuration is of importance. The mean electric field over a distance d between two conductors with a potential difference of V_{12} is

$$E_{av} = \frac{V_{12}}{d} \qquad (1.17)$$

In field configurations of non-uniform fields, the maximum electric field E_m is always higher than the average value. For some common field configurations, the maximum value of E_m and the field enhancement factor f given by E_m/E_{av} are presented in Table 1.1.

Table 1.1 Some Geometrical Configurations and the Field Factors

Geometrical configuration	Maximum electric field E_m	Field enhancement factor $f = E_m/E_{av}$
Parallel plates	$\dfrac{V}{r}$	1.0
Concentric cylinders	$\dfrac{V}{r \ln \dfrac{R}{r}}$	$\dfrac{(R - r)}{r \ln \dfrac{R}{r}}$
Figure same as above Concentric spheres	$\dfrac{VR}{r(R - r)}$	$\dfrac{R}{r}$
Parallel cylinders of equal diameter	$\dfrac{V \sqrt{D^2 - 4r^2}}{2r(D - r)\cosh^{-1}(D/2r)}$ $\approx \dfrac{V}{2r} \ln \dfrac{D}{r}$ if $D \gg r$	$\approx \dfrac{d}{2r \ln \dfrac{d}{r}}$ if $d \gg r$
Equal spheres with dimensions as above	$\approx \dfrac{V}{d} f$ $\approx \dfrac{V}{2r}$, if $d \gg r$	$f = \dfrac{\left(\dfrac{d}{r} + 1\right) + \sqrt{\left(\dfrac{d}{r} + 1\right)^2 + 8}}{4}$ $\approx \dfrac{d}{2r}$, if $d \gg r$

For other configurations like sphere-plane and cylinder-plane f is approximately given by

$$f = 0.94 \frac{d}{r} + 0.8 \text{ (sphere-plane)}$$

$$f = 0.25 \frac{d}{r} + 1.0 \text{ (cylinder-plane)}$$

Many electric conductors are normally either plane, cylindrical, or spherical in shape or can be approximated to these shapes. In other situations the conductors may be approximated into spheroidal, elliptical, toroidal, and other geometrical shapes, and thus estimation of E_m can be made.

1.6 SURGE VOLTAGES, THEIR DISTRIBUTION AND CONTROL

The design of power apparatus particularly at high voltages is governed by their transient behaviour. The transient high voltages or surge voltages originate in power systems due to lightning and switching operations. The effect of the surge voltages is severe in all power apparatuses. The response of a power apparatus to the impulse or surge voltage depends on the capacitances between the coils of windings and between the different phase windings of the multi-phase machines. The transient voltage distribution in the windings as a whole are generally very non-uniform and are complicated by travelling wave voltage oscillations set up within the windings. In the actual design of an apparatus, it is, of course, necessary to consider the maximum voltage differences occurring, in each region, at any instant of time after the application of an impulse, and to take into account their durations especially when they are less than one microsecond.

An experimental assessment of the dielectric strength of insulation against the power frequency voltages and surge voltages, on samples of basic materials, on more or less complex assemblies, or on complete equipment must involve high voltage testing. Since the design of an electrical apparatus is based on the dielectric strength, the design cannot be completely relied upon, unless experimentally tested. High voltage testing is done by generating the voltages and measuring them in a laboratory.

When high voltage testing is done on component parts, elaborate insulation assemblies, and complete full-scale prototype apparatus (called development testing), it is possible to build up a considerable stock of design information; although expensive, such data can be very useful. However, such data can never really be complete to cover all future designs and necessitates use of large factors of safety. A different approach to the problem is the exact calculation of dielectric strength of any insulation arrangement. In an ideal design each part of the dielectric would be uniformly stressed at the maximum value which it will safely withstand. Such an ideal condition is impossible to achieve in practice, for dielectrics of different electrical strengths, due to the practical limitations of construction. Nevertheless it provides information on stress concentration factors — the ratios of maximum local voltage gradients to the mean value in the adjacent regions of relatively uniform stress. A survey of typical power apparatus designs suggests that factors ranging from 2 to 5 can occur in practice; when this factor is high, considerable quantities of insulation must be used. Generally, improvements can be effected in the following ways:

(*i*) by shaping the conductors to reduce stress concentrations,
(*ii*) by insertion of higher dielectric strength insulation at high stress points, and
(*iii*) by selection of materials of appropriate permittivities to obtain more uniform voltage gradients.

The properties of different insulating media and their applications are presented in Chapters 2, 3, 4 and 5. The generation and measurement of high voltages and currents are discussed in Chapters 6 and 7, and high voltage test methods and the design of high voltage laboratories are detailed in Chapters 9, 10 and 11. The various aspects of insulation co-ordination in high voltage power systems are discussed in Chapter 8.

REFERENCES

1. Alston, L.L., *High Voltage Technology*, Oxford University Press, Oxford (1967).
2. Seely, S., *Electromagnetic Fields*, McGraw-Hill, New York (1960).
3. Kuffel, E. and Zaengl, W.S., *High Voltage Engineering Fundamentals*, Pergamon Press, Oxford (1984).

2
Conduction and Breakdown in Gases

2.1 GASES AS INSULATING MEDIA

The simplest and the most commonly found dielectrics are gases. Most of the electrical apparatus use air as the insulating medium, and in a few cases other gases such as nitrogen (N_2), carbon dioxide (CO_2), freon (CCl_2F_2) and sulphur hexafluoride (SF_6) are also used.

Various phenomena occur in gaseous dielectrics when a voltage is applied. When the applied voltage is low, small currents flow between the electrodes and the insulation retains its electrical properties. On the other hand, if the applied voltages are large, the current flowing through the insulation increases very sharply, and an electrical breakdown occurs. A strongly conducting spark formed during breakdown practically produces a short circuit between the electrodes. The maximum voltage applied to the insulation at the moment of breakdown is called the breakdown voltage. In order to understand the breakdown phenomenon in gases, a study of the electrical properties of gases and the processes by which high currents are produced in gases is essential.

The electrical discharges in gases are of two types, i.e. (*i*) non-sustaining discharges, and (*ii*) self-sustaining types. The breakdown in a gas, called spark breakdown is the transition of a non-sustaining discharge into a self-sustaining discharge. The build-up of high currents in a breakdown is due to the process known as ionization in which electrons and ions are created from neutral atoms or molecules, and their migration to the anode and cathode respectively leads to high currents. At present two types of theories, viz. (*i*) Townsend theory, and (*ii*) Streamer theory are known which explain the mechanism for breakdown under different conditions. The various physical conditions of gases, namely, pressure, temperature, electrode field configuration, nature of electrode surfaces, and the availability of initial conducting particles are known to govern the ionization processes.

2.2 IONIZATION PROCESSES

A gas in its normal state is almost a perfect insulator. However, when a high voltage is applied between the two electrodes immersed in a gaseous medium, the gas becomes a conductor and an electrical breakdown occurs.

The processes that are primarily responsible for the breakdown of a gas are ionization by collision, photo-ionization, and the secondary ionization processes. In insulating gases (also called electron-attaching gases) the process of attachment also plays an important role.

2.2.1 Ionization by Collision

The process of liberating an electron from a gas molecule with the simultaneous production of a positive ion is called ionisation. In the process of ionisation by collision, a free electron collides with a neutral gas molecule and gives rise to a new electron and a positive ion. If we consider a low pressure gas column in which an electric field E is applied across two plane parallel electrodes, as shown in Fig. 2.1 then, any electron starting at the cathode will be accelerated more and more between collisions with other gas molecules during its travel towards the anode. If the energy (ε) gained during this travel between collisions exceeds the ionisation potential, V_i, which is the energy required to dislodge an electron from its atomic shell, then ionisation takes place. This process can be represented as

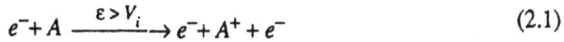

$$e^- + A \xrightarrow{\ \varepsilon > V_i\ } e^- + A^+ + e^- \qquad (2.1)$$

Where, A is the atom, A^+ is the positive ion and e^- is the electron.

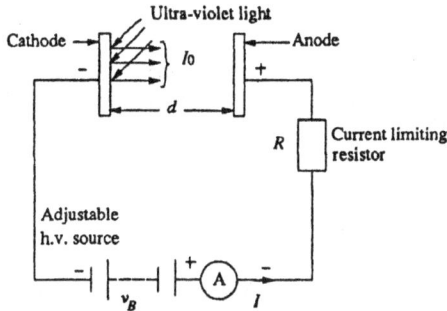

Fig. 2.1   Arrangement for study of a Townsend discharge

A few of the electrons produced at the cathode by some external means, say by ultra-violet light falling on the cathode, ionise neutral gas particles producing positive ions and additional electrons. The additional electrons, then, themselves make 'ionising collisions' and thus the process repeats itself. This represents an increase in the electron current, since the number of electrons reaching the anode per unit time is greater than those liberated at the cathode. In addition, the positive ions also reach the cathode and on bombardment on the cathode give rise to secondary electrons.

2.2.2 Photo-Ionization

The phenomena associated with ionisation by radiation, or photo-ionisation, involves the interaction of radiation with matter. Photo-ionisation occurs when the amount of radiation energy absorbed by an atom or molecule exceeds its ionisation potential.

There are several processes by which radiation can be absorbed by atoms or molecules. They are

(a) excitation of the atom to a higher energy state

(b) continuous absorption by direct excitation of the atom or dissociation of diatomic molecule or direct ionisation etc.

Just as an excited atom emits radiation when the electron returns to the lower state or to the ground state, the reverse process takes place when an atom absorbs radiation. This reversible process can be expressed as

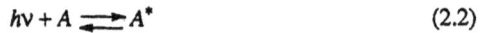

$$h\nu + A \rightleftharpoons A^*$$ (2.2)

Ionisation occurs when

$$\lambda \leq c \cdot \frac{h.}{V_i}$$ (2.3)

where, h is the Planck's constant , c is the velocity of light, λ is the wavelength of the incident radiation and V_i is the ionisation energy of the atom. Substituting for h and c, we get

$$\lambda \leq \left(\frac{1.27}{V_i}\right) \times 10^{-6} \text{cm}$$

where V_i is in electron volts (eV). The higher the ionisation energy, the shorter will be the wavelength of the radiation capable of causing ionisation. It was observed experimentally that a radiation having a wavelength of 1250 $A°$ is capable of causing photo-ionisation of almost all gases.

2.2.3 Secondary Ionisation Processes

Secondary ionisation processes by which secondary electrons are produced are the one which sustain a discharge after it is established due to ionisation by collision and photo-ionization.

They are briefly described below.

(a) Electron Emission due to Positive Ion Impact

Positive ions are formed due to ionisation by collision or by photo-ionisation, and being positively charged, they travel towards the cathode.

A positive ion approaching a metallic cathode can cause emission of electrons from the cathode by giving up its kinetic energy on impact. If the total energy of the positive ion, namely, the sum of its kinetic energy and the ionisation energy, is greater than twice the work function of the metal, then one electron will be ejected and a second electron will neutralise the ion. The probability of this process is measured as γ_i which is called the Townsend's secondary ionisation coefficient due to positive ions and is defined as the net yield of electrons per incident positive ion. γ_i increases with ion velocity and depends on the kind of gas and electrode material used.

(b) Electron Emission due to Photons

To cause an electron to escape from a metal, it should be given enough energy to overcome the surface potential barrier. The energy can also be supplied in the form of a photon of ultraviolet light of suitable frequency. Electron emission from a metal surface occurs at the critical condition (see Eq. 2.3)

$$h.v \geq \varphi$$

where φ is the work function of the metallic electrode. The frequency (v) is given by the relationship

$$v = \frac{\varphi}{h} \tag{2.4}$$

is known as the threshold frequency. For a clean nickel surface with $\varphi = 4.5$ eV, the threshold frequency will be that corresponding to a wavelength $\lambda = 2755 \overset{\circ}{A}$. If the incident radiation has a greater frequency than the threshold frequency, then the excess energy goes partly as the kinetic energy of the emitted electron and partly to heat the surface of the electrode. Since φ is typically a few electron volts, the threshold frequency lies in the far ultra-violet region of the electromagnetic radiation spectrum.

(c) Electron Emission due to Metastable and Neutral Atoms

A metastable atom or molecule is an excited particle whose lifetime is very large (10^{-3}s) compared to the lifetime of an ordinary particle (10^{-8}s). Electrons can be ejected from the metal surface by the impact of excited (metastable) atoms, provided that their total energy is sufficient to overcome the work function. This process is most easily observed with metastable atoms, because the lifetime of other excited states is too short for them to reach the cathode and cause electron emission, unless they originate very near to the cathode surface. Therefore, the yields can also be large nearly 100%, for the interactions of excited *He* atom with a clean surface of molybdenum, nickel or magnesium. Neutral atoms in the ground state also give rise to secondary electron emission if their kinetic energy is high ($\simeq 1000$ eV). At low energies the yield is considerably less.

2.2.4 Electron Attachment Process

The types of collisions in which electrons may become attached to atoms or molecules to form negative ions are called attachment collisions. Electron attachment process depends on the energy of the electron and the nature of the gas and is a very important process from the engineering point of view. All electrically insulating gases, such as O_2, CO_2, Cl_2, F_2, $C_2 F_6$, $C_3 F_8$, $C_4 F_{10}$, $CCl_2 F_2$, and SF_6 exhibit this property. An electron attachment process can be represented as:

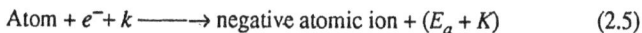

$$\text{Atom} + e^- + k \longrightarrow \text{negative atomic ion} + (E_a + K) \tag{2.5}$$

The energy liberated as a result of this process is the kinetic energy K plus the electron affinity E_a. In the attaching or insulating gases, the atoms or molecules have vacancies in their outermost shells and, therefore, have an affinity for electrons. The attachment process plays a very important role in the removal of free electrons from an ionised gas when arc interruption occurs in gas-insulated switchgear. The effect of attachment on breakdown in gases is discussed in sec. 2.7 of this chapter.

2.3 TOWNSEND'S CURRENT GROWTH EQUATION

Referring to Fig. 2.1 let us assume that n_0 electrons are emitted from the cathode. When one electron collides with a neutral particle, a positive ion and an electron are formed. This is called an ionizing collision. Let α be the average number of ionizing collisions made by an electron per centimetre travel in the direction of the field (α depends on gas pressure p and E/p, and is called the Townsend's first ionization coefficient). At any distance x from the cathode, let the number of electrons be n_x. When these n_x electrons travel a further distance of dx they give rise to ($\alpha n_x dx$) electrons.

At
$$x = 0, n_x = n_0 \tag{2.6}$$

Also,
$$\frac{dn_x}{dx} = \alpha n_x; \text{ or } n_x = n_0 \exp(\alpha x) \tag{2.7}$$

Then, the number of electrons reaching the anode ($x = d$) will be
$$n_d = n_0 \exp(\alpha d) \tag{2.8}$$

The number of new electrons created, on the average, by each electron is
$$\exp(\alpha d) - 1 = \frac{n_d - n_0}{n_0} \tag{2.9}$$

Therefore, the average current in the gap, which is equal to the number of electrons travelling per second will be
$$I = I_0 \exp(\alpha d) \tag{2.10}$$

where I_0 is the initial current at the cathode.

2.4 CURRENT GROWTH IN THE PRESENCE OF SECONDARY PROCESSES

The single avalanche process described in the previous section becomes complete when the initial set of electrons reaches the anode. However, since the amplification of electrons [$\exp(\alpha d)$] is occurring in the field, the probability of additional new electrons being liberated in the gap by other mechanisms increases, and these new electrons create further avalanches. The other mechanisms are

(i) The positive ions liberated may have sufficient energy to cause liberation of electrons from the cathode when they impinge on it.

(*ii*) The excited atoms or molecules in avalanches may emit photons, and this will lead to the emission of electrons due to photo-emission.

(*iii*) The metastable particles may diffuse back causing electron emission.

The electrons produced by these processes are called secondary electrons. The secondary ionization coefficient γ is defined in the same way as α, as the net number of secondary electrons produced per incident positive ion, photon, excited particle, or metastable particle, and the total value of γ is the sum of the individual coefficients due to the three different processes, i.e., $\gamma = \gamma_1 + \gamma_2 + \gamma_3$. γ is called the Townsend's secondary ionization coefficient and is a function of the gas pressure p and E/p.

Following Townsend's procedure for current growth, let us assume

$n_0' =$ number of secondary electrons produced due to secondary (γ) processes.

Let $\qquad n_0'' =$ total number of electrons leaving the cathode.

Then $\qquad\qquad\qquad\qquad n_0'' = n_0 + n_0 \qquad\qquad\qquad\qquad (2.11)$

The total number of electrons n reaching the anode becomes,

$$n = n_0'' \exp(\alpha d) = (n_0 + n_0') \exp(\alpha d);$$

and $\qquad\qquad\qquad\qquad n_0' = \gamma [n - (n_0 + n_0')]$

Eliminating n_0', $\qquad\qquad n = \dfrac{n_0 \exp(\alpha d)}{1 - \gamma [\exp(\alpha d) - 1]}$

or $\qquad\qquad\qquad\qquad I = \dfrac{I_0 \exp(\alpha d)}{1 - \gamma [\exp(\alpha d) - 1]} \qquad\qquad (2.12)$

2.5 TOWNSEND'S CRITERION FOR BREAKDOWN

Equation (2.12) gives the total average current in a gap before the occurrence of breakdown. As the distance between the electrodes d is increased, the denominator of the equation tends to zero, and at some critical distance $d = d_s$.

$$1 - \gamma [\exp(\alpha d) - 1] = 0 \qquad\qquad (2.13)$$

For values of $d < d_s$, I is approximately equal to I_0, and if the external source for the supply of I_0 is removed, I becomes zero. If $d = d_s$, $I \to \infty$ and the current will be limited only by the resistance of the power supply and the external circuit. This condition is called Townsend's breakdown criterion and can be written as

$$\gamma [\exp(\alpha d) - 1] = 1$$

Normally, $\exp(\alpha d)$ is very large, and hence the above equation reduces to

$$\gamma \exp(\alpha d) = 1 \qquad\qquad (2.14)$$

For a given gap spacing and at a give pressure the value of the voltage V which gives the values of α and γ satisfying the breakdown criterion is called the spark breakdown voltage V_s and the corresponding distance d_s is called the sparking distance.

The Townsend mechanism explains the phenomena of breakdown only at low pressures, corresponding to $p \times d$ (gas pressure $\times$ gap distance) values of 1000 torr-cm and below.

2.6 EXPERIMENTAL DETERMINATION OF COEFFICIENTS α AND γ

The experimental arrangement is shown in Fig. 2.2. The electrode system consists of two uniform field electrodes. The high voltage electrode is connected to a variable high voltage d.c. source (of 2 to 10 kV rating). The low voltage electrode consists of a central electrode and a guard electrode. The central electrode is connected to the ground through the high resistance of an electrometer amplifier having an input resistance of 10^9 to 10^{13} ohms. The guard electrode is directly earthed. The electrometer amplifier measures currents in the range 10^{-14} to 10^{-8} A.

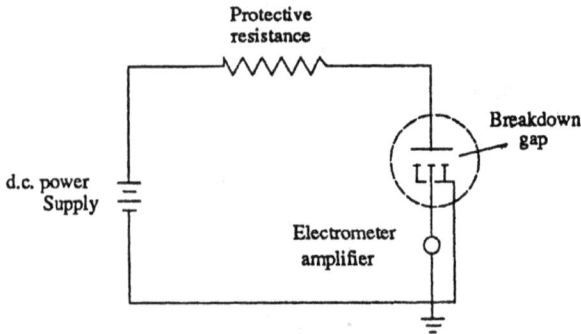

Fig. 2.2 Experimental arrangement to measure ionization coefficients α and η

The electrode system is placed in an ionization chamber which is either a metal chamber made of chromium plated mild steel or stainless steel, or a glass chamber. The electrodes are usually made of brass or stainless steel. The chamber is evacuated to a very high vacuum of the order of 10^{-4} to 10^{-6} torr. Then it is filled with the desired gas and flushed several times till all the residual gases and air are removed. The pressure inside the chamber is adjusted to a few torr depending on the gap separation and left for about half an hour for the gas to fill the chamber uniformly.

The cathode is irradiated using an ultra-violet (U.V.) lamp kept outside the chamber. The U.V. radiation produces the initiatory electrons (n_0) by photo-electric emission.

When the d.c. voltage is applied and when the voltage is low, the current pulses start appearing due to electrons and positive ions as shown in Figs. 2.3a and 2.3b. These records are obtained when the current is measured using a cathode ray oscillograph.

When the applied voltage is increased, the pulses disappear and an average d.c. current is obtained as shown in Fig. 2.4. In the initial portion (T_0), the current increases slowly but unsteadily with the voltage applied. In the regions T_1 and T_2, the current increases steadily due to the Townsend mechanism. Beyond T_2 the current rises very sharply, and a spark occurs.

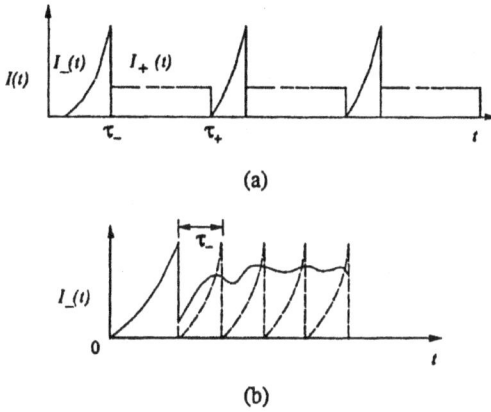

Fig. 2.3 Current as a function of time

(a) When secondary electrons are produced at the cathode by positive ions.

(b) When secondary electrons are produced by photons at the cathode.

– – – ideal, —— actual.

$I(t)$ is the total current and I_- and I_+ are electron ion currents. τ_- and τ_+ are the electron and ion transit times.

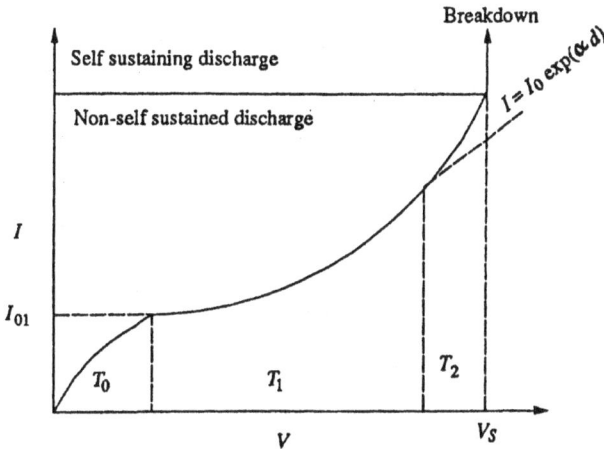

Fig. 2.4 Typical current growth curve in a townsend discharge

For determining the α and γ coefficients, the voltage-current characteristics for different gap settings are obtained. From these results, a log I/I_0 versus gap distance plot is obtained under constant field (E) conditions as shown in Fig. 2.5. The slope of the initial portion of the curves gives the value of α. Knowing α, γ can be found from Eq. (2.12) using points on the upcurving portion of the graphs. The experiment can be repeated for different pressures.

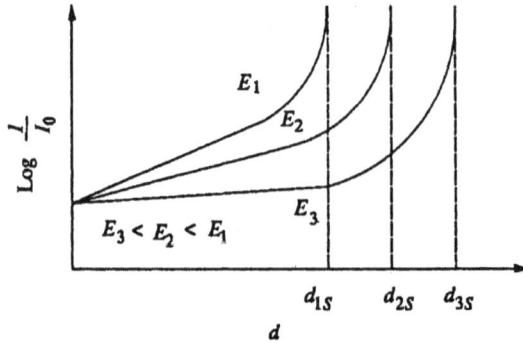

Fig. 2.5 Townsend type log (I/I_0) vs. d plot

It can be easily seen that α/p and γ are functions of E/p. The spark-over voltage for any gap length d_s is $V_s = Ed_s$ where d_s is the critical gap length for that field strength as obtained from the graph. It may be noted that if I_0, the initial current, is more, the average anode current I will also be more, and the relation log I/I_0 versus remains the same. Typical values of α and γ are shown in Figs. 2.6 and 2.7.

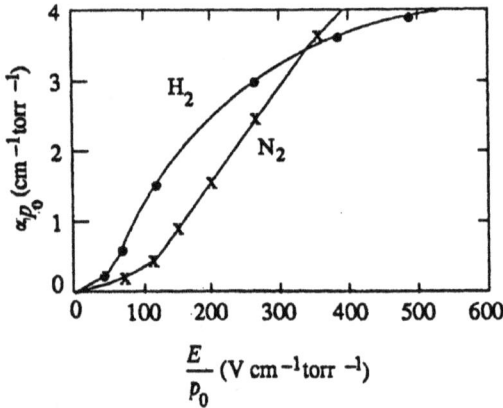

Fig. 2.6 The variation of α/p with E/p in hydrogen and nitrogen, p_0 in both x and y axes refers to values of pressure reduced to 0°C

2.7 BREAKDOWN IN ELECTRONEGATIVE GASES

It has been recognised that one process that gives high breakdown strength to a gas is the electron attachment in which free electrons get attached to neutral atoms or molecules to form negative ions. Since negative ions like positive ions are too massive to produce ionization due to collisions, attachment represents an effective way of removing electrons which otherwise would have led to current growth and breakdown at low voltages. The gases in which attachment plays an active role are called electronegative gases.

The most common attachment processes encountered in gases are (a) the direct attachment in which an electron directly attaches to form a negative ion, and (b) the dissociative attachment in which the gas molecules split into their constituent atoms and the electronegative atom forms a negative ion. These processes may be symbolically represented as:

(a) Direct attachment

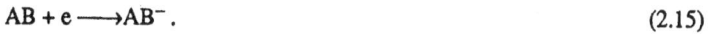

$$AB + e \longrightarrow AB^- .$$ (2.15)

(b) Dissociative attachment

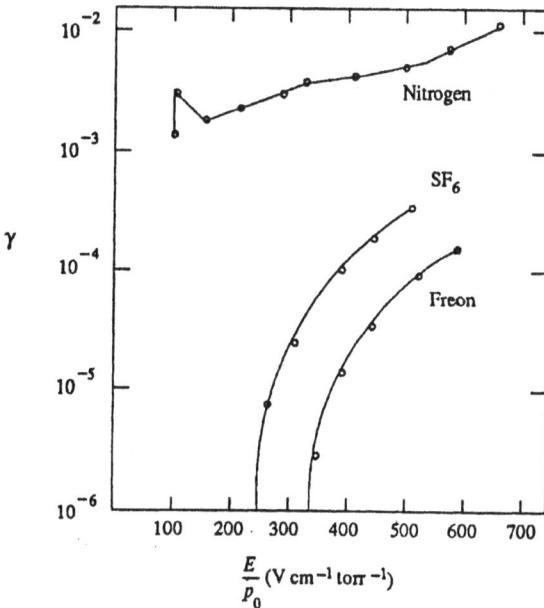

$$AB + e \longrightarrow A + B^-$$ (2.16)

Fig. 2.7   Secondary ionization coefficient (γ) as a function of E/p in nitrogen, freon and SF_6 gases

A simple gas of this type is oxygen. Other gases are sulphur hexafluoride, freon, carbon dioxide, and fluorocarbons. In these gases, 'A' is usually sulphur or carbon atom, and 'B' is oxygen atom or one of the halogen atoms or molecules.

With such gases, the Townsend current growth equation is modified to include ionization and attachment. An attachment coefficient (η) is defined, similar to α, as the number of attaching collisions made by one electron drifting one centimetre in the direction of the field. Under these conditions the current reaching the anode, can be written as

$$I = I_0 \frac{[\{\alpha/(\alpha - \eta)\} \exp{(\alpha - \eta)d}] - [\eta/(\alpha - \eta)]}{1 - \left\{ \gamma \frac{\alpha}{(\alpha - \eta)} [\{ \exp{(\alpha - \eta)d} \} - 1] \right\}} \qquad (2.17)$$

The Townsend breakdown criterion for attaching gases can also be deduced by equating the denominator in Eq. (2.17) to zero, i.e.

$$\gamma \frac{\alpha}{(\alpha - \eta)} [\exp(\alpha - \eta)d - 1] = 1 \qquad (2.18)$$

This shows that for $\alpha > \eta$, breakdown is always possible irrespective of the values of α, η, and γ. If on the other hand, $\eta > \alpha$ Eq. (2.18) approaches an asymptotic form with increasing value of d, and

$$\gamma \frac{\alpha}{(\alpha - \eta)} = 1 ; \text{ or } \alpha = \frac{\eta}{(1 - \gamma)} \qquad (2.19)$$

Normally, γ is very small ($\leq 10^{-4}$) and the above equation can be written as $\alpha = \eta$. This condition puts a limit for E/p below which no breakdown is possible irrespective of the value of d, and the limit value is called the critical E/p. Critical E/p for SF_6 is 117 V cm^{-1} torr^{-1}, and for CCl_2F_2 it is 121 V cm^{-1} torr^{-1} (both at 20°C). η values are also experimentally determined as described in Sec. 2.6. Typical values of η in a few gases are shown in Fig. 2.8.

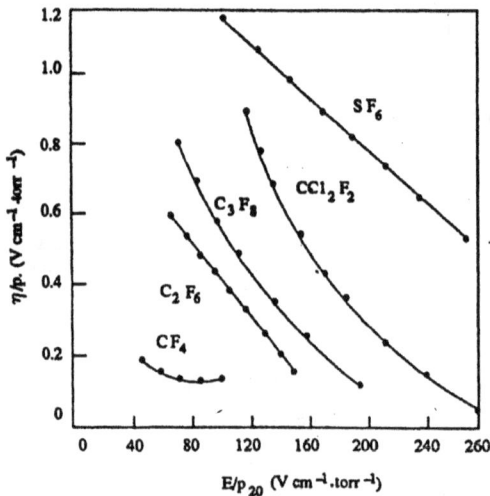

Fig. 2.8   The variation of η/p with E/p in some insulating gases. p_{20} refers to values of pressure reduced to 20°C

2.8 TIME LAGS FOR BREAKDOWN

In the previous section, the mechanism of spark breakdown is considered as a function of ionization processes under uniform field conditions. But in practical engineering designs, the breakdown due to rapidly changing voltages or impulse voltages is of great importance. Actually, there is a time difference between the application of a voltage sufficient to cause breakdown and the occurrence of breakdown itself. This time difference is called the time lag.

The Townsend criterion for breakdown is satisfied, only if at least one electron is present in the gap between the electrodes. In the case of applied d.c. or slowly varying (50 Hz a.c) voltages, there is no difficulty in satisfying this condition. However, with rapidly varying voltages of short duration ($\approx 10^{-6}$ s), the initiatory electron may not be present in the gap, and in the absence of such an electron breakdown cannot occur. The time t which lapses between the application of the voltage sufficient to cause breakdown and the appearance of the initiating electron is called a statistical time lag (t_s) of the gap. The appearance of electrons is usually statistically distributed. After the appearance of the electron, a time t_t is required for the ionization processes to develop fully to cause the breakdown of the gap, and this time is called the formative time lag (t_t). The total time $t_s + t_t = t$ is called the total time lag.

Time lags are of considerable practical importance. For breakdown to occur the applied voltage V should be greater than the static breakdown voltage V_s as shown in Fig. 2.9. The difference in voltage $\Delta V = V - V_s$ is called the overvoltage, and the ratio V/V_s is called the impulse ratio. The variation of t_t with overvoltage (ΔV) is shown in Fig. 2.10. The volt-time characteristics of different electrical apparatus, which are very important in insulation co-ordination, are shown in Fig. 2.11. It can be seen from the Fig. 2.11 that a rod gap will protect a bushing, whereas a sphere gap is required for the complete protection of a transformer against high voltage surges (see also Chapter 8).

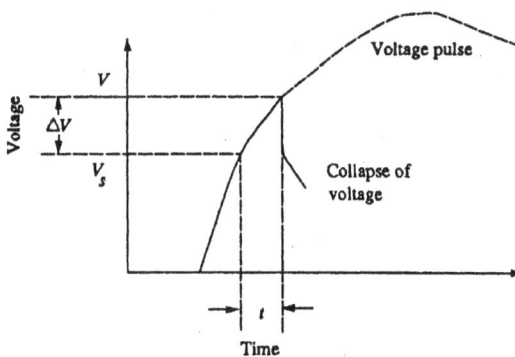

Fig. 2.9   Breakdown on the front of the applied impulse voltage wave

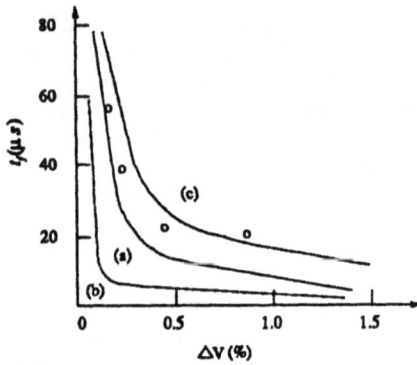

Fig. 2.10 Formative time lag (*t*) as a function of Δ*V*. *a, b, c* are for
different gap spacings
0 experimental point, — calculated
ref: F. Llewellyn Jones, ionization and breakdown in gases,
Methuen, London (1957)

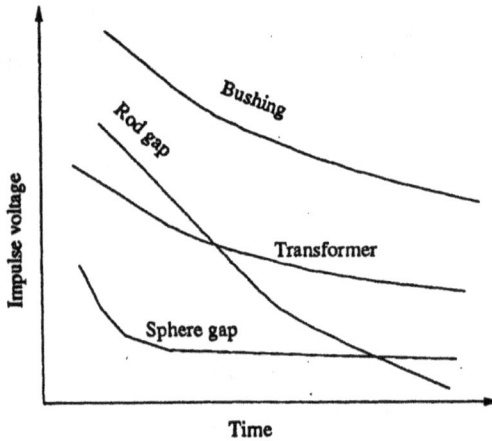

Fig. 2.11 Impulse voltage-time to flashover characteristics

2.9 STREAMER THEORY OF BREAKDOWN IN GASES

Townsend mechanism when applied to breakdown at atmospheric pressure was found
to have certain drawbacks. Firstly, according to the Townsend theory, current growth
occurs as a result of ionization processes only. But in practice, breakdown voltages
were found to depend on the gas pressure and the geometry of the gap. Secondly, the

mechanism predicts time lags of the order of 10^{-5}s, while in actual practice break-down was observed to occur at very short times of the order of 10^{-8}s. Also, while the Townsend mechanism predicts a very diffused form of discharge, in actual practice, discharges were found to be filamentary and irregular. The Townsend mechanism failed to explain all these observed phenomena and as a result, around 1940, Raether and, Meek and Loeb independently proposed the Streamer theory.

The theories predict the development of a spark discharge directly from a single avalanche in which the space charge developed by the avalanche itself is said to transform the avalanche into a plasma streamer. Consider Fig. 2.12. A single electron starting at the cathode by ionization builds up an avalanche that crosses the gap. The electrons in the avalanche move very fast compared with the positive ions. By the time the electrons reach the anode the positive ions are virtually in their original positions and form a positive space charge at the anode. This enhances the field, and the secondary avalanches are formed from the few electrons produced due to photo-ionization in the space charge region. This occurs first near the anode where the space charge is maximum. This results in a further increase in the space charge. This process is very fast

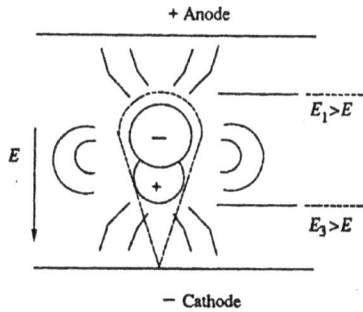

Fig. 2.12 Effect of space charge produced by an avalanche on the applied electric field

and the positive space charge extends to the cathode very rapidly resulting in the formation of a streamer. Comparatively narrow luminous tracks occurring at break-down at high pressures are called streamers. As soon as the streamer tip approaches the cathode, a cathode spot is formed and a stream of electrons rush from the cathode to neutralize the positive space charge in the streamer; the result is a spark, and the spark breakdown has occurred. The three successive stages in the development of the streamer are shown diagrammatically in Fig. 2.13 in which (a) shows the stage when avalanche has crossed the gap, (b) shows that the streamer has crossed half the gap length, and (c) shows that the gap has been bridged by a conducting channel.

Meek proposed a simple quantitative criterion to estimate the electric field that transforms an avalanche into a streamer. The field E_r produced by the space charge, at the radius r, is given by

$$E_r = 5.27 \times 10^{-7} \frac{\alpha \exp(\alpha x)}{(x/p)^{1/2}} \text{ V/cm} \qquad (2.20)$$

where α is Townsend's first ionization coefficient, p is the gas pressure in torr, and x is the distance to which the streamer has extended in the gap. According to Meek, the minimum breakdown voltage is obtained when $E_r = E$ and $x = d$ in the above equation.

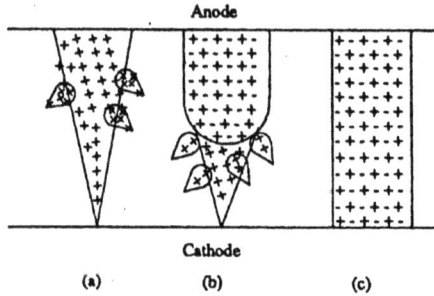

Fig. 2.13 Cathode directed streamer

The equation simplifies into,

$$\alpha d + \ln\left(\frac{\alpha}{p}\right) = 14.5 + \ln\left(\frac{E}{p}\right) + \frac{1}{2}\ln\left(\frac{d}{p}\right) \tag{2.21}$$

This equation is solved between α/p and E/p at which a given p and d satisfy the equation. The breakdown voltage is given by the corresponding product of E and d.

The above simple criterion enabled an agreement between the calculated and the measured breakdown voltages. This theory also neatly fits in with the observed filamentary, crooked channels and the branching of the spark channels, and cleared up many ambiguities of the Townsend mechanism when applied to breakdown in a high pressure gas across a long gap.

It is still controversial as to which mechanism operates in uniform field conditions over a given range of pd values. It is generally assumed that for pd values below 1000 torr-cm and gas pressures varying from 0.01 to 300 torr, the Townsend mechanism operates, while at higher pressures and pd values the Streamer mechanism plays the dominant role in explaining the breakdown phenomena. However, controversies still exist on these statements.

2.10 PASCHEN'S LAW

It has been shown earlier (refer Sec. 2.5) that the breakdown criterion in gases is given as

$$\gamma\left[\exp\left(\alpha d\right) - 1\right] = 1 \tag{2.22}$$

where the coefficients α and γ are functions of E/p, i.e

$$\frac{\alpha}{p} = f_1\left(\frac{E}{p}\right)$$

and

$$\gamma = f_2\left(\frac{E}{p}\right)$$

Also

$$E = \frac{V}{d}$$

Substituting for E in the expressions for α and γ and rewriting Eq. (2.18) we have

$$f_2\left(\frac{V}{pd}\right)[\exp\left\{pd f_1\left(\frac{V}{pd}\right)\right\} - 1] = 1 \tag{2.23}$$

This equation shows a relationship between V and pd, and implies that the breakdown voltage varies as the product pd varies. Knowing the nature of functions f_1 and f_2 we can rewrite Eq. (2.22) as,

$$V = f(pd) \tag{2.24}$$

This equation is known as Paschen's law and has been experimentally established for many gases, and it is a very important law in high voltage engineering.

The Paschen's curve, the relationship between V and pd is shown in Fig. 2.14 for three gases CO_2, air and H_2. It is seen that the relationship between V and pd is not linear and has a minimum value for any gas. The minimum breakdown voltages for various gases are given in Table 2.1

Table 2.1 Minimum Sparking Potential For Various Gases

Gas	V_s min (V)	pd at V_s min (torr-cm)
Air	327	0.567
Argon	137	0.9
H_2	273	1.15
Helium	156	4.0
CO_2	420	0.51
N_2	251	0.67
N_2O	418	0.5
O_2	450	0.7
SO_2	457	0.33
H_2S	414	0.6

The existence of a minimum sparking potential in Paschen's curve may be explained as follows:

For values of $pd > (pd)_{min}$, electrons crossing the gap make more frequent collisions with gas molecules than at $(pd)_{min}$, but the energy gained between collisions is lower. Hence, to maintain the desired ionization more voltage has to be applied. For $pd < (pd)_{min}$, electron may cross the gap without even making a collision or making only less number of collisions. Hence, more voltage has to be applied for breakdown to occur.

However, in some gases Paschen's law is not strictly obeyed, and sparking potentials at larger spacings for a given value of pd are higher than at lower spacings for the same pd value. This is attributed to the loss of electrons from the gap due to diffusion.

The sparking potentials for uniform field gaps in air, CO_2 and H_2 at 20°C are shown in Fig. 2.14. It has been observed that the cathode materials also affect the breakdown values. This is shown in Fig. 2.15 for cathodes made of barium, magnesium and aluminium.

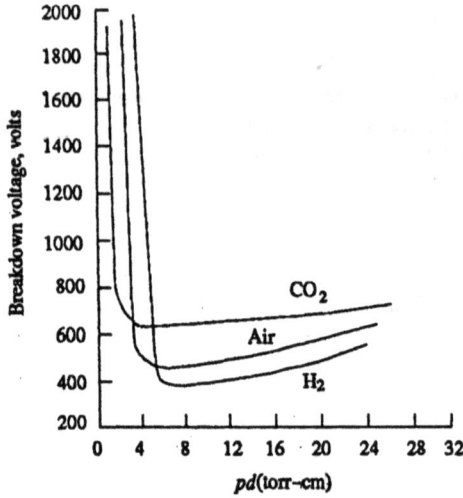

Fig. 2.14 Breakdown voltage-*pd* characteristics for air, CO_2 and hydrogen

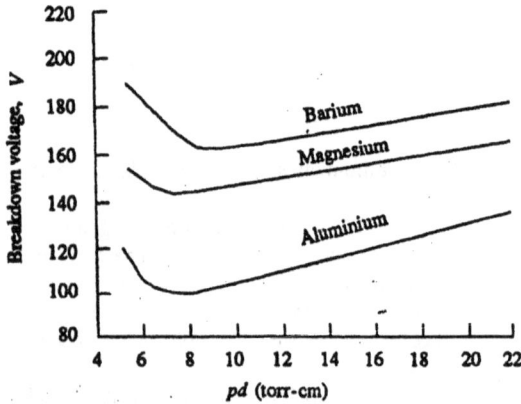

Fig. 2.15 Dependence of breakdown voltage on the cathode materials

In order to account for the effect of temperature, the Paschen's law is generally stated as $V = f(Nd)$ where N is the density of the gas molecules. This is necessary, because the pressure of the gas changes with temperature according to the gas law pv = NRT, where v is the volume of the gas, T is the temperature, and R is a constant.

Based on the experimental results, the breakdown potential of air is expressed as a power function in pd as

$$V = 24.22 \left[\frac{293\, pd}{760T} \right] + 6.08 \left[\frac{293\, pd}{760T} \right]^{1/2} \tag{2.25}$$

It may be noted from the above formula that the breakdown voltage at constant pressure and temperature is not constant.

At 760 torr and $293°K$.

$$E = V/d = 24.22 + \left[\frac{6.08}{\sqrt{d}} \right] \text{kV/cm} \tag{2.26}$$

This equation yields a limiting value for E of 24 kV/cm for long gaps and a value of 30 kV/cm for $\left(\frac{293pd}{760T} \right) = 1$, which means a pressure of 760 torr at 20°C with 1 cm gap. This is the usually quoted breakdown strength of air at room temperature and atmospheric pressure.

2.11 BREAKDOWN IN NON-UNIFORM FIELDS AND CORONA DISCHARGES

2.11.1 Corona Discharges

If the electric field is uniform, a gradual increase in voltage across a gap produces a breakdown of the gap in the form of a spark without any preliminary discharges. On the other hand, if the field is non-uniform, an increase in voltage will first cause a discharge in the gas to appear at points with highest electric field intensity, namely at sharp points or where the electrodes are curved or on transmission lines. This form of discharge is called a corona discharge and can be observed as a bluish luminiscence. This phenomenon is always accompanied by a hissing noise, and the air surrounding the corona region becomes converted into ozone. Corona is responsible for considerable loss of power from high voltage transmission lines, and it leads to the deterioration of insulation due to the combined action of the bombardment of ions and of the chemical compounds formed during discharges. Corona also gives rise to radio interference.

The voltage gradient required to produce visual a.c. corona in air at a conductor surface, called the corona inception field, can be approximately given for the case of parallel wires of radius r as

$$E_{v_c} = 30md \left[1 + \frac{0.301}{\sqrt{dr}} \right] \tag{2.27}$$

For the case of coaxial cylinders, whose inner cylinder has a radius r the equation becomes

$$E_c = 31md\left[1 + \frac{0.308}{\sqrt{dr}}\right] \tag{2.28}$$

where m is the surface irregularity factor which becomes equal to unity for highly polished smooth wires; d is the relative air density correction factor given by,

$$d = \frac{0.392b}{(273 + t)} \tag{2.29}$$

where b is the atmospheric pressure in torr, and t is the temperature in °C, $d = 1$ at 760 torr and 25°C. The expressions were found to hold good from atmospheric pressure down to a pressure of several torr.

On the high voltage conductors at high pressures there is a distinct difference in the visual appearance of the corona under positive and negative polarities of the applied voltage. When the voltage is positive, corona appears as a uniform bluish white sheath over the entire surface of the conductor. On the other hand, when the voltage is negative, the corona will appear like reddish glowing spots distributed along the length of the wire. Investigations with point-plane gaps in air showed that when point is negative, corona appears as current pulses called Trichel pulses, and the repetition frequency of these pulses increases as the applied voltage is increased and decreases with decrease in pressure. On the other hand, observations when the point is positive in air showed that the corona current increases steadily with voltage. At sufficiently high voltage, current amplification increases rapidly with voltage, up to a current of about 10^{-7}A, after which the current becomes pulsed with repetition frequency of about 1 kHz composed of small bursts. This form of corona is called burst corona. The average current then increases steadily with applied voltage leading to breakdown.

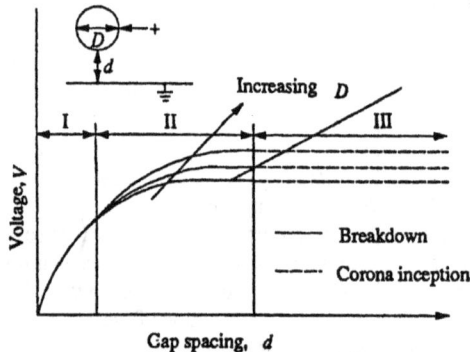

Fig. 2.16 Breakdown and corona inception characteristics for spheres of different diameters in sphere-plane gap geometry

The corona inception and breakdown voltages of the sphere-plane arrangement are shown in Fig. 2.16. From this figure it can be seen that

    (a) at small spacings (region I), the field is uniform, and the breakdown voltage mainly depends on the spacing;

(b) at fairly large spacings (region II), the field is non-uniform, and the break-down voltage depends both on the sphere diameter and the spacing; and

(c) at large spacings (region III), the field is non-uniform, and the breakdown is preceded by corona and is controlled only by the spacing. The corona inception voltage mainly depends on the sphere diameter.

The actual breakdown characteristics of the sphere-plane gap in air is shown in Fig. 2.17. It may be summarized that the study of corona and non-uniform field breakdown is very complicated and investigations are still under progress.

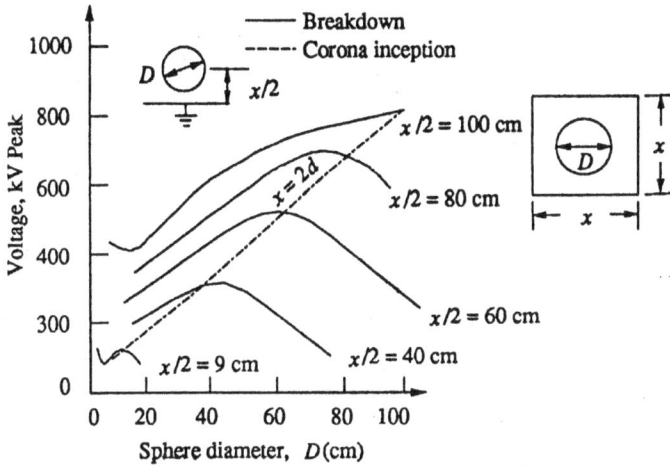

Fig. 2.17 Breakdown and corona inception characteristics of sphere-plane geometry in air, x is the fixed overall dimension of a system as shown

ref: B.L. Goodlet, F.S. Edwards and F.R. Perry, Journal of the Institution of Electrical Engineers; vol. 69, 695 (1931)

2.11.2 Breakdown in Non-uniform Fields

In non-uniform fields, such as coaxial cylinders, point-plane and sphere-plane gaps, the applied field varies across the gap. Similarly, Townsend's first ionization coefficient (α) also varies with the gap. Hence αd in Townsend's criterion [refer to Eq. (2.14)] is rewritten by replacing αd by $\int_0^d \alpha \, dx$. Townsend's criterion for breakdown now becomes

$$\gamma \left\{ \exp[\int_0^d \alpha dx] - 1 \right\} = 1 \qquad (2.30)$$

Meek and Raether also discussed the non-uniform field breakdown process as applied to their Streamer theory, and the Meek's equation [Eq. (2.19)] for the radial field at the head of an avalanche when it has crossed a distance x is modified as

$$E_r = \frac{5.27 \times 10^{-7} \alpha_x \exp \left(\int_0^x \alpha dx \right)}{(x/p)^{1/2}} \text{ V/cm} \qquad (2.31)$$

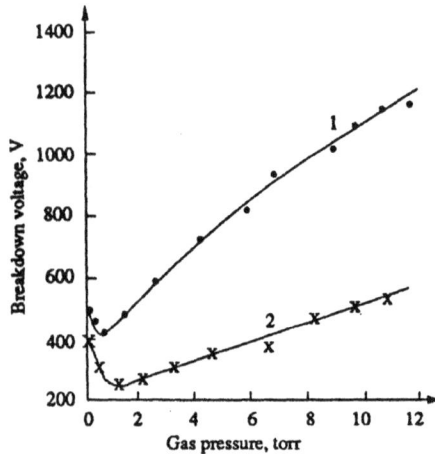

Fig. 2.18 Breakdown characteristics for nitrogen between a wire and a coaxial cylinder of radii 0.083 and 2.3 cm. 1-wire positive, 2-wire negative

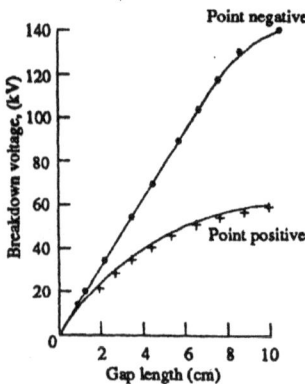

Fig. 2.19 d.c. breakdown characteristics for air between 30° conical point and a plane

where α_x is the value of α at the head of the avalanche, and p is the gas pressure. The criterion for the formation of the streamer is reached when the space charge field E_r approaches a value equal to the applied field at the head of the avalanche.

This equation has been successfully used for determining the corona onset voltages of many non-uniform geometries. However, the condition for the advancement of streamers has not been arrived at so far. Figures 2.18 and 2.19 show the d.c. breakdown characteristics for a wire-coaxial cylinder geometry in nitrogen and for a point-plane geometry in air, respectively.

From the practical engineering point of view, rod-rod gap and sphere-sphere gap are of great importance, as they are used for the measurement of high voltages and for the protection of electrical apparatus such as transformers. The breakdown characteristics of rod-rod gaps are shown n Fig. 2.20. From this figure it can be seen that the breakdown voltages are higher for negative polarity. The breakdown voltages were also observed to depend on humidity in air. In the case of rod gaps the field is non-uniform, while in the case of sphere gaps field is uniform, if the gap is small compared with the diameter. In the case of sphere gaps, the breakdown voltages do not depend on humidity and are also independent of the voltage waveform. The formative time lag is quite small ($\sim 0.5 \, \mu$ s) even with 5% over voltage. Hence sphere gaps are used for breakdown voltage (peak value) measurements. These are further discussed in Chapter 7 (Sec. 7.2.6).

2.12 POST-BREAKDOWN PHENOMENA AND APPLICATIONS

This is the phenomenon which occurs after the actual breakdown has taken place and is of technical importance. Glow and arc discharges are the post-breakdown phenomena, and there are many devices that operate over these regions. In a Townsend discharge (see Fig. 2.21) the current increases gradually as a function of the applied voltage. Further to this point (B) only the current increases, and the discharge changes from the Townsend type to Glow type (BC). Further increase in current results in a very small reduction in voltage across the gap (CD) corresponding to the normal glow region. The gap voltage again increases (DE), when the current is increased more, but eventually leads to a considerable drop in the applied voltage. This is the region of the arc discharge (EG). The phenomena that occur in the region CG are the post-breakdown phenomena consisting of glow discharge (CE) and the arc discharge (EG):

Glow Discharge

A glow discharge is characterized by a diffused luminous glow. The colour of the glow discharge depends on the cathode material and the gas used. The glow discharge covers the cathode partly and the space between the cathode, and the anode will have intermediate dark and bright regions. This is called normal glow. If the current in the normal glow is increased such that the discharge covers the entire cathode surface, then it becomes abnormal glow. In a glow discharge, the voltage drop between the electrodes is substantially constant, ranging from 75 to 300 V over a current range of 1 mA to 100 mA depending on the type of the gas. The properties of the glow discharge are used in many practical applications, such as cold cathode gaseous voltage stabilized tubes (voltage regulation tubes or VR tubes), for rectification, as a relaxation oscillator, and as an amplifier.

Arc Discharge

If the current in the gap is increased to about 1 A or more, the voltage across the gap suddenly reduces to a few volts (20-50 V). The discharge becomes very luminous and noisy (region EG in Fig. 2.21). This phase is called the arc discharge and the current

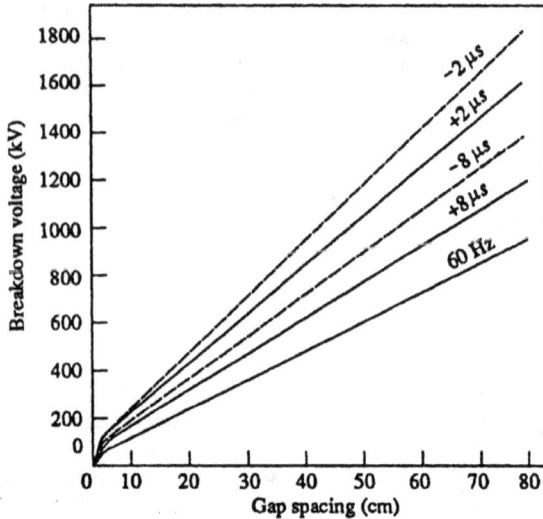

Fig. 2.20 Power frequency (60 Hz) and impulse breakdown voltage
curves for a rod-rod gap in air at n.t.p. One rod is earthed.
Absolute humidity is 6.5 gms/ft². impulse breakdown curves
are for various times of breakdown on the wave tail
ref.: B.S.S. 171: 1959, power transformers

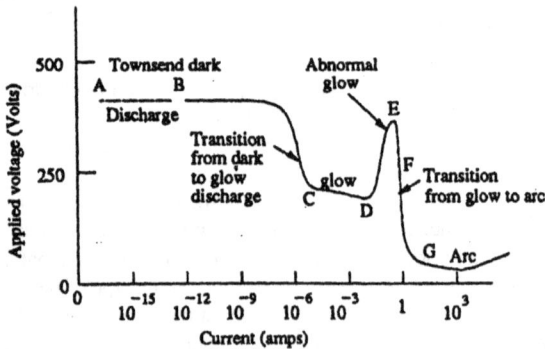

Fig. 2.21 d.c. voltage-current characteristic of an electrical discharge
with electrodes having no sharp points or edges

density over the cathode region increases to very high values of 10^3 to 10^7 A/cm². Arcing is associated with high temperatures, ranging from 1000°C to several thousand degrees celsius. The discharge will contain a very high density of electrons and positive ions, called the arc plasma. The study of arcs is important in circuit breakers and other switch contacts. It is a convenient high temperature high intensity light source. It is used for welding and cutting of metals. It is the light source in lamps such as carbon arc lamp. High temperature plasmas are used for generation of electricity through magneto-hydro dynamic (MHD) or nuclear fusion processes.

2.13 PRACTICAL CONSIDERATIONS IN USING GASES FOR INSULATION PURPOSES

In recent years, considerable amount of work has been done to adopt a specific gas for practical use. Before adopting a particular gas for a practical purpose, it is useful to gain a knowledge of what the gas does, what its composition is, and what the factors that influence its performance are. The greater the versatility of the operating performance demanded from an insulating gas, the more rigorous would be the requirements which the gas should meet. These requirements needed by a good dielectric gas do not exist in a majority of the gases. Generally, the preferred properties of a gaseous dielectric for high voltage applications are:

(a) high dielectric strength,
(b) thermal stability and chemical inactivity towards materials of construction,
(c) non-flammability and physiological intertness,
(d) low temperature of condensation,

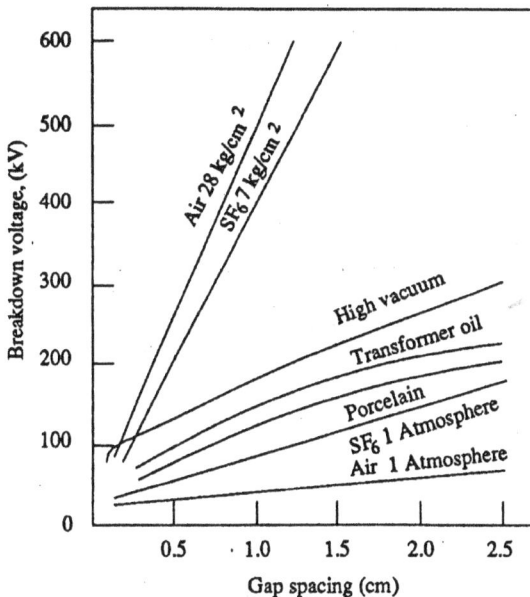

Fig. 2.22 d.c. breakdown strength of typical solid, liquid, gas and vacuum insulations in uniform fields

Table 2.2 Properties of Insulating Gases

Name of the gas	Formula	Molecular weight	Melting point at 760 torr °C	Boiling point at 760 torr °C	Relative dielectric strength ($N_2 = 1$)	Dielectric constant	Specific gravity (Air = 1)	Flammability	Toxicity
Air	—	29	—	-194	1	1.00059	1.00000	No	Physio-inert
Nitrogen	N_2	28	-210	-196	1	1.00058	0.96724	No	-do-
Hydrogen	H_2	2	-259	-253	—	1.00026	0.06952	Yes	-do-
Carbon tetrafluoride	CF_4	88	-183	-128	1.01	1.00050	—	No	-do-
Hexafluoro-ethane	C_2F_6	138	-101	-78	2.02	1.00200	—	No	-do-
Perfluoro-propane	C_3F_8	188	-160	-37	2.2	—	—	No	-do-
Perfluoro-butane	C_4F_{10}	238	-80	-2	2.6	—	—	No	-do-
Perfluoro-n-butane	C_4F_8	200		+2	3.6	1.00340	7.3323	No	-do-
Sulphur hexafluoride	SF_6	146		-63	2.5	1.00191	5.1900	No	-do-
30% SF_6 + 70% Air (Vol.)	(Vol.)				2.0			No	-do-
Freon-12	CCl_2F_2	121	-158	-30	2.46	1.00160	—	No	P*

*P: Physio-inert for durations of 2 hr or less with 20% concentration.

Table 2.3

Set 1:

Gap distance (mm)	0.5	1.0	1.5	2.0	2.5	3.0	3.5	4.0	5.0
Applied voltage V (volts)	1000	2000	3000	4000	5000	6000	7000	8000	10000
Observed current I (A)	10^{-13}	3×10^{-13}	6×10^{-13}	10^{-12}	4×10^{-12}	10^{-11}	10^{-10}	10^{-9}	5×10^{-7}

Set 2:

V (volts)	500	1000	1500	2000	2500	3000	3500	4000	4500
I (A)	5×10^{-14}	1.5×10^{-13}	3×10^{-13}	6×10^{-13}	10^{-12}	5×10^{-12}	5×10^{-11}	3×10^{-10}	10^{-8}

The minimum current observed when 150 V was applied was 5×10^{-14} A.

Table 2.4

Gap (mm)	0.5	1.0	1.5	2.0	2.5	3.0	3.5	4.0	5.0
I/I_0 for $E_1 = 20$ k V/cm	2	6	12	20	80	200	2×10^3	2×10^4	5×10^7
log I/I_0	0.3010	0.7181	1.0792	1.3010	1.9031	2.3010	3.3010	4.3010	7.6990
I/I_0 for $E_2 = 10$ k V/cm	1	3	6	12	20	100	1000	6000	2×10^5
log I/I_0	0	0.4771	0.7781	1.0792	1.3010	2.0	3.0	3.7781	5.3010

(e)   good heat transfer, and

(f)   ready availability at moderate cost.

Sulphur hexafluoride (SF_6) which has received much study in recent years has been found to possess most of the above requirements.

Of the above properties, dielectric strength is the most important property of a gaseous dielectric for practical use. The dielectric strength of gases is comparable with those of solid and liquid dielectrics (see Fig. 2.22). In recent years, the dielectric properties of many complex chlorinated and flourinated molecular compounds have also been studied. These are shown in Fig. 2.23. This feature of high dielectric strength of gases is attributed to the molecular complexity and the high rates of electron attachment (see Sec. 2.7). The relative dielectric strengths and the chemical and physical properties of some of the commercially important gases are shown in Table 2.2.

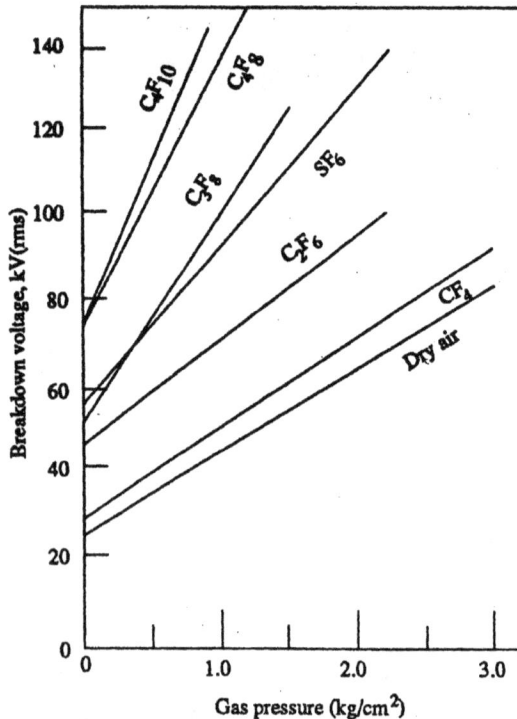

Fig. 2.23 Breakdown strength of insulating gases for 75 cm diameter uniform field electrodes having 12 mm gap

From the figures and the table, it can be seen that SF_6 has high dielectric strength and low liquification temperature, and it can be used over a wide range of operating conditions. SF_6 was also found to have excellent arc-quenching properties. Therefore, it is widely used as an insulating as well as arc-quenching medium in high voltage apparatus such as high voltage cables, current and voltage transformers, circuit-breakers and metal enclosed substations. It can also be seen from the table that addition of 30% SF_6 to air (by volume) increases the dielectric strength of air by 100%. One of the qualitative effects of mixing SF_6 to air is to reduce the overall cost of the gas, and at the same time attaining relatively high dielectric strength or simply preventing the onset of corona at desired operating voltages. In addition to the use of SF_6 gas in recent times, everyone knows of the essential quality of air as an insulating medium for overhead power transmission lines and in air blast circuit-breakers.

2.14 VACUUM INSULATION

2.14.1 Introduction

The idea of using vacuum for insulation purposes is very old. According to the Townsend theory, the growth of current in a gap depends on the drift of the charged particles. In the absence of any such particles, as in the case of perfect vacuum, there should be no conduction and the vacuum should be a perfect insulating medium. However, in practice, the presence of metallic electrodes and insulating surfaces within the vacuum complicate the issue and, therefore, even in vacuum, a sufficiently high voltage will cause a breakdown.

In recent years a considerable amount of work has been done to determine the electrical properties of high vacuum. This is mainly aimed at adopting such a medium for a wide range of applications in devices such as vacuum contractors and interrupters, high frequency capacitors and relays, electrostatic generators, microwave tubes, etc. The contractors and circuit breakers using vacuum as insulation are finding increasing applications in power systems.

2.14.2 What Is Vacuum?

A vacuum system which is used to create vacuum is a system in which the pressure is maintained at a value much below the atmospheric pressure. In vacuum systems the pressure is always measured in terms of millimetres of mercury, where one standard atmosphere is equal to 760 millimetres of mercury at a temperature of 0°C. The term "millimetres of mercury" has been standardised as "Torr" by the International Vacuum Society, where one millimetre of mercury is taken as equal to one Torr. Vacuum may be classified as

High vacuum   : 1×10^{-3} to 1×10^{-6} Torr

Very high vacuum : 1×10^{-6} to 1×10^{-8} Torr

Ultra high vacuum : 1×10^{-9} torr and below.

For electrical insulation purposes, the range of vacuum generally used is the "high vacuum", in the pressure range of 10^{-3} Torr to 10^{-6} Torr.

2.14.3 Vacuum Breakdown

In the Townsend type of discharge in a gas described earlier, electrons get multiplied due to various ionisation processes and an electron avalanche is formed. In a high vacuum, even if the electrodes are separated by, say, a few centimetres, an electron crosses the gap without encountering any collisions. Therefore, the current growth prior to breakdown cannot be due to the formation of electron avalanches. However, if a gas is liberated in the vacuum gap, then, breakdown can occur in the manner described by the Townsend process. Thus, the various breakdown mechanisms in high vacuum aim at establishing the way in which the liberation of gas can be brought about in a vacuum gap.

During the last 70 years or so, many different mechanisms for breakdown in vacuum have been proposed. These can be broadly divided into three categories

(a) Particle exchange mechanism

(b) Field emission mechanism

(c) Clump theory

(a) Particle Exchange Mechanism

In this mechanism it is assumed that a charged particle would be emitted from one electrode under the action of the high electric field, and when it impinges on the other electrode, it liberates oppositely charged particles. These particles are accelerated by the applied voltage back to the first electrode where they release more of the original type of particles. When this process becomes cumulative, a chain reaction occurs which leads to the breakdown of the gap.

The particle-exchange mechanism involves electrons, positive ions, photons and the absorbed gases at the electrode surfaces. Qualitatively, an electron present in the vacuum gap is accelerated towards the anode, and on impact releases A positive ions

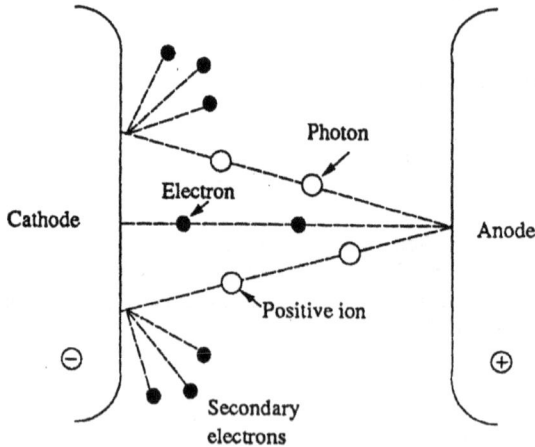

Fig. 2.24 Particle exchange mechanism of vacuum breakdown

and C photons. These positive ions are accelerated towards the cathode, and on impact each positive ion liberates B electrons and each photon liberates D electrons. This is shown schematically in Fig. 2.24. The breakdown will occur if the coefficients of production of secondary electrons exceeds unity. Mathematically, the condition for breakdown can be written as

$$(AB + CD) > 1 \qquad (2.32)$$

Later, Trump and Van de Graaff measured these coefficients and showed that they were too small for this process to take place. Accordingly, this theory was modified to allow for the presence of negative ions and the criterion for breakdown then becomes

$$(AB + EF) > 1 \qquad (2.33)$$

Where A and B are the same as before and E and F represent the coefficients for negative and positive ion liberation by positive and negative ions. It was experimentally found that the values of the product EF were close enough to unity for copper, aluminium and stainless steel electrodes to make this mechanism applicable at voltages above 250 kV.

(b) Field Emission Theory

(i) Anode Heating Mechanism

This theory postulates that electrons produced at small micro-projections on the cathode due to field emission bombard the anode causing a local rise in temperature and release gases and vapours into the vacuum gap. These electrons ionise the atoms of the gas and produce positive ions. These positive ions arrive at the cathode, increase the primary electron emission due to space charge formation and produce secondary electrons by bombarding the surface. The process continues until a sufficient number of electrons are produced to give rise to breakdown, as in the case of a low pressure Townsend type gas discharge. This is shown schematically in Fig. 2.25.

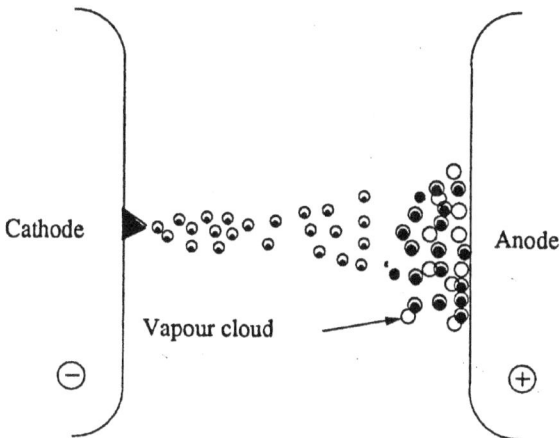

Fig. 2.25 Electron beam anode heating mechanism of vacuum breakdown

(ii) Cathode Heating Mechanism

This mechanism postulates that near the breakdown voltages of the gap, sharp points on the cathode surface are responsible for the existence of the pre-breakdown current, which is generated according to the field emission process described below.

This current causes resistive heating at the tip of a point and when a critical current density is reached, the tip melts and explodes, thus initiating vacuum discharge. This mechanism is called field emission as shown schematically in Fig. 2.26. Thus, the initiation of breakdown depends on the conditions and the properties of the cathode surface. Experimental evidence shows that breakdown takes place by this process when the effective cathode electric field is of the order of 10^6 to 10^7 V/cm.

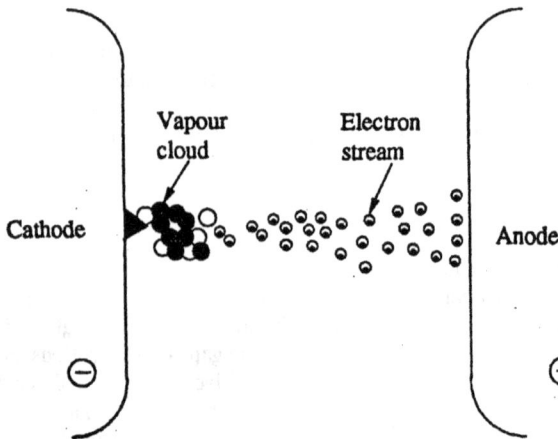

Fig. 2.26 Breakdown in vacuum caused by the heating of a microprojection on the cathode

(c) Clump Mechanism

Basically this theory has been developed on the following assumptions (Fig. 2.27):
   (i) A loosely bound particle (clump) exists on one of the electrode surfaces.
   (ii) On the application of a high voltage, this particle gets charged, subsequently gets detached from the mother electrode, and is accelerated across the gap.
   (iii) The breakdown occurs due to a discharge in the vapour or gas released by the impact of the particle at the target electrode.

Cranberg was the first to propose this theory. He initially assumed that breakdown will occur when the energy per unit area, W, delivered to the target electrode by a clump exceeds a value C', a constant, characteristic of a given pair of electrodes. The quantity W is the product of gap voltage (V) and the charge density on the clump. The latter is proportional to the electric field E at the electrode of origin. The criterion for breakdown, therefore, is

$$VE = C' \tag{2.34}$$

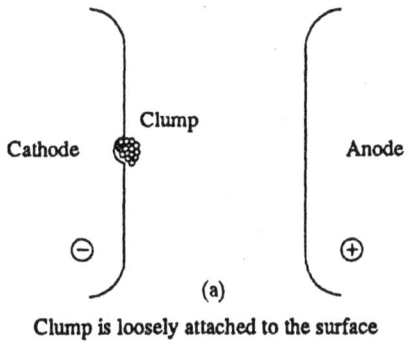

(a)

Clump is loosely attached to the surface

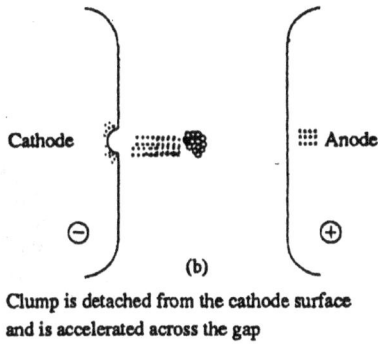

(b)

Clump is detached from the cathode surface
and is accelerated across the gap

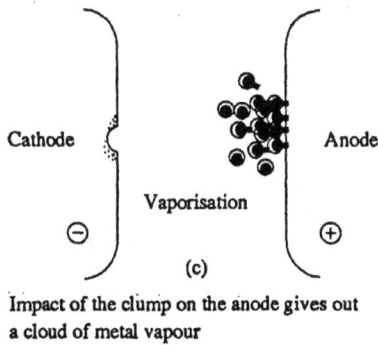

(c)

Impact of the clump on the anode gives out
a cloud of metal vapour

Fig. 2.27 (a, b, c) Clump mechanism of vacuum breakdown

In case of parallel plane electrodes the field $E = V/d$, where d is the distance between the electrodes. So the generalised criterion for breakdown becomes

$$V = (C\,d)^{1/2} \tag{2.35}$$

where C is another constant involving C' and the electrode surface conditions.

Cranberg presented a summary of the experimental results which satisfied this breakdown criterion with reasonable accuracy. He stated that the origin of the clump was the cathode and obtained a value for the constant C as 60×10^{10} V^2/cm (for iron particles). However the equation was later modified as $V = C\,d^{\alpha}$, where α varies between 0.2 and 1.2 depending on the gap length and the electrode material, with a maximum at 0.6. The dependence of V on the electrode material, comes from the observations of markings on the electrode surfaces. Craters were observed on the anode and melted regions on the cathode or vice-versa after a single breakdown.

(d) Summary

Although there has been a large amount of work done on vacuum breakdown phenomena, so far, no single theory has been able to explain all the available experimental measurements and observations. Since experimental evidence exists for all the postulated mechanisms, it appears that each mechanism would depend, to a great extent, on the conditions under which the experiments were performed. The most significant experimental factors which influence the breakdown mechanism are: gap length, geometry and material of the electrodes, surface uniformity and treatment of the surface, presence of extraneous particles and residual gas pressure in the vacuum gap. It was observed that the correct choice of electrode material, and the use of thin insulating coatings in long gaps can increase the breakdown voltage of a vacuum gap. On the other hand, an increase of electrode area or the presence of particles in the vacuum gap will reduce the breakdown voltage.

QUESTIONS

Q.2.1 Explain the difference between photo-ionisation and photo-electric emission.

Q.2.2 Explain the term "electron attachment". Why are electron attaching gases useful for practical use as insultants when compared to non-attaching gases.

Q.2.3 Describe the current growth phenomenon in a gas subjected to uniform electric fields.

Q.2.4 Explain the experimental set-up for the measurement of pre-breakdown currents in a gas.

Q.2.5 Define Townsend's first and second ionization coefficients. How is the condition for breakdown obtained in a Townsend discharge?

Q.2.6 What are electronegative gases? Why is the breakdown strength higher in these gases compared to that in other gases?

Q.2.7 Derive the criterion for breakdown in electronegative gases.

Q.2.8 Explain the Streamer theory of breakdown in air at atmospheric pressure.

Q.2.9 What are the anode and the cathode streamers? Explain the mechanism of their formation and development leading to breakdown.

Q.2.10 What is Paschen's law? How do you account for the minimum voltage for breakdown under a given '$p \times d$' condition?

Q.2.11 Describe the various factors that influence breakdown in a gas.

Q.2.12 What is vacuum? How is it categorised? What is the usual range of vacuum used in high voltage apparatus?

Q.2.13 Describe how vacuum breakdown is different from normal breakdown of a gas.

Q.2.14 Discuss the various mechanisms of vacuum breakdown.

WORKED EXAMPLES

Example 2.1 : Table 2.3 gives the sets of observations obtained while studying the Townsend phenomenon in a gas. Compute the values of the Townsend's primary and secondary ionization coefficients from the data given.

Solution : The current at minimum applied voltage, I_0, is taken as 5×10^{-14} A, and the graph of d versus log I/I_0 is plotted as shown in Fig. E.2.1. The values of log I/I_0 versus d for two values of electric field, $E_1 = 20$ kV/cm and $E_2 = 10$ kV/cm are given in Table 2.4.

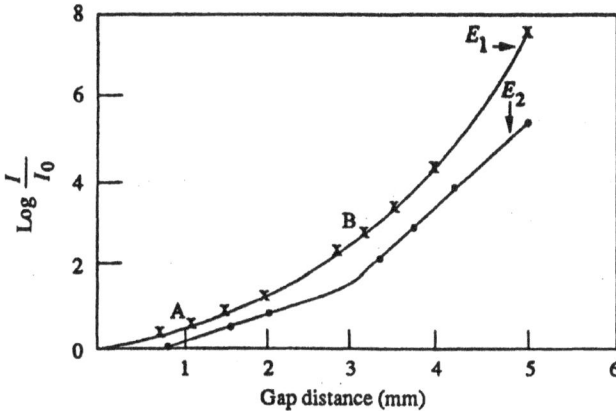

Fig. E.2.1 Log I/I_0 as a function of gap distance

Value of α at $E_1 (= 20$ kV/cm) i.e. $\alpha_2 =$ slope of curve E_1

$$= \frac{2.9}{2.5 \times 10^{-1}}$$

$$= 11.6 \text{ cm}^{-1} \text{torr}^{-1}$$

Value of α at $E_2 (= 10$ kV/cm) i.e. $\alpha_1 =$ slope of curve E_2

$$= \frac{13}{2 \times 10^{-1}}$$

$$= 6.5 \text{ cm}^{-1} \text{torr}^{-1}$$

As the sparking potential and the critical gap distance are not known, the last observations will be made use in determining the values of γ.

For a gap distance of 5 mm, at $E_1 = 20$ kV/cm,

$$I = \frac{I_0 \exp(\alpha d)}{1 - \gamma[\exp(\alpha d)-1]}$$

$$\frac{I}{I_0} = \frac{\exp(\alpha d)}{1 - \gamma[\exp(\alpha d)-1]}$$

Substituting $\alpha_1 = 11.6$, $d = 0.5$ cm, and $I/I_0 = 5 \times 10^7$

$$5 \times 10^7 = \frac{\exp(5.8)}{1 - \gamma[\exp(5.8) - 1]}$$

$$= \frac{330.3}{1 - \gamma(330.3 - 1)}$$

or $\qquad \gamma = 3.0367 \quad 10^{-3}/\text{cm . torr, at } E_1 = 20 \text{ kV/cm}$

(Check this value with other observations also.)

For $\qquad\qquad E_2 = 10$ kV/cm

$\qquad\qquad\qquad \alpha_2 = 6.5/\text{cm . torr}$

$\qquad\qquad\qquad d = 0.5$ cm

and $\qquad\qquad\qquad I/I_0 = 2 \times 10^5$

Substituting these values in the same equation,

$$2 \times 10^5 = \frac{\exp(3.25}{1\gamma[\exp(3.25) - 1]}$$

$$= \frac{25.79}{1 - \gamma(25.79 - 1)}$$

or, $\qquad\qquad \gamma = 4.03 \times 10^{-2}/\text{cm . torr, at } E_2 = 10 \text{ kV/cm}$

Example 2.2 : A glow discharge tube is to be designed such that the breakdown occurs at the Paschen minimum voltage. Making use of Fig. 2.14 suggest the suitable gap distance and pressure in glow discharge tube when the gas in it is (a) hydrogen, (b) air.

Solution : In the case of hydrogen, the Paschen minimum voltage occurs at a pd (product of pressure and gap spacing) of 7.5 torr-cm, and in the case of air the corresponding value of pd is 4.5 torr-cm (see Fig. 2.14).

Since the usual gap distance used for glow discharge tubes of smaller sizes is about 3 mm, the gas pressure used in case of hydrogen will be

$$\frac{7.5}{0.3} = 25 \text{ torr}$$

and in the case of air it will be $\dfrac{4.5}{0.3} = 15$ torr.

Example 2.3 : What will the breakdown strength of air be for small gaps (1 mm) and large gaps (20 cm) under uniform field conditions and standard atmospheric conditions?

Solution : The breakdown strength of air under uniform field conditions and standard atmospheric conditions is approximately given by

$$E = \frac{V}{d} = \left(24.22 + \frac{6.08}{d^{1/2}}\right) kV/cm$$

Substituting for 1 mm gap,

$$E = 24.22 + \frac{6.08}{(0.1)^{1/2}} = 43.45 \; kV/cm$$

for 20 cm gap,

$$E = 24.22 + \frac{6.08}{(20.1)^{1/2}} = 25.58 \; kV/cm$$

Example 2.4 : In an experiment in a certain gas it was found that the steady state current is 5.5×10^{-8} A at 8 kV at a distance of 0.4 cm between the plane electrodes. Keeping the field constant and reducing the distance to 0.1 cm results in a current of 5.5×10^{-9}A. Calculate Townsend's primary ionization coefficient α.

Solution: The current at the anode I is given by

$$I = I_0 \exp (\alpha \, d)$$

where I_0 is the initial current and d is the gap distance.
Given,

$$d_1 = 0.4 \; cm \qquad d_2 = 0.1 \; cm$$
$$I_1 = 5.5 \times 10^{-8}A \quad I_2 = 5.5 \times 10^{-9}A$$
$$\frac{I_1}{I_2} = \exp \alpha(d_1 - d_2)$$

i.e., $$10 = \exp (\alpha \times 0.3)$$
i.e., $$0.3\alpha = \ln (10)$$
$\therefore$ $$\alpha = 7.676/cm \,.\, torr$$

REFERENCES

1. Meek, J.M. and Craggs, J.D., *Electrical Breakdown of Gases*, John Wiley, New York (1978).
2. Llewellyn Jones, F., *Ionization and Breakdown in Gases*, Methuen, London (1957).
3. Cobine, J.D., *Gaseous Conductors*, Dover Publications, New York (1958).
4. Raether, H., *Electron Avalanches and Breakdown in Gases*, Butterworth, London (1964).
5. Naidu, M.S. and Maller, V.N., *Advances in High Voltage Breakdown and Arc Interruption in SF₆ and Vacuum*, Pergamon Press, Oxford (1981).
6. Nasser, E., *Fundamentals of Gaseous Ionization and Plasma Electronics*, John Wiley, New York (1974).

7. Alston, L.L., *High Voltage Technology*, Oxford University Press, Oxford (1968).
8. Kuffel, E. and Abdullah, M., *High Voltage Engineering*, Pergamon Press, Oxford (1970).
9. Hawley, R. *High Voltage Technology—Chapter on Vacuum Breakdown*, (edt. L.L. Alston), Oxford University Press, Oxford, p. 58 (1969).
10. Hawley, R. and Zaky, A.A. *Progress in Dielectrics* (edt. J.B. Birks), vol. 7, Haywood, London, p. 115 (1967).

3

Conduction and Breakdown in Liquid Dielectrics

3.1 LIQUIDS AS INSULATORS

Liquid dielectrics, because of their inherent properties, appear as though they would be more useful as insulating materials than either solids or gases. This is because both liquids and solids are usually 10^3 times denser than gases and hence, from Paschen's law it should follow that they possess much higher dielectric strength of the order of 10^7 V/cm. Also, liquids, like gases, fill the complete volume to be insulated and simultaneously will dissipate heat by convection. Oil is about 10 times more efficient than air or nitrogen in its heat transfer capability when used in transformers. Although liquids are expected to give very high dielectric strength of the order of 10 MV/cm, in actual practice the strengths obtained are only of the order of 100 kV/cm.

Liquid dielectrics are used mainly as impregnants in high voltage cables and capacitors, and for filling up of transformers, circuit breakers etc. Liquid dielectrics also act as heat transfer agents in transformers and as arc quenching media in circuit breakers. Petroleum oils (Transformer oil) are the most commonly used liquid dielectrics. Synthetic hydrocarbons and halogenated hydrocarbons are also used for certain applications. For very high temperature application, silicone oils and fluorinated hydrocarbons are also employed. In recent times, certain vegetable oils and esters are also being tried. However, it may be mentioned that some of the isomers of poly-chlorinated diphenyls (generally called askerels) have been found to be very toxic and poisonous, and hence, their use has been almost stopped. In recent years, a synthetic ester fluid with the trade name 'Midel' has been developed as a replacement for askerels.

Liquid dielectrics normally are mixtures of hydrocarbons and are weakly polarised. When used for electrical insulation purposes they should be free from moisture, products of oxidation and other contaminants. The most important factor that affects the electrical strength of an insulating oil is the presence of water in the form of fine droplets suspended in the oil. The presence of even 0.01% water in transformer oil reduces its electrical strength to 20% of the dry oil value. The dielectric strength of oil reduces more sharply, if it contains fibrous impurities in addition to water. Table 3.1 shows the properties of some dielectrics commonly used in electrical equipment.

Table 3.1 Dielectric Properties of Some Liquid Dielectrics

Property	Transformer oil	Cable oil	Capacitor oil	Askerels	Silicone oils
Breakdown strength at 20°C on 2.5 mm standard sphere gap	15 kV/mm	30 kV/mm	20 kV/mm	20-25 kV/mm	30-40 kV/mm
Relative permittivity (50 Hz)	2.2-2.3	2.3-2.6	2.1	4.8	2-73
Tan δ (50 Hz)	0.001	0.002	0.25×10^{-3}	0.60×10^{-3}	10^{-3}
(1 kHz)	0.0005	0.0001	0.10×10^{-3}	0.50×10^{-3}	10^{-4}
Resistivity (ohm-cm)	$10^{12}\text{-}10^{13}$	$10^{12}\text{-}10^{13}$	$10^{13}\text{-}10^{14}$	2×10^{12}	3×10^{14}
Specific gravity at 20°C	0.89	0.93	0.88-0.89	1.4	1.0-1.1
Viscosity at 20°C (CS)	30	30	30	100-150	10-1000
Acid value (mg/gm of KOH)	Nil	Nil	Nil	Nil	Nil
Refractive index	1.4820	1.4700	1.4740	1.6000	1.5000-1.6000
Saponification (mg of KOH/gm of oil)	0.01	0.01	0.01	<0.01	<0.01
Expansion (20 – 100°C)	$7 \times 10^{-4}/°C$	$7 \times 10^{-4}/°C$	$7 \times 10^{-4}/°C$	$7 \times 10^{-1}/°C$	$5 \times 10^{-4}/°C$
Maximum permissible water content (in ppm)	50	50	50	<30 (negligible)	<30 (negligible)

Of the insulating liquids shown in table 3.1, transformer oils are the cheapest and the most commonly used. The electrical properties of transformer oil are given in the above table. Oils used in the capacitors are similar to transformer oil but they are subjected to a very high degree of purification. Various kinds of oils are used in cables as impregnants for paper insulation and to improve their heat transfer capability. Table 3.1 gives the dielectric properties of various liquid dielectrics used in cables, capacitors and in other special applications.

In practice, the choice of a liquid dielectric for a given application is made mainly on the basis of its chemical stability. Other factors such as saving of space, cost, previous usage, and susceptibility to the environmental influences are also considered.

3.1.1 Transformer Oil

As already mentioned, transformer oil is the most commonly used liquid dielectric in power apparatus. It is an almost colourless liquid consisting a mixture of hydrocarbons which include paraffins, iso-paraffins, naphthalenes and aromatics. When in service, the liquid in a transformer is subjected to prolonged heating at high temperatures of about 95°C, and consequently it undergoes a gradual ageing process. With time the oil becomes darker due to the formation of acids and resins, or sludge in the liquid. Some of the acids are corrosive to the solid insulating materials and metal parts in the transformer. Deposits of sludge on the transformer core, on the coils and inside the oil ducts reduce circulation of oil and thus its heat transfer capability gets considerably reduced. Complete specifications for the testing of transformer oils are given in IS 1866 (1983), IEC 296 (1969) and IEC 474 (1974).

3.1.2 Electrical Properties

The electrical properties that are essential in determining the dielectric performance of a liquid dielectric are
- (a) its capacitance per unit volume or its relative permittivity
- (b) its resistivity
- (c) its loss tangent (tan δ) or its power factor which is an indication of the power loss under a.c. voltage application
- (d) its ability to withstand high electric stresses.

Permittivities of most of the petroleum oils vary from 2.0 to 2.6 while those of askerels vary between 4.5 and 5.0 and those of silicone oils from 2.0 to 73 (see Table 3.1). In case of the non-polar liquids, the permittivity is independent of frequency but in the case of polar liquids, such as water, it changes with frequency. For example, the permittivity of water is 78 at 50 Hz and reduces to about 5.0 at 1 MHz.

Resistivities of insulating liquids used for high voltage applications should be more than 10^{16} ohm-metre and most of the liquids in their pure state exhibit this property.

Power Factor of a liquid dielectric under a.c. voltage will determine its performance under load conditions. Power factor is a measure of the power loss and is an important parameter in cable and capacitor systems. However, in the case of transformers, the dielectric loss in the oil is negligible when compared to copper and iron losses. Pure and dry transformer oil will have a very low power factor varying between 10^{-4} at 20°C and 10^{-3} at 90°C at a frequency of 50 Hz.

Dielectric Strength is the most important parameter in the choice of a given liquid dielectric for a given application. The dielectric strength depends on the atomic and molecular properties of the liquid itself. However, under practical conditions the dielectric strength depends on the material of the electrodes, temperature, type of applied voltage, gas content in the liquid etc., which change the dielectric strength by changing the molecular properties of the liquid. The above factors which control the breakdown strength and lead to electrical breakdown of the liquid dielectrics are discussed in subsequent sections.

3.2 PURE LIQUIDS AND COMMERCIAL LIQUIDS

Pure liquids are those which are chemically pure and do not contain any other impurity even in traces of 1 in 10^9, and are structurally simple. Examples of such simple pure liquids are n-hexane (C_6H_{14}), n-heptane (C_7H_{16}) and other paraffin hydrocarbons. By using simple and pure liquids, it is easier to separate out the various factors that influence conduction and breakdown in them. On the other hand, the commercial liquids which are insulating liquids like oils which are not chemically pure, normally consist of mixtures of complex organic molecules which cannot be easily specified or reproduced in a series of experiments.

3.2.1 Purification

The main impurities in liquid dielectrics are dust, moisture, dissolved gases and ionic impurities. Various methods employed for purification are filtration (through mechanical filters, spray filters, and electrostatic filters), centrifuging, degassing and distillation, and chemical treatment (adding ion exchange materials such as alumina, fuller's earth, etc. and filtering). Dust particles when present become charged and reduce the breakdown strength of the liquid dielectrics, and they can be removed by careful filtration. Liquid will normally contain moisture and dissolved gases in small quantities. Gases like oxygen and carbon dioxide significantly affect the breakdown strength of the liquids, and hence it is necessary to control the amount of gas present. This is done by distillation and degassing. Ionic impurity in liquids, like water vapour which easily dissociates, leads to very high conductivity and heating of the liquid depending on the applied electric field. Water is removed using drying agents or by vacuum drying. Sometimes, liquids are shaken with concentrated sulphuric acid to remove wax and residue and washed with caustic soda and distilled water. A commonly used closed-cycle liquid purification system to prepare liquids as per the above requirements is shown in Fig. 3.1. This system provides for cycling the liquid. The liquid from the reservoir flows through the distillation column where ionic impurities are removed. Water is removed by drying agents or frozen out in the low-temperature bath. The gases dissolved in the liquid are removed by passing them through the cooling tower and/or pumped out by the vacuum pumps. The liquid then passes through the filter where dust particles are removed. The liquid thus purified is then used in the test cell. The used liquid then flows back into the reservoir. The vacuum system thus helps to remove the moisture and other gaseous impurities.

3.2.2 Breakdown Tests

Breakdown tests are normally conducted using test cells. For testing pure liquids, the test cells used are small so that less quantity of liquid is used during testing. Also, test cells are usually an integral part of the purification system as shown in Fig. 3.1. The electrodes used for breakdown voltage measurements are usually spheres of 0.5 to 1 cm in diameter with gap spacings of about 100-200 μm. The gap is accurately controlled by using a micrometer. Sometimes parallel plane uniform-field electrode systems are also used. Electrode separation is very critical in measurements with liquids, and also the electrode surface smoothness and the presence of oxide films have a marked influence on the breakdown strength. The test voltages required for these tests are usually low, of the order of 50-100 kV, because of small electrode spacings. The breakdown strengths and d.c. conductivities obtained in pure liquids are very high, of the order of 1 MV/cm and 10^{-18} - 10^{-20} mho/cm respectively, the conductivity being measured at electric fields of the order of 1 kV/cm. However, the corresponding values in commercial liquids are relatively low, as can be seen from Table 3.1.

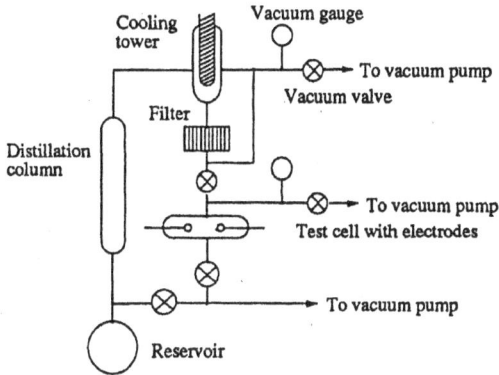

Fig. 3.1 Liquid purification system with test cell

3.3 CONDUCTION AND BREAKDOWN IN PURE LIQUIDS

When low electric fields less than 1 kV/cm are applied, conductivities of 10^{-18} - 10^{-20} mho/cm are obtained. These are probably due to the impurities remaining after purification. However, when the fields are high (> 100 kV/cm) the currents not only increase rapidly, but also undergo violent fluctuations which will die down after some time. A typical mean value of the conduction current in hexane is shown in Fig. 3.2.

This is the condition nearer to break-down. However, if this figure is redrawn starting from very small currents, a current-electric field characteristic as shown in Fig. 3.3, can be obtained. This curve will have three distinct regions as shown. At very low fields the current is due to the dissociation of ions. With intermediate fields ithe current reaches a saturation value, and at high fields the current generated because of the field-aided electron emission from the cathode gets multiplied in the liquid medium by a Townsend type of mechanism (see Chapter 2). The current multiplication also occurs from the electrons generated at the interfaces of liquid and impurities. The increase in current by these processes continues till breakdown occurs.

Fig. 3.2 Conduction current-electric field characteristic in hexane at high fields

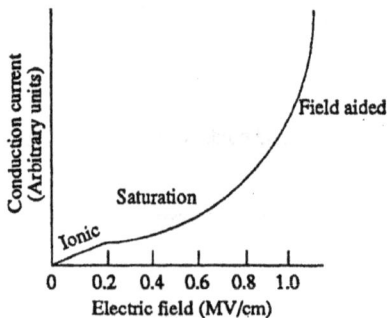

The exact mechanism of current growth is not known; however, it appears that the electrons are generated from the cathode by field emission of electrons. The electrons so liberated get multiplied by a process similar to Townsend's primary and secondary ionization in gases (see Chapter 2, Secs. 2.2, 2.3, 2.4). As the breakdown field is approached, the current increases rapidly due to a process similar to the primary ionization process and also the positive ions reaching the cathode generate secondary electrons, leading to breakdown. The breakdown voltage depends on the field, gap separation, cathode work-function, and the temperature of the cathode. In addition, the liquid viscosity, the liquid temperature, the density, and the molecular structure of the liquid also influence the breakdown strength of the liquid. Typical maximum

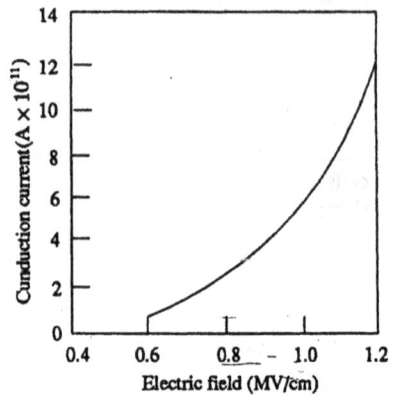

Fig. 3.3 Conduction current-electric field characteristic in a hydrocarbon liquid

breakdown strengths of some highly purified liquids and liquefied gases are given in
Table 3.2.

Table 3.2 Maximum Breakdown Strengths of Some Liquids

Liquid	Maximum breakdown strength (MV/cm)
Hexane	1.1-1.3
Benzene	1.1
Transformer oil	1.0
Silicone	1.0-1.2
Liquid Oxygen	2.4
Liquid Nitrogen	1.6-1.9
Liquid Hydrogen	1.0
Liquid Helium	0.7
Liquid Argon	1.10-1.42

It has been observed that the increase in breakdown strength is more, if the
dissolved gases are electronegative in character (like oxygen).
Similarly, the increase in the liquid hydrostatic pressure increases the breakdown
strength. These properties are shown in Figs. 3.4 and 3.5.

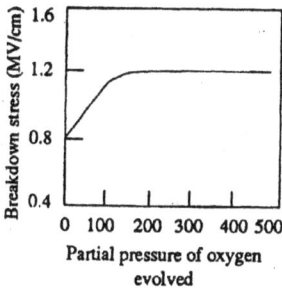

Fig. 3.4 Effect of oxygen gas evolved on
the breakdown stress in *n*-hex-
ane.

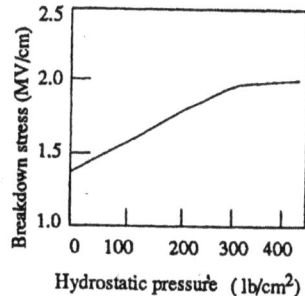

Fig. 3.5 Effect of hydrostatic
pressure on breakdown
stress in *n*-hexane

To sum up, this type of breakdown process in pure liquids, called the electronic
breakdown, involves emission of electrons at fields greater than 100 kV/cm. This
emission occurs either at the electrode surface irregularities or at the interfaces of
impurities and the liquid. These electrons get further multiplied by Townsend's type
of primary and secondary ionisation processes, leading to breakdown.

3.4 CONDUCTION AND BREAKDOWN IN COMMERCIAL LIQUIDS

As already mentioned, commercial insulating liquids are not chemically pure and have impurities like gas bubbles, suspended particles, etc. These impurities reduce the breakdown strength of these liquids considerably. The breakdown mechanisms are also considerably influenced by the presence of these impurities. In addition, when breakdown occurs in these liquids, additional gases and gas bubbles are evolved and solid decomposition products are formed. The electrode surfaces become rough, and at times explosive sounds are heard due to the generation of impulsive pressure through the liquid.

The breakdown mechanism in commercial liquids is dependent, as seen above, on several factors, such as, the nature and condition of the electrodes, the physical properties of the liquid, and the impurities and gases present in the liquid. Several theories have been proposed to explain the breakdown in liquids, and they are classified as follows:

(a) Suspended Particle Mechanism
(b) Cavitation and Bubble Mechanism
(c) Stressed Oil Volume Mechanism
These are explained briefly below.

3.4.1 Suspended Particle Theory

In commercial liquids, the presence of solid impurities cannot be avoided. These impurities will be present as fibres or as dispersed solid particles. The permittivity of these particles (ε_2) will be different from the permittivity of the liquid (ε_1). If we consider these impurities to be spherical particles of radius r, and if the applied field is E, then the particles experience a force F, where

$$F = \frac{1}{2r^3} \frac{(\varepsilon_2 - \varepsilon_1)}{2\varepsilon_1 + \varepsilon_2} \text{grad } E^2 \tag{3.1}$$

This force is directed towards areas of maximum stress, if $\varepsilon_2 > \varepsilon_1$, for example, in the case of the presence of solid particles like paper in the liquid. On the other hand, if only gas bubbles are present in the liquid. i.e. $\varepsilon_2 < \varepsilon_1$, the force will be in the direction of areas of lower stress. If the voltage is continuously applied (d.c.) or the duration of the voltage is long (a.c.), then this force drives the particles towards the areas of maximum stress. If the number of particles present are large, they becomes aligned due to these forces, and thus form a stable chain bridging the electrode gap causing a breakdown between the electrodes.

If there is only a single conducting particle between the electrodes, it will give rise to local field enhancement depending on its shape. If this field exceeds the breakdown strength of the liquid, local breakdown will occur near the particle, and this will result in the formation of gas bubbles which may lead to the breakdown of the liquid.

The vales of the breakdown strength of liquids containing solid impurities was found to be much less than the vlaues for pure liquids. The impurity particles reduce the breakdown strength, and it was also observed that the larger the size of the particles the lower were the breakdown strengths.

3.4.2 Cavitation and the Bubble Theory

It was experimentally observed that in many liquids, the breakdown strength depends strongly on the applied hydrostatic pressure, suggesting that a change of phase of the medium is involved in the breakdown process, which in other words means that a kind of vapour bubble formed is responsible for breakdown. The following processes have been suggested to be responsible for the formation of the vapour bubbles:

- (a) Gas pockets at the surfaces of the electrodes;
- (b) electrostatic repulsive forces between space charges which may be sufficient to overcome the surface tension;
- (c) gaseous products due to the dissociation of liquid molecules by electron collisions; and
- (d) vapourization of the liquid by corona type discharge from sharp points and irregularities on the electrode surfaces.

Once a bubble is formed it will elongate in the direction of the electric field under the influence of electrostatic forces. The volume of the bubble remains constant during elongation. Breakdown occurs when the voltage drop along the length of the bubble becomes equal to the minimum value on the Paschen's curve (see Chapter 2, Sec. 2.10) for the gas in the bubble. The breakdown field is given as

$$E_0 = \frac{1}{(\varepsilon_1 - \varepsilon_2)} \left[\frac{2\pi\sigma(2\varepsilon_1 + \varepsilon_2)}{r} \left\{ \frac{\pi}{4} \sqrt{\left(\frac{V_b}{2rE_0} \right)} - 1 \right\} \right]^{\frac{1}{2}} \tag{3.2}$$

where σ is the surface tension of the liquid, ε_1 is the permittivity of the liquid, ε_2 is the permittivity of the gas bubble, r is the initial radius of the bubble assumed as a sphere and V_b is the voltage drop in the bubble (corresponding to minimum on the Paschen's curve). From this equation, it can be seen that the breakdown strength depends on the initial size of the bubble which in turn is influenced by the hydrostatic pressure and temperature of the liquid.

This theory does not take into account the production of the initial bubble and hence the results given by this theory do not agree well with the experimental results. This is shown in Fig. 3.6.

Later this theory was modified, and it was suggested that only incompressible bubbles like water globules can elongate at constant volume, according to the simple gas law $pv = RT$. Under the influence of the applied electric field the shape of the globule is assumed to be approximately a prolate spheroid. The incompressible bubbles reach the condition of instability when β, the ratio of the longer to the shorter diameter of the spheriod, is about 1.85, and the critical field producing the instability will be:

$$E_c = 600 \sqrt{\frac{\pi \sigma}{\varepsilon_1 R}} \left[\frac{\varepsilon_1}{\varepsilon_1 - \varepsilon_2} - G \right] H \tag{3.3}$$

where σ = surface tension,

R = initial radius of the bubble,

ε_1 = permittivity of the liquid dielectric,

ε_2 = permittivity of the globule,

$$G = \frac{1}{\beta^2 - 1}\left[\frac{\beta \cosh^{-1}\beta}{(\beta^2 - 1)^{1/2}} - 1\right]$$

and $H^2 = 2\beta^{\frac{1}{3}}\left(2\beta - 1 - \frac{1}{\beta^2}\right)$

For a water globule having $R = 1$ μm with $\sigma = 43$ dyne/cm and $\varepsilon_1 = 2.0$ (transformer oil), the above equation gives a critical field $E_c = 226$ kV/cm which is approximately the maximum strength obtained for commercial oils.

In the case of gas bubbles the equation for the critical field is rewritten as

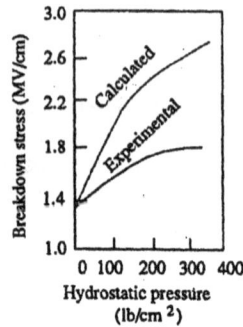

Fig. 3.6 Theoretical and experimental breakdown stresses in n-hexane

$$E_c = 600\left[\frac{\pi\sigma}{\varepsilon_1 R}\right]^{1/2}\left[\frac{\varepsilon_1}{\varepsilon_1 - \varepsilon_2} - G\right]\left[\frac{8A^2B}{3\beta(\varepsilon_1 - \varepsilon_2)}\right]^{1/4}(\cosh\theta)^{1/2} \qquad (3.4)$$

where,

$$A = \frac{2}{\beta} - 1 - \frac{1}{\beta^2}$$

$$B = 2\varepsilon_1\beta^3 - \varepsilon_2(1 - \beta^2)$$

G, σ and R are as above for liquid globules, and

$$\theta = \frac{1}{3}\cosh^{-1}\left[\frac{PR}{\sigma}\left\{\frac{27\beta^5(\varepsilon_1 - \varepsilon_2)^3}{2B^3}\right\}^{1/2}\right]$$

where P is the hydrostatic pressure (Equations 3.2-3.4 are in c.g.s. units). The expressions are quite complicated, and the breakdown voltages were obtained using a computer. Results thus obtained showed good agreement with the experimental results in n-hexane. This theory suggests that sub-microscopic particles (diameter 100-250 Å) and bubbles greatly influence the maximum electrical strength attainable in commercial liquids.

This theory can also be extended to the study of pure liquids. The critical condition is reached when cavities are formed due to zero pressure conditions given by

$$P_c = P_{vp} = P_{es} + P_s + P_h, \qquad (3.5)$$

where, P_c = coulombic pressure,

P_{vp} = vapour pressure inside the cavity,

P_{es} = electrostatic pressure,

P_s = pressure due to surface tension, and

P_h = hydrostatic pressure

From this condition, an expression has been obtained for the maximum breakdown strength of pure liquids which was found to be in good agreement with the experimental results.

In general, the cavitation and bubble theories try to explain the highest breakdown strengths obtainable, considering the cavities or bubbles formed in the liquid dielectrics.

3.4.3 Thermal Mechanism of Breakdown

Another mechanism proposed to explain breakdown under pulse conditions is thermal breakdown. This mechanism is based on the experimental observations of extremely large currents just before breakdown. This high current pulses are believed to originate from the tips of the microscopic projections on the cathode surface with densities of the order of 1 A/cm^3. These high density current pulses give rise to localised heating of the oil which may lead to the formation of vapour bubbles. The vapour bubbles are formed when the energy exceeds 10^7 W/cm^2. When a bubble is formed, breakdown follows, either because of its elongation to a critical size or when it completely bridges the gap between the electrodes. In either case, it will result in the formation of a spark. According to this mechanism, the breakdown strength depends on the pressure and the molecular structure of the liquid. For example, in n-alkanes the breakdown strength was observed to depend on the chain length of the molecule. This theory is only applicable at very small lengths ($\leq 100 \mu$ m) and does not explain the reduction in breakdown strength with increased gap lengths.

3.4.4 Stressed Oil Volume Theory

In commercial liquids where minute traces of impurities are present, the breakdown strength is determined by the "largest possible impurity" or "weak link". On a statistical basis it was proposed that the electrical breakdown strength of the oil is defined by the weakest region in the oil, namely, the region which is stressed to the maximum and by the volume of oil included in that region. In non-uniform fields, the stressed oil volume is taken as the volume which is contained between the maximum stress (E_{max}) contour and 0.9 E_{max} contour. According to this theory the breakdown strength is inversely proportional to the stressed oil volume.

The breakdown voltage is highly influenced by the gas content in the oil, the viscosity of the oil, and the presence of other impurities. These being uniformly distributed, increase in the stressed oil volume consequently results in a reduction in the breakdown voltage. The variation of the breakdown voltage stress with the stressed oil volume is shown in Fig. 3.7.

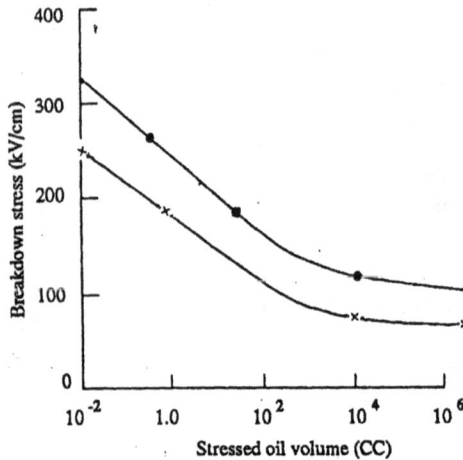

Fig. 3.7 Power frequency (50 Hz) a.c. breakdown stress as a
function of the stressed oil volume
- With steady voltage rise
× One minute withstand voltage

3.4.5 Conclusions

All the theories discussed above do not consider the dependence of breakdown
strength on the gap length. They all try to account for the maximum obtainable
breakdown strength only. However, the experimental evidence showed that the
breakdown strength of a liquid depends on the gap length, given by the following
expression,

$$V_b = Ad^n \qquad (3.6)$$

where, d = gap length,

 A = constant, and

 n = constant, always less than 1.

The breakdown voltage also depends on the nature of the voltage, the mode in
which the voltage is applied, and the time of application. The above relationship is of
practical importance, and the electrical stress of a given oil used in design is obtained
from this. During the last ten years, research work is directed on the measurements
of discharge inception levels in oil and the breakdown strengths of large volumes of
oil under different conditions.

It may be summarized that the actual mechanism of breakdown in oil is not a
simple phenomenon and the breakdown voltages are determined by experimental
investigations only. Electrical stresses obtained for small volumes should not be
used in the case of large volumes. As a general guideline, the properties of good
dielectric oils for electrical purpose are tabulated (see Table 3.1), and the designer
should satisfy himself on all the properties before acceptance.

Table 3.3 gives the typical breakdown strengths for highly purified liquids and the design stresses actually used. A factor of safety of about 10 is used as can be seen from the data in the table. The reasons for such an approach can be understood from the various factors considered in the breakdown theories discussed.

Table 3.3 Power Frequency Design Field Strengths and Breakdown Field Strengths for Highly Purified Dielectric Liquids—a Comparison

Dielectric liquid	Used in (equipment)	Design field strength (MV/m)	Breakdown Strength (MV/m)
Transformer oil	Transformers	2-5	100
n-hexane	Cables	13-20	132
Polybutane (Synthetic hydrocarbon)	Capacitors	10-25	109

In the next 20 to 30 years, transformer oils, which are derivatives of petroleum crude, may be in short supply. Therefore, various synthetic insulating oils are being examined as potential insultants in high voltage apparatus. Among the synthetic oils, polybutane liquids have been used for some years in cables and paper capacitors. They are superior to transformer oils in various electrical properties including dielectric strength. Fluorocarbon and silicone liquids are also used in special apparatus. However, their use will require further evaluation of their properties. Further details can be found in refeences 7, 9 and 10 cited at the end of this chapter.

QUESTIONS

Q.3.1 Explain the phenomena of electrical conduction in liquids. How does it differ from that in gases?

Q.3.2 What are commercial liquid dielectrics, and how are they different from pure liquid dielectrics?

Q.3.3 What are the factors that influence conduction in pure liquid dielectrics and in commercial liquid dielectrics?

Q.3.4 Explain the various theories that explain breakdown in commercial liquid dielectrics.

Q.3.5 What is "stressed oil volume theory", and how does it explain breakdown in large volumes of commercial liquid dielectrics?

WORKED EXAMPLES

Example 3.1: In an experiment for determining the breakdown strength of transformer oil, the following observations were made. Determine the power law deendence between the gap spacing and the applied voltage of the oil.

Gap spacing (mm) :	4	6	10	12
Voltage at breakdown (kV) :	90	140	210	255

The relationship between the breakdown voltage and gap is normally given as

$$V = Kd^n$$

or,
$$\log V = \log K + n \log d$$

i.e.,
$$\log V - \log K = n \log d$$

or,
$$n = \frac{\log V - \log K}{\log d}$$

$$= \text{slope of the straight line as shown in Fig. E.3.1.}$$

$$= 0.947$$

From Fig. E.3.1.,

$$K = 24.5$$

∴ Relationship between the breakdown voltage and the gap spacing for the transformer oil studied is

$$V = 24.5 \, d^{0.947}$$

where V is in kV and d is in mm.

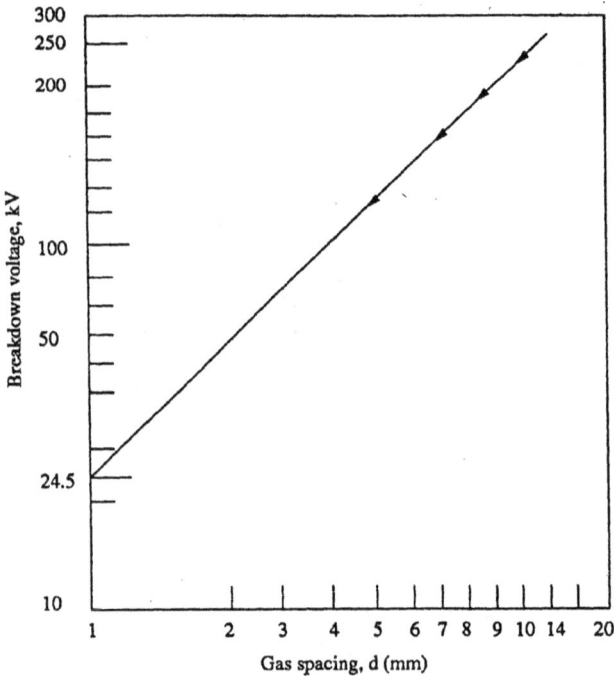

Fig. E.3.1   Breakdown voltage as a function of gap spacing

REFERENCES

1. Adam Czewski, I., *Ionization, Conduction and Breakdown in Dielectric Liquids*, Taylor and Francis, London (1969).
2. Gallager, T.J., *Simple Dielectric Liquids*, Clarendon Press, Oxford (1975).
3. Hawley, R. and Zaky, A.A., *Conduction and Breakdown in Mineral Oil*, Peter Peregrinus, London (1973).
4. Alston, L.L., *High Voltage Technology*, Oxford University Press, Oxford (1968).
5. Lewis, T.J., *Progress in Dielectrics*, Vol. 1, Heywood, London (1959), pp. 97-140.
6. Sharbough, A. and Watson, P.K., *Progress in Dielectrics*, vol. 4, Heywood, London (1962), pp. 199-248.
7. Code of practice for maintenance and supervision of insulating oil in equipment, *IS: 1866-1983*; also *IS: 6262-1971* (for tan δ measurement) and *IS: 6792-1972* (for measurement of electrical strength).
8. "Specifications for new insulating oils for transformers and switchgear", *IEC No. 269*, 1969.
9. "Oxidation tests for inhibited oils", *IEC No. 474*, 1974.
10. Wilson, A.C.M., *Insulating liquids: Their uses, manufacture and properties*, Peter Peregrinus and IEE, London (1980).
11. "IEE Colloquim on New Dielectric Fluids for Power Engineering" IEE, London (1980).

4

Breakdown in Solid Dielectrics

4.1 INTRODUCTION

Solid dielectric materials are used in all kinds of electrical circuits and devices to insulate one current carrying part from another when they operate at different voltages. A good dielectric should have low dielectric loss, high mechanical strength, should be free from gaseous inclusions, and moisture, and be resistant to thermal and chemical deterioration. Solid dielectrics have higher breakdown strength compared to liquids and gases.

Studies of the breakdown of solid dielectrics are of extreme importance in insulation studies. When breakdown occurs, solids get permanently damaged while gases fully and liquids partly recover their dielectric strength after the applied electric field is removed.

The mechanism of breakdown is a complex phenomena in the case of solids, and varies depending on the time of application of voltage as shown in Fig. 4.1. The various breakdown mechanisms can be classified as follows:

(a) intrinsic or ionic breakdown,
(b) electromechanical breakdown,
(c) failure due to treeing and tracking,
(d) thermal breakdown,
(e) electrochemical breakdown, and
(f) breakdown due to internal discharges.

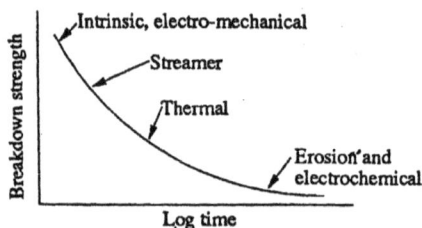

Fig. 4.1 Variation of breakdown strength with time after application of voltage

4.2 INTRINSIC BREAKDOWN

When voltages are applied only for short durations of the order of 10^{-8}s the dielectric strength of a solid dielectric increases very rapidly to an upper limit called the intrinsic electric strength. Experimentally, this highest dielectric strength can be obtained only under the best experimental conditions when all extraneous influences have been isolated and the value depends only on the structure of the material and the temperature. The maximum electrical strength recorded is 15 MV/cm for polyvinyl-alcohol at $-196°C$. The maximum strength usually obtainable ranges from 5 MV/cm to 10 MV/cm.

Intrinsic breakdown depends upon the presence of free electrons which are capable of migration through the lattice of the dielectric. Usually, a small number of conduction electrons are present in solid dielectrics, along with some structural imperfections and small amounts of impurities. The impurity atoms, or molecules or both act as traps for the conduction electrons up to certain ranges of electric fields and temperatures. When these ranges are exceeded, additional electrons in addition to trapped electrons are released, and these electrons participate in the conduction process. Based on this principle, two types of intrinsic breakdown mechanisms have been proposed.

4.2.1 Electronic Breakdown

As mentioned earlier, intrinsic breakdown occurs in time of the order of 10^{-8}s and therefore is assumed to be electronic in nature. The initial density of conduction (free) electrons is also assumed to be large, and electron-electron collisions occur. When an electric field is applied, electrons gain energy from the electric field and cross the forbidden energy gap from the valency to the conduction band. When this process is repeated, more and more electrons become available in the conduction band, eventually leading to breakdown.

4.2.2 Avalanche or Streamer Breakdown

This is similar to breakdown in gases due to cumulative ionization. Conduction electrons gain sufficient energy above a certain critical electric field and cause liberation of electrons from the lattice atoms by collisions. Under uniform field conditions, if the electrodes are embedded in the specimen, breakdown will occur when an electron avalanche bridges the electrode gap.

An electron within the dielectric, starting from the cathode will drift towards the anode and during this motion gains energy from the field and loses it during collisions. When the energy gained by an electron exceeds the lattice ionization potential, an additional electron will be liberated due to collision of the first electron. This process repeats itself resulting in the formation of an electron avalanche. Breakdown will occur, when the avalanche exceeds a certain critical size.

In practice, breakdown does not occur by the formation of a single avalanche itself, but occurs as a result of many avalanches formed within the dielectric and extending step by step through the entire thickness of the material as shown in Fig. 4.2. This can be readily demonstrated in a laboratory by applying an impulse voltage between point-plane electrodes with point embedded in a transparent solid dielectric such as perspex.

4.3 ELECTROMECHANICAL BREAKDOWN

When solid dielectrics are subjected to high electric fields, failure occurs due to electrostatic compressive forces which can exceed the mechanical compressive strength. If the thickness of the specimen is d_0 and is compressed to a thickness d under an applied voltage V, then the electrically developed compressive stress is in equilibrium if

$$\varepsilon_0\, \varepsilon_r \frac{V^2}{2d^2} = Y \ln\left[\frac{d_0}{d}\right] \quad (4.1)$$

where Y is the Young's modulus.
From Eq. (4.1)

$$V^2 = d^2 \left[\frac{2Y}{\varepsilon_0\,\varepsilon_r}\right] \ln\left[\frac{d_0}{d}\right] \quad (4.2)$$

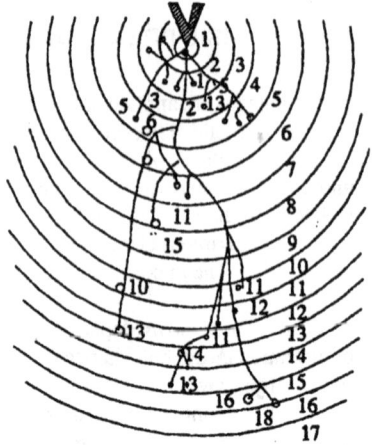

Fig. 4.2 Breakdown channels in perspex between point-plane electrodes. Radius of point 0.01 in, thickness 0.19 in. Total number of impulses 190. Number of channels produced 16; (n) point indicates end of nth channel. Radii of circles increases in units of 10^{-2} in.

Source: R. Cooper, *International Journal of Elec. Engg. Education.* vol. 1, 241 (1963)

Usually, mechanical instability occurs when

$$d/d_0 = 0.6 \text{ or } d_0/d = 1.67$$

Substituting this in Eq. 4.2, the highest apparent electric stress before breakdown,

$$E_{max} = \frac{V}{d_0} = 0.6 \left[\frac{Y}{\varepsilon_0\,\varepsilon_r}\right]^{\frac{1}{2}} \quad (4.3)$$

The above equation is only approximate as Y depends on the mechanical stress. Also when the material is subjected to high stresses the theory of elasticity does not hold good, and plastic deformation has to be considered.

4.4 THERMAL BREAKDOWN

In general, the breakdown voltage of a solid dielectric should increase with its thickness. But this is true only up to a certain thickness above which the heat generated in the dielectric due to the flow of current determines the conduction.

When an electric field is applied to a dielectric, conduction current, however small it may be, flows through the material. The current heats up the specimen and the temperature rises. The heat generated is transferred to the surrounding medium by conduction through the solid dielectric and by radiation from its outer surfaces. Equilibrium is reached when the heat used to raise the temperature of the dielectric, plus the heat radiated out, equals the heat generated. The heat generated under d.c. stress E is given as

$$W_{d.c.} = E^2\sigma \qquad \text{W/cm}^3 \qquad (4.4)$$

where, σ is the d.c. conductivity of the specimen.
Under a.c. fields, the heat generated

$$W_{a.c.} = \frac{E^2 f \varepsilon_r \tan \delta}{1.8 \times 10^{12}} \qquad \text{W/cm}^3 \qquad (4.5)$$

where, $f =$ frequency in Hz,

$\delta =$ loss angle of the dielectric material, and

$E =$ rms value.

The heat dissipated (W_T) is given by

$$W_T = C_V \frac{dT}{dt} + \text{div} (K \text{ grad } T) \qquad (4.6)$$

where, $C_V =$ specific heat of the specimen,

$T =$ temperature of the specimen,

$K =$ thermal conductivity of the specimen, and

$t =$ time over which the heat is dissipated.

Equilibrium is reached when the heat generated ($W_{d.c.}$ or $W_{a.c.}$) becomes equal to the heat dissipated (W_T). In actual practice there is always some heat that is radiated out.

Breakdown occurs when $W_{d.c.}$ or $W_{a.c.}$ exceeds W_T. The thermal instability condition is shown in Fig. 4.3. Here, the heat lost is shown by a straight line, while the heat generated at fields E_1 and E_2 are shown by separate curves. At field E_2 breakdown occurs both at temperatures T_A and T_B. In the temperature region of T_A and T_B heat generated is less than the heat lost for the field E_2, and hence the breakdown will not occur.

This is of great importance to practising engineers, as most of the insulation failures in high voltage power apparatus occur due to thermal breakdown. Thermal breakdown sets up an upper limit for increasing the breakdown voltage when the thickness of the insulation is increased. For a given loss angle and applied stress, the heat generated is proportional to the frequency and hence thermal breakdown is more serious at high frequencies. Table 4.1 gives the thermal breakdown voltages of various materials under d.c. and a.c. fields.

It can be seen from this table that since the power loss under a.c. fields is higher, the heat generation is also high, and hence the thermal breakdown stresses are lower under a.c. conditions than under d.c. conditions.

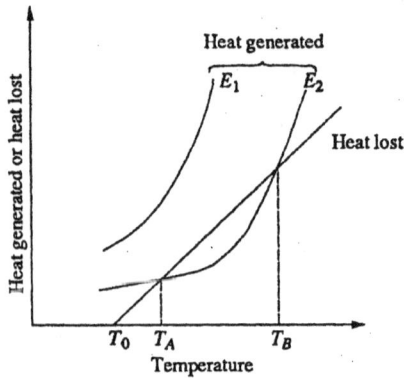

Fig. 4.3 Thermal instability in solid dielectrics

Table 4.1 Thermal Breakdown Stresses in Dielectrics

Material	Maximum thermal breakdown stress in MV/cm	
	d.c.	a.c.
Muscovite mica	24	7.18
Rock salt	38	1.4
High grade porcelain	—	2.8
H.V. Steatite	—	9.8
Quartz—perpendicular to axis	1200	—
—parallel to axis	66	—
Capacitor paper	—	3.4–4.4
Polythene	—	3.5
Polystryene	—	5.0

4.5 BREAKDOWN OF SOLID DIELECTRICS IN PRACTICE

There are certain types of breakdown which do not come under either intrinsic breakdown or thermal breakdown, but actually occur after prolonged operation. These are, for example, breakdown due to tracking in which dry conducting tracks are formed on the surface of the insulation. These tracks act as conducting paths on the insulator surfaces leading to gradual breakdown along the surface of the insulator. Another type of breakdown in this category is the electrochemical breakdown caused by chemical transformations such as electrolysis, formation of ozone, etc. In addition, failure also occurs due to partial discharges which are brought about in the air pockets inside the insulation. This type of breakdown is very important in the impregnated paper insulation used in high voltage cables and capacitors.

4.5.1 Chemical and Electrochemical Deterioration and Breakdown

In the presence of air and other gases some dielectric materials undergo chemical changes when subjected to continuous electrical stresses. Some of the important chemical reactions that occur are:

Oxidation: In the presence of air or oxygen, materials such as rubber and polyethylene undergo oxidation giving rise to surface cracks.

Hydrolysis: When moisture or water vapour is present on the surface of a solid dielectric, hydrolysis occurs and the materials lose their electrical and mechanical properties. Electrical properties of materials such as paper, cotton tape, and other cellulose materials deteriorate very rapidly due to hydrolysis. Plastics like polyethylene undergo changes, and their service life considerably reduces.

Chemical Action: Even in the absence of electric fields, progressive chemical degradation of insulating materials can occur due to a variety of processes such as chemical instability at high temperatures, oxidation and cracking in the presence of air and ozone, and hydrolysis due to moisture and heat. Since different insulating materials come into contact with each other in any practical apparatus, chemical reactions occur between these various materials leading to reduction in electrical and mechanical strengths resulting in failure.

The effects of electrochemical and chemical deterioration could be minimized by carefully studying and examining the materials. High soda content glass insulation should be avoided in moist and damp conditions, because sodium, being very mobile, leaches to the surface giving rise to the formation of a strong alkali which will cause deterioration. It was observed that this type of material will lose its mechanical strength within 24 hrs, when it is explosed to atmospheres having 100% relative humidity at 70°C. In paper insulation, even if partial discharges are prevented completely, breakdown can occur due to chemical degradation. The chemical and electrochemical deterioration increases very rapidly with temperature, and hence high temperatures should be avoided.

4.5.2 Breakdown Due to Treeing and Tracking

When a solid dielectric subjected to electrical stresses for a long time fails, normally two kinds of visible markings are observed on the dielectric materials. They are :
 (a) the presence of a conducting path across the surface of the insulation;
 (b) a mechanism whereby leakage current passes through the conducting path finally leading to the formation of a spark. Insulation deterioration occurs as a result of these sparks.

The spreading of spark channels during *tracking*, in the form of the branches of a tree is called *treeing*.

Consider a system of a solid dielectric having a conducting film and two electrodes on its surface. In practice, the conducting film very often is formed due to moisture. On application of voltage, the film starts conducting, resulting in generation of heat, and the surface starts becoming dry. The conducting film becomes separate due to drying, and so sparks are drawn damaging the dielectric surface. With organic insulating materials such as paper and bakelite, the dielectric carbonizes at the region

of sparking, and the carbonized regions act as permanent conducting channels resulting in increased stress over the rest of the region. This is a cumulative process, and
insulation failure occurs when carbonized tracks bridge the distance between the
electrodes. This phenomena, called tracking is common between layers of bakelite,
paper and similar dielectrics built of laminates.

On the other hand treeing occurs due to the erosion of material at the tips of the
spark. Erosion results in the roughening of the surfaces, and hence becomes a source
of dirt and contamination. This causes increased conductivity resulting either in the
formation of a conducting path bridging the electrodes or in a mechanical failure of
the dielectric.

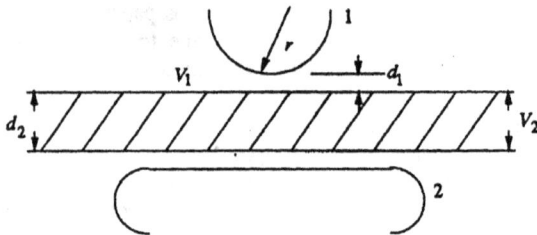

Fig. 4.4 Arrangement for study of treeing phenomena. 1 and 2
are electrodes

When a dielectric material lies between two electrodes as shown in Fig. 4.4, there
is a possibility for two different dielectric media, the air and the dielectric, to come in
series. The voltages across the two media are as shown (V_1 across the air gap, and V_2
across the dielectric). The voltage V_1 across the air gap is given as,

$$V_1 = \frac{V d_1}{d_1 + \left(\dfrac{\varepsilon_1}{\varepsilon_2}\right) d_2} \qquad (4.7)$$

where V is the applied voltage.

Since $\varepsilon_2 > \varepsilon_1$, most of the voltage appears across d_1, the air gap. Sparking will
occur in the air gap and, charge accumulation takes place on the surface of the
insulation. Sometimes the spark erodes the surface of the insulation. As time passes,
breakdown channels spread through the insulation in an irregular "tree" like fashion
leading to the formation of conducting channels. This kind of channelling is called
treeing.

Under a.c. voltage conditions treeing can occur in a few minutes or several hours.
Hence, care must be taken to see that no series air gaps or other weaker insulation
gaps are formed.

Usually, tracking occurs even at very low voltages of the order of about 100 V,
whereas treeing requires high voltage. For testing of tracking, low and medium
voltage tracking tests are specified. These tests are done at low voltages but for times
of about 100 hr or more. The insulation should not fail. Sometimes the tests are done
using 5 to 10 kV with shorter durations of 4 to 6 hr. The numerical value of voltage

that initiates or causes the formation of a track is called the "tracking index" and this is used to qualify the surface properties of dielectric materials.

Treeing can be prevented by having clean, dry, and undamaged surfaces and a clean environment. The materials chosen should be resistant to tracking. Sometimes moisture repellant greases are used. But this needs frequent cleaning and regreasing. Increasing creepage distances should prevent tracking, but in practice the presence of moisture films defeat the purpose.

Usually, treeing phenomena is observed in capacitors and cables, and extensive work is being done to investigate the real nature and causes of this phenomenon.

4.5.3 Breakdown Due to Internal Discharges

Solid insulating materials, and to a lesser extent liquid dielectrics contain voids or cavities within the medium or at the boundaries between the dielectric and the electrodes. These voids are generally filled with a medium of lower dielectric strength, and the dielectric constant of the medium in the voids is lower than that of the insulation. Hence, the electric field strength in the voids is higher than that across the dielectric. Therefore, even under normal working voltages the field in the voids may exceed their breakdown value, and breakdown may occur.

Let us consider a dielectric between two conductors as shown in Fig. 4.5a. If we divide the insulation into three parts, an electrical network of C_1, C_2 and C_3 can be formed as shown in Fig. 4.5b. In this C_1 represents the capacitance of the void or cavity, C_2 is the capacitance of the dielectric which is in series with the void, and C_3 is the capacitance of the rest of the dielectric. When the applied voltage is V, the voltage across the void, V_1 is given by the same equation as (4.7)

$$V_1 = \frac{V d_1}{d_1 + \left(\dfrac{\varepsilon_0}{\varepsilon_1}\right) d_2}$$

where d_1 and d_2 are the thickness of the void and the dielectric, respectively, having permittivities ε_0 and ε_1. Usually $d_1 \ll d_2$, and if we assume that the cavity is filled with a gas, then

$$V_1 = V \varepsilon_r \left(\frac{d_1}{d_2}\right) \tag{4.8}$$

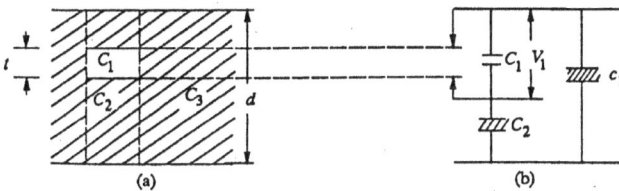

Fig. 4.5 Electrical discharge in a cavity and its equivalent circuit

where ε_r is the relative permittivity of the dielectric.

When a voltage V is applied, V_1 reaches the breakdown strength of the medium in the cavity (V_i) and breakdown occurs. V_i is called the "discharge inception voltage". When the applied voltage is a.c., breakdown occurs on both the half cycles and the number of discharges will depend on the applied voltage. The voltage and the discharge current waveforms are shown in Fig. 4.6. When the first breakdown across the cavity occurs the breakdown voltage across it becomes zero. When once the voltage V_1 becomes zero, the spark gets extinguished and again the voltage rises till breakdown occurs again. This process repeats again and again, and current pulses as shown, will be obtained both in the positive and negative half cycles.

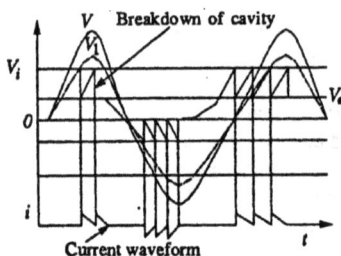

Fig. 4.6 Sequence of cavity breakdown under alternating voltages

These internal discharges (also called partial discharges) will have the same effect as "treeing" on the insulation. When the breakdown occurs in the voids, electrons and positive ions are formed. They will have sufficient energy and when they reach the void surfaces they may break the chemical bonds. Also, in each discharge there will be some heat dissipated in the cavities, and this will carbonize the surface of the voids and will cause erosion of the material. Channels and pits formed on the cavity surfaces increase the conduction. Chemical degradation may also occur as a result of the active discharge products formed during breakdown.

All these effects will result in a gradual erosion of the material and consequent reduction in the thickness of insulation leading to breakdown. The life of the insulation with internal discharges depends upon the applied voltage and the number of discharges. Breakdown by this process may occur in a few days or may take a few years.

4.6 BREAKDOWN IN COMPOSITE DIELECTRICS

4.6.1 Introduction

It is difficult to imagine a complete insulation system in an electrical equipment which does not consist of more than one type of insulation. If an insulation system as a whole is considered, it will be found that more than one insulating material is used. These different materials can be in parallel with each other, such as air or SF_6 gas in parallel with solid insulation or in series with one another. Such insulation systems are called composite dielectrics.

The composite nature of an insulation system arises from the mechanical requirements involved in separating electrical conductors which are at different potentials. Also, parts of a single system that are normally composed of a single material are in fact composite in nature. In actual practice, these single materials will normally have small volumes of another material present in their bulk. For example, a solid will contain gas pockets or voids, while a liquid or gas will contain metallic or dust particles, gas bubbles etc. A commonly encountered composite dielectric is the solid/liquid combination or liquid impregnated flexible solid like thin sheets of paper or plastic. This type of composite dielectric is widely used in a variety of low and high voltage apparatus such as cables, capacitors, transformers, oil-filled switchgear, bushings etc. In recent years solid/SF_6 gas technology has become more acceptable.

All the desirable properties of composite dielectrics cannot be realised to the fullest extent owing to the presence of impurities in them. For example, in a solid-liquid system, the presence of gas-bubbles in the liquid phase and cavities in the solid phase will give rise to a number of processes, both physical and chemical, which will reduce the dielectric strength of the system.

In the practical system, in order to reduce the undesirable effects mentioned above, composite insulation is used by combining different dielectrics either in series or in parallel such that it is possible to obtain superior dielectric properties than that possible for a single material of the same thickness.

4.6.2 Properties of Composite Dielectrics

A composite dielectric generally consists of a large number of layers arranged one over the other. This is called "the layered construction" and is widely used in cables, capacitors and transformers. Three properties of composite dielectrics which are important to their performance are given below.

(a) Effect of Multiple Layers

The simplest composite dielectric consists of two layers of the same material. Here, advantage is taken of the fact that two thin sheets have a higher dielectric strength than a single sheet of the same total thickness. The advantage is particularly significant in the case of materials having a wide variation in dielectric strength values measured at different points on its surface.

(b) Effect of Layer Thickness

Increase in layer thickness normally gives increased breakdown voltage. In a layered construction, breakdown channels occur at the interfaces only and not directly through another layer. Also, a discharge having penetrated one layer cannot enter the next layer until a part of the interface also attains the potential which can produce an electric field stress comparable to that of the discharge channel.

The use of layered construction is very important in the case of insulating paper since the paper thickness itself varies from point to point and consequently the dielectric strength across its surface is not homogeneous. The differences in the thickness impart a rough surface to the paper which can produce an electric field stress comparable to that of the discharge channel. The rough surface of the paper also helps in better impregnation when tightly wound. On the other hand, the existence of areas with lower thickness in the paper can cause breakdown at these points at considerably lower voltages.

Various investigations on composite dielectrics have shown that

(i)   the discharge inception voltage depends on the thickness of the solid dielectric, as well as on the dielectric constant of both the liquid and solid dielectric, and

(ii) the difference in the dielectric constants between the liquid and solid dielectrics does not significantly affect the rate of change of electric field at the electrode edge with the change in the dielectric thickness.

(c) Effect of Interfaces

The interface between two dielectric surfaces in a composite dielectric system plays an important role in determining its pre-breakdown and breakdown strengths. Discharges usually occur at the interfaces and the magnitude of the discharge depends on the associated surface resistance and capacitance. When the surface conductivity increases, the discharge magnitude also increases, resulting in damage to the dielectric.

In a composite dielectric, it is essential to maintain low dielectric losses because they normally operate at high electric stresses. However, even in an initially pure dielectric liquid, when used under industrial conditions for impregnating solid dielectrics, impurities arise, resulting in increased dielectric losses. The effect of various impurities in causing the breakdown of composite dielectrics is discussed in the next section.

4.6.3 Mechanisms of Breakdown in Composite Dielectrics

As mentioned in the earlier section, if dielectric losses are low the cumulative heat produced will be low and thermal breakdown will not occur. However, several other factors can cause short and long time breakdown.

(a) Short-Term Breakdown

If the electric field stresses are very high, failure may occur in seconds or even faster without any substantial damage to the insulating surface prior to breakdown. It has been observed that breakdown results from one or more discharges when the applied voltage is close to the observed breakdown value. There exists a critical stress in the volume of the dielectric at which discharges of a given magnitude can enter the insulation from the surface and propagate rapidly into its volume to cause breakdown. Experiments with single discharges on an insulating material have shown that breakdown occurs more rapidly when the electric field in the insulation is such that it assists the charged particles in the discharge to penetrate into the insulation, than when the field opposes their entry. Breakdown was observed to occur more readily when the bombarding particles are electrons, rather than positive ions. In addition, there are local field intensifications due to the presence of impurities and variations in the thickness of solid insulation and these local field intensifications play a very important role in causing breakdown under high field conditions; the actual effect being dependent on the field in the insulation before the discharge impinges on it. A more detailed description is given in the book by Bradley (Reference 7).

(b) Long-Term Breakdown

Long-term breakdown is also called the ageing of insulation. The principal effects responsible for the ageing of the insulation which eventually leads to breakdown arise from the thermal processes and partial discharges. Partial discharges normally occur within the volume of the composite insulation systems. In addition, the charge accumulation and conduction on the surface of the insulation also contributes significantly towards the ageing and failure of insulation.

(i) *Ageing and breakdown due to partial discharges*
During the manufacture of composite insulation, gas filled cavities will be present within the dielectric or adjacent to the interface between the conductor and the dielectric. When a voltage is applied to such a system, discharges occur within the gas-filled cavities. These discharges are called the "partial discharges" and involve the transfer of electric charge between the two points in sufficient quantity to cause the discharge of the local capacitance. At a given voltage, the impact of this charge on the dielectric surface produces a deterioration of the insulating properties, in many ways, depending on the geometry of the cavity and the nature of the dielectric. The study of breakdown by partial discharges is very important in industrial systems.

The degree of ageing depends on the discharge inception voltage, V_i and the discharge magnitude. It has been shown that V_i is strongly dependent on the permittivity of the dielectric ε_r and the thickness of the cavity g. V_i can be estimated approximately by

$$V_i = \left(\frac{E_g}{\varepsilon_r}\right)(t + \varepsilon_r\, g) \tag{4.9}$$

where E_g is the breakdown stress of the cavity air gap of thickness g and t is the thickness of the dielectric in series with the cavity. For any given arrangement $(g + t)$ will be constant. Let us call this constant as C. Then, the above equation can be written as

$$V_i = \left(\frac{E_g}{\varepsilon_r}\right)[(\varepsilon_r - 1)g + C\,] \tag{4.10}$$

Differentiating, we get

$$\frac{dV_i}{dg} = \frac{\varepsilon_r - 1}{\varepsilon_r}\left[E_g + \left\{g + \frac{C}{\varepsilon_r - 1}\right\}\frac{dE}{dg}\right] \tag{4.11}$$

Here, E_g is always positive and dE/dg is always negative or zero.

In obtaining the above expressions, two assumptions have been made. One is that in the cavity the stress $E_g = \varepsilon_r \cdot E$, where E is the electric field across the dielectric. The other is that within the cavity $E_{g(max)}/E_g = 1$. When these assumptions are valid, the Paschen law can be used to explain the breakdown of the gas gap (cavity).

However, the validity of the above assumptions depends upon the shape and dimensions of the cavity.

For the breakdown of the gas in the cavity to occur, the discharge has to start at one end and progress to the other end. As the discharge progresses, the voltage across the cavity drops due to charge accumulation on the cavity surface towards which it is progressing, and often the discharge gets extinguished. The discharge extinction voltage depends on the conditions inside the cavity. The discharge causes a rise in the temperature and pressure of the gas in the cavity and gaseous deterioration products are also formed. At high frequencies, when the discharges occur very rapidly, these may cause the extinction voltage levels to reach lower values in spite of the erosion of the cavity walls.

From the above analysis the following conclusions can be arrived at

    (i) for very small cavities, V_i decreases as the cavity depth increases, following the Paschen curve of gas breakdown.

    (ii) in spite of the erosion in the cavity walls, breakdown will not occur and the life of the insulation is very long if the applied voltage is less than $2V_i$.

    (iii) for applied voltages greater than $2V_i$, erosion is faster and therefore ageing of the insulation is quicker.

    (iv) the total capacitance of the cavity is not discharged as a single event but as a result of many discharges, each discharge involving only a small area of the cavity wall determined by the conductivity of the cavity surface in the region of the discharge.

Further details on ageing and breakdown due to partial discharges can be had from the book edited by A. Bradley (1984).

(ii) *Ageing and breakdown due to accumulation of charges on insulator surfaces*

During discharges at the solid or liquid or solid-gas or solid-vacuum interfaces, certain quantity of charge (electrons or positive ions) gets deposited on the solid insulator surface. The charge thus deposited can stay there for very long durations, lasting for days or even weeks. The presence of this charge increases the surface conductivity thereby increasing the discharge magnitude in subsequent discharges. Increased discharge magnitude in subsequent discharges causes damage to the dielectric surface. Experiments using electro-photography and other methods have shown that transverse discharges occur on the faces of the dielectric, and these discharges cause a large area to be discharged instantaneously. Charges that exist in surface conductivity are due to the discharges themselves such that changes in discharge magnitude will occur spontaneously during the life of a dielectric.

It has been generally observed that the discharge characteristics change with the life of the insulation. This can be explained as follows: for clean surfaces, at the discharge inception voltage V_i, the discharge characteristic depends on the nature of the dielectric, its size and shape. The discharge normally consists of a large number of comparatively small discharges originating from sites on the insulator surface where the necessary discharge condition exists. After some time, erosion at these sites causes the discharges to decrease in number as well as in magnitude, and consequently total extinction may occur. With the passage of time, the phenomena involved become complex because the charges from the surface-induced conductivity add to the charge accumulation in the bulk due to partial discharges.

4.7 SOLID DIELECTRICS USED IN PRACTICE

The majority of the insulating systems used in practice are solids. They can be broadly classified into three groups: organic materials, inorganic materials and synthetic polymers. Some of these materials are listed in Table 4.2 below.

Table 4.2 Classification of Solid Insulating Materials

Organic materials	Inorganic materials	Synthetic polymers	
		Thermoplastic	Thermosetting
Amber	Asbestos	Polyethylene	Bakelite
Cotton	Ceramics	Perspex	Epoxy resins
Paper	Glass	Polypropylene	
Pressboard	Mica	Polystyrene	
Rubber		Polyvinyl chloride	
Wax			
Wood			

Organic materials are those which are produced from vegetable or animal matter and all of them have similar characteristics. They are good insulators and can be easily adopted for practical applications. However, their mechanical and electrical properties always deteriorate rapidly when the temperature exceeds 100°C. Therefore, they are generally used after treating with a varnish or impregnation with an oil. Examples are paper and press board used in cables, capacitors and transformers.

Inorganic materials, unlike the organic materials, do not show any appreciable reduction (< 10%) in their electrical and mechanical properties almost up to 250°C. Important inorganic materials used for electric applications are glasses and ceramics. They are widely used for the manufacture of insulators, bushings etc., because of their resistance to atmospheric pollutants and their excellent performance under varying conditions of temperature and pressure.

Synthetic polymers are the polymeric materials which possess excellent insulating properties and can be easily fabricated and applied to the apparatus. These are generally divided into two groups, the thermoplastic and the thermosetting plastic types. Although they have low melting temperatures in the range 100–120°C, they are very flexible and can be moulded and extruded at temperatures below their melting points. They are widely used in bushings, insulators etc. Their electrical use depends on their ability to prevent the absorption of moisture.

Some of the important dielectric properties of the above materials are discussed below.

4.7.1 Paper

The kind of paper normally employed for insulation purposes is a special variety known as tissue paper or Kraft paper. The thickness and density of paper vary depending on the application. Low-density paper (0.8 gms/cm^3) is preferred in high frequency capacitors and cables, while medium density paper is used in power capacitors. High-density papers are preferable in d.c. and energy storage capacitors and for the insulation of d.c. machines.

Paper is hygroscopic. Therefore, it has to be dried and impregnated with impregnants, such as mineral oil, chlorinated diphenyl and vegetable oils. The relative dielectric constant of impregnated paper depends upon the permittivity of cellulose of which the paper is made, and permittivity of the impregnant and the density of the paper. Table 4.3 gives the dielectric constants for different densities of paper impregnated with different oils.

Table 4.3 Classification of Solid Insulating Materials

Impregnant	Density (g/cm^3)		
	0.8	1.0	1.2
Trichlorodiphenyl at 20°C $\varepsilon = 6.1$	6.28	6.30	6.40
Trichlorodiphenyl at 50°C $\varepsilon = 5.6$	6.0	6.14	6.24
Pentachlorodiphenyl at 20°C $\varepsilon = 5.7$	5.71	5.88	6.06
Transformer oil $\varepsilon = 2.2$	3.26	3.72	4.30

When very thin (thickness 8-20 μm) paper is used, it is very essential to see that the number of conducting particles on the surface of the paper is minimum. The conventional method of detecting conducting particles is by means of using a roller and plate, the conduction being indicated by means of headphones.

4.7.2 Fibres

Fibres when used for electrical purposes will have the ability to combine strength and durability with extreme fineness and flexibility. The fibres used are both natural and man-made. They include cotton, jute, flax, wool, silk (natural fibres), rayon, nylon, terylene, teflon and fibreglass.

The properties of fibrous materials depend on the temperature and humidity. Figures 4.7 and 4.8 show the variation of ε_r and tan δ of various fibrous materials as a function of the frequency. It can be observed from these figures that ε_r decreases with frequency, while tan δ is higher at lower frequencies. Most of the perfectly-dried fibres have a dielectric constant between 3 and 8. The presence of ionic impurities (e.g., salt) considerably reduces the electrical resistance of the fibre. Artificial fibres, such as terylene and fibreglass absorb very little water and hence have very high resistance. Table 4.4 gives the density, ε_r and tan δ of various fibres.

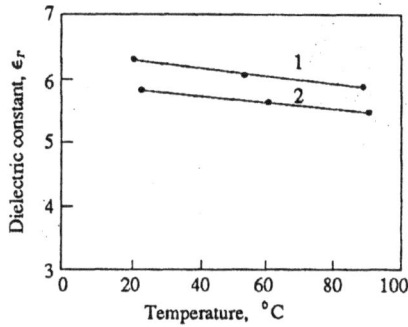

Fig. 4.7   Variation of dielectric constant, ε_r with temperature for paper
1. Trichlorodiphenyl impregnated paper
2. Pentachlorodiphenyl impregnated paper

Table 4.4   Electrical Properties of Fibrous Dielectrics

Fibres	Density	ε_r	tan δ
Vegetable Fibre—Natural			
Cotton	1.53	4.4-7.3	0.120
Flax	1.53	4.4-7.3	0.120
Jute	1.53	4.4-7.3	0.120
Animal Fibre—Natural			
Wool	1.31	1.52	0.016
Silk	1.30	3.4	0.016
		(4.4)†	
Man-made Fibres			
Rayon	1.52	2.03	0.031
Acetate (cellulose acetate)	1.33	2.2	0.015
Nylon	1.14	2.51	0.053
(Polyamide)		(3.5-4.2)*	
Terylene (Decrom)	1.38	1.97	0.030
Teflon (P.T.F.E.)	2.30	1.9-2.2	0.001-0.003
Fibreglass	2.54	5-7	0.001-0.0025

*Dielectric constant when the material contains no air voids.

4.7.3 Mica and Its Products

Mica is the generic name of a class of crystalline mineral silicates of alumina and potash. It can be classified into four main groups: (i) muscovite, (ii) phlogopite, (iii) fibiolite, and (iv) lipidolite. The last two groups are hard and brittle and hence are

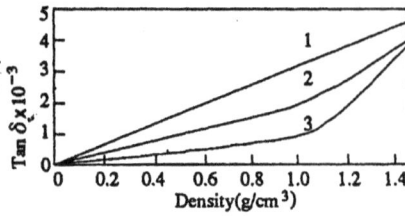

Fig. 4.8 Variation of tan δ with the density of paper
1. Trichlorodiphenyl impregnated paper
2. Mineral oil impregnated paper
3. Dry paper

unsuitable for electrical insulation purposes. Mica can be split into very thin flat laminae. It has got a unique combination of electrical properties, such as high dielectric strength, low dielectric losses, resistance to high temperatures and good mechanical strength. These have made it possible for it to be used in many electrical apparatus. Very pure mica is used for high frequency applications. Spotted mica is used for low voltage insulation, such as for commutator segment separators, armature windings, switchgear and in electrical heating and cooling equipments. Table 4.5 shows the electrical properties of mica.

Table 4.5 Electrical Properties of Muscovite and Phlogopite Mica

Property	Muscovite	Phlogopite
Chemical formula	$K_2O.3Al_2O_3.6SiO_2.2H_2O$	$K_2O.6MgO.Al_2O_3.6SiO_2.2H_2O$
Hardness (shore)	76–106	77–101
Tensile strength (0.02–0.05 mm thick)	17–36 kgf/mm^2	16–26 kgf/mm^2
Compressive strength (20 × 20 mm sample, 10–20 mm thick)	2000–5800 kgf/cm^2	1500–4600 kgf/cm^2
Dielectric strength (up to 30°C)	1000 kV/mm	700 kV/mm
Dielectric constant (1 kHz–3000 MHz)	6–7.5	6–7.5
Loss tangent	0.03 (50 Hz) 0.001 (1 MHz)	0.03 (50 Hz) 0.001 (1MHz)
Surface resistivity (60% humidity)	10^{11}–10^{12} ohm–cm	10^{10}–10^{11} ohm–cm
Volume resistivity (constant up to 200°C)	10^{14}–10^{15} ohm–cm	10^{13}–10^{14} ohm–cm

Mica is built into sheet form by bounding together with a suitable resin or varnish. Depending on the type of application, mica can be mixed with the required type of resin to meet the operating temperature requirements. Micanite is another form of mica which is extensively used for insulation purposes. Mica splittings and mica powder are used as filters in insulating materials, such as glass and phenolic resins. The use of mica as a filler results in improved dielectric strength, reduced dielectric loss and improved heat resistance and hardness of the material.

4.7.4 Glass

Glass is a thermoplastic inorganic material comprising complex systems of oxides (SiO_2). The dielectric constant of glass varies from 3.7 to 10 and the density varies from 2.2 to 6 g/cm^3. At room temperature, the volume resistivity of glass varies from 10^{12} to 10^{20} ohm-cm. The dielectric loss of glass varies from 0.004 to 0.020 depending on the frequency. The losses are highest at lowest frequencies. The dielectric strength of glass varies from 3000 to 5000 kV/cm and decreases with increase in temperature, reaching half the value at 100°C. Glass is used as a cover and for internal supports in electric bulbs, electronic valves, mercury arc switches, x-ray equipment, capacitors and as insulators in telephones.

4.7.5 Ceramics

Ceramics are inorganic materials produced by consolidating minerals into monolithic bodies by high temperature heat treatment. Ceramics can be divided into two groups depending on the dielectric constant. Low permittivity ceramics ($\varepsilon_r < 12$) are used as insulators, while the high permittivity ceramics ($\varepsilon_r > 12$) are used in capacitors and transducers.

Tables 4.6 and 4.7 give the various dielectric properties of some ceramics commonly used for electrical insulation purposes.

Table 4.6 Properties of Low Permittivity Ceramics

Property	H.T. porcelain	L.T. porcelain	Low loss steatite	Alumina	Forsterite
Chemical composition	50% Clay, 25% Feldspar, 25% Flint	50% Clay, 25% Feldspar, 25% Flint	3 MgO, 4 SiO$_2$, H$_2$O	95%	2 MgO, SiO,
Water absorption (p.p.m.)	0	0.5-2	0	0	0
Safe temperature (°C)	1000	900	1050	1600	1050
Dielectric strength (kV/mm)	25	3	8-25	16	8-12
ε_r	5-7	5-7	6	9	6
tan $\delta \times 10^4$	50-100	100-200	10	5	3-4

Table 4.7 Properties of High Permittivity Ceramics

Ceramics	Chemical composition	ε_r	$\tan \delta \times 10^4$
Magnesium metatitanate	$MgTiO_3$	16	2
Strantium zirconate	$SrZrO_3$	38	3
Titanium oxide	TiO_2	90	5
Calcium titanate	$CaTiO_3$	150	3
Strantium titanate	$SrTiO_3$	200	5
Barium titanate	$BaTiO_3$	1500	150

4.7.6 Rubber

Rubber is a natural or synthetic vulcanizable high polymer having high elastic properties. Electrical properties of rubber depend on the degree of compounding and vulcanizing. General impurities, chemical changes due to ageing, moisture content and variations in temperature and frequency have substantial effects on the electrical properties of rubber. Some important electrical properties and applications of different types of rubber are given in Table 4.8.

Table 4.8 Properties and Applications of Rubber

Type of rubber	Water absorption (per cent)	ε_r (50 Hz)	$\tan\delta$ (50 Hz)	a.c. break-down voltage kV/cm	Application and limitations
Natural rubber	0–4.8	2.9–6.6	0.02–0.1	100–390	Inexpensive, flexible, good electrical properties, resistant to corona, maximum operating temperature 60°C.
Polylsar kryflex rubber and Styrene butadine rubber	0–4.5	3.8–6.2	0.02–0.09	80–380	Used for low voltage (<11 kV) cables. Not good under weathering, water, oils, etc.
Butyl rubber and Polysar butyl rubber	0–2	2.2–3.2	0.003–0.03	80–200	Used in aerial, submarine and underground cables. Good electrical properties at low temperatures. Excellent resistance to tear, abrasion, acids, alkalies and chemicals.

(Contd.)

Type of rubber	Water absorption (per cent)	ε_r (50 Hz)	tanδ (50 Hz)	a.c. break-down voltage kV/cm	Application and limitations
Silicone rubber	0–3	2.6–3.4	0.006–0.02	90–390	Used in shipping and aircraft cables, transformers, lightning equipment, etc. High service temperature (150°C). Easily attacked by acids, alkalies and chlorinated compounds.

4.7.7 Plastics

Plastics are very widely used as insulating materials because of their excellent dielectric properties. Many new developments in electrical engineering and electronics would not have been possible without the development of plastics. Plastics are made by combining large numbers of small molecules into a few big ones. When small molecules link to form the bigger molecules of the plastics, many different types of structures result. Most thermoplastic resins approximate to a structure in which several thousand atoms are tied together in one direction. The thermosetting resins on the other hand, form a three-dimensional network. In view of the large number of plastics available, it will not be possible to deal with all of them, and only materials which are commonly used for insulation purposes are described.

Polyethylene

Polyethylene is a thermoplastic material which combines unusual electrical properties, high resistance to moisture and chemicals, easy processability, and low cost. It has got high resistivity and good dielectric properties at high frequencies, and therefore, is widely used for power and coaxial cables, telephone cables, multi-conductor control cables, TV lead-in wires, etc.

Table 4.9 Electrical Properties of Polyethylene

Property	Low density polyethylene	Medium density polyethylene	High density polyethylene	Irradiated polyethylene
Volume resistivity (ohm–cm)	$> 10^{16}$	$> 10^{16}$	$> 10^{16}$	$> 10^{16}$
Dielectric strength (kV/cm)	170–280	200–280	180–240	720–1000
Dielectric constant (50 Hz–10^6 Hz)	2.3	2.3	2.35	2.3
tan (50 Hz–10^6 Hz)	0.0002	0.0002	0.0002	0.0005
Arc resistance	melts	melts	melts	melts

By varying the methods of manufacture different types of polyethylene are made with specific properties for different applications. They may have low density, medium density, high density or may be irradiated types. The dielectric properties of these are summarized in Table 4.9.

Fluorocarbon Plastics

Polytetrafluoroethylene (P.T.F.E.), polychlorotrifluoroethylene (P.C.T.F.E.) and polyvinylidene (P.V.F$_2$) plastics come under this category. P.T.F.E. is the most thermally stable and chemically resistant of all the three. It is considered as one of the best plastics used for insulation because of its excellent electrical and mechanical properties. It can be used without decomposition up to temperatures of 327°C. It is widely used in almost all applications. P.C.T.F.E. has higher dielectric constant and loss factor than P.T.F.E., but melts at 190°C. P.V.F$_2$ can be worked in the temperature range –30 °C to 150°C. It is used as thin wall insulation, as jacketing for computer wires and special control wires, and for tubing and sleeving for capacitors, resistors, terminal junctions, and solder sleeves. The electrical properties of fluorocarbons are tabulated in Table 4.10.

Table 4.10 Properties of Fluorocarbon Plastics

Property	P.T.F.E	P.C.T.F.E.	P.V.F$_2$
Volume resistivity (ohm–cm)	$> 10^{18}$	1.2×10^{18}	2×10^{14}
Dielectric strength (kV/cm)	200	210	104–512
Dielectric constant (50 Hz–10^6Hz)	2.0	2.3–2.8	8.4–6.49
tan δ (50 Hz–10^6 Hz)	< 0.0002	0.0012–0.0036	0.0491–0.15

Nylon

Nylon is a thermoplastic which possesses high impact, tensile, and flexural strengths over a wide range of temperature (0 to 300°C). It also has high dielectric strength and good surface and volume resistivities even after lengthy exposure to high humidity. It is also resistant to chemical action, and can be easily moulded, extruded and machined. It is generally recommended for high frequency low loss applications. In electrical engineering, nylon mouldings are used to make coil forms, fasteners, connectors, washers, cable clamps, switch housings, etc.

There are three different types of nylon commonly used. They are nylon 6/6, nylon 6 and nylon 6/10. The dielectric properties of these three types are given in Table 4.11.

Table 4.11 Dielectric Properties of Nylon

Property	Nylon 6/6	Nylon 6	Nylon 6/10
Volume resistivity (ohm–cm)	4.5×10^{13}	$10^{13} \times 10^{15}$	$4. \times 10^{14}$
Dielectric strength (kV/cm)	154	176–204	190
Dielectric constant			
50 Hz	4.1	5.0–14.0	4.6
10^3 Hz	4.0	4.9–10.1	4.5
10^6 Hz	3.4	4.0–4.7	3.5
tan δ			
50 Hz	0.014	0.06–0.10	0.04
10^3 Hz	0.02	0.06–0.10	0.04
10^6 Hz	0.04	0.04–0.13	0.03

Polyvinyl Chloride

Polyvinyl chloride or P.V.C. is used commercially in various forms. It is available as an unplasticized, tough, and rigid sheet material and can be easily shaped to any required form. It is chemically resistant to strong acids and alkalis and is insoluble in water, alcohol and organic solvents like benzene. The upper temperature limit of operation is about 60°C. The dielectric strength, volume resistivity and surface resistivity are relatively high. The dielectric constant and loss tangent are 3.0-3.3 and 0.015-0.02 respectively, at all frequencies up to 1 MHz.

P.V.C. is also available as a highly plasticized flexible material, which is used extensively for wire covering, insulated sleeving, and cable sheathing in preference to natural rubber because of its resistance to the action of sunlight, water and oxygen.

Polyesters

Polyesters have excellent dielectric properties and superior surface hardness, and are highly resistant to most chemicals. They represent a whole family of thermosetting plastics produced by the condensation of dicarboxylic acids and dihydric alcohols, and are classified as either saturated or unsaturated types. Unsaturated polyesters are used in glass laminates and glass fibre reinforced mouldings, both of which are widely used for making small electrical components to very large structures. Saturated polyesters are used in producing fibres and film. Polyester fibre is used to make paper, mat and cloth for electrical applications. The film is used for insulating wires and cables in motors, capacitors and transformers. The dielectric properties of polyester compounds are given in Table 4.12.

Table 4.12 Dielectric Properties of Polyesters

Property	Glass reinforced type		Cast resins	
	Premixed	Preformed	Rigid	Flexible
Volume resistivity (ohm–cm)	10^{12}–10^{15}	10^{14}	10^{12}–10^{14}	—
Dielectric constant				
50 Hz	5.3–7.3	3.8–6.0	3.3–4.3	4.4–8.1
10^3 Hz	4.68	4.0–6.0	3.2–4.3	4.5–7.1
10^6 Hz	5.6–6.4	3.5–5.5	3.2–4.3	4.1–5.7
tan δ				
50 Hz	0.01–0.04	0.01–0.04	0.006–0.05	0.026–0.031
10^3 Hz	—	0.01–0.05	0.006–0.04	0.016–0.050
10^6 Hz	0.008–0.022	0.01–0.03	0.017–0.019	0.020–0.060

Mylor polyester film is being largely used in preference to paper insulation. At power frequencies, its dissipation factor is very low, and it decreases as the temperature increases. It has got a dielectric strength of 2000 kV/cm, and its volume resistivity is better than 10^{15} ohm-cm at 100°C. Its high softening point enables it to be used at temperatures above the operating limit of paper insulation. It has got high resistance to weathering and can be buried under the soil also. Therefore, this can be used for motor and transformer insulation at power frequencies and also for high frequency applications which are subjected to varying weather conditions.

Polystyrenes

Polystyrenes are obtained when styrene is polymerized with itself or with other polymers or monomers producing a variety of thermoplastic materials with varying properties in different colours. Electrical grade polystyrenes have a dielectric strength comparable to that of mica, and have low dielectric losses which are independent of the frequency. Their volume resistivity is about 10^{19} ohm-cm and the dielectric strength is 200-350 kV/cm. The dielectric constant at 20°C is 2.55, and the loss tangent is 0.0002 at all frequencies up to 10,000 MHz.

Polystyrene films are extensively used in the manufacture of low loss capacitors, which will have a very stable capacitance and extremely high insulation resistance. Films and drawn threads of polystyrene are also used for high frequency and cable insulations.

4.7.8 Epoxy Resins

Epoxy resins are thermosetting types of insulating materials. They possess excellent dielectric and mechanical properties. They can be easily cast into desired shapes even at room temperature. They are very versatile, and their basic properties can be modified either by the selection of a curing agent or by the use of modifiers or fillers. They are highly elastic; samples tested under very high pressures, up to about 180,00 psi (12,000 atm) returned to their original shape after the load was removed, and the sample showed no permanent damage. Resistance to weathering and chemicals is also very good. The tensile strength of araldite CT 200 and hardner HV 901 is in the range 5.5–8.5 kg/mm^2, and the compressive strength is 11-13 kg/mm^2. The dielectric constant varies between 2.5 and 3.8. The dielectric loss factor is very small under power frequency conditions lying in the range 0.003-0.03. The dielectric strength is 75 kV/mm, when the specimen thickness is 0.025 mm or 1 mil. The volume resistivity of the material is of the order of 10^{13} ohm-cm.

Epoxy resin can be formed into an insulator of any desired shape for almost any type of high voltage application. Insulators, bushings, apparatus, etc. can be made out of epoxy resin. It can also be used for encapsulation of electronic components, generator windings and transformers. It is used for bonding of very diverse materials such as porcelain, wood, metals, plastics, etc. It is a very important adhesive used for sealing of high vacuum joints. In any laboratory or industry in which electrical or electronic components or equipments are handled or manufactured, numerous occasions arise wherein epoxy resins can be used with an advantage saving time, labour and money.

In the previous sections details are given of a variety of insulating materials, commonly used for electrical insulation purposes. A good insulating material should have good dielectric strength, high mechanical strength, high thermal conductivity, very low loss factor, and high insulation resistance. The specific application of these materials in various power apparatus, electronic equipments, capacitors and cables are discussed in Chapter 5.

QUESTIONS

Q.4.1 What do you understand by "intrinsic strength" of a solid dielectric? How does breakdown occur due to electrons in a solid dielectric?

Q.4.2 What is "thermal breakdown" in solid dielectrics, and how is it practically more significant than other mechanisms?

Q.4.3 Explain the different mechanisms by which breakdown occurs in solid dielectrics in practice.

Q.4.4 How does the "internal discharge" phenomena lead to breakdown in solid dielectrics?

Q.4.5 What is a composite dielectric and what are its properties?

Q.4.6 Describe the mechanism of short-term breakdown of composite insulation.

Q.4.7 How do the temperature and moisture affect the breakdown strength of solid dielectrics?

Q.4.8 What are the advantages of using plastic film insulation over the paper insulation?

Q.4.9 What are the properties that make plastics more suitable as insulating materials?

Q.4.10 What are the special features of epoxy resin insulation?

WORKED EXAMPLES

Example 4.1: A solid specimen of dielectric has a dielectric constant of 4.2, and tan $\delta = 0.001$ at a frequency of 50 Hz. If it is subjected to an alternating field of 50 kV/cm, calculate the heat generated in the specimen due to the dielectric loss.

Solution: Dielectric heat loss at any electric stress E [Eq. (4.5)]

$$= \frac{E^2 f \varepsilon_r \tan \delta}{1.8 \times 10^{12}} \text{ W/cm}^3$$

For the specimen under study, the heat loss will be

$$= \frac{50 \times 50 \times 10^6 \times 50 \times 4.2 \times .001}{1.8 \times 10^{12}}$$

$$= 0.291 \text{ mW/cm}^3$$

Example 4.2: A solid dielectric specimen of dielectric constant of 4.0 shown in the figure has an internal void of thickness 1 mm. The specimen is 1 cm thick and is subjected to a voltage of 80 kV (rms). If the void is filled with air and if the breakdown strength of air can be taken as 30 kV (peak)/cm, find the voltage at which an internal discharge can occur.

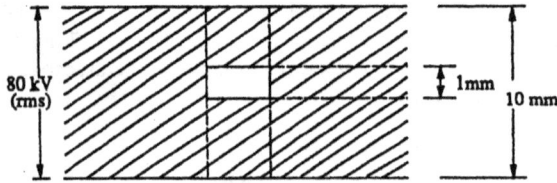

Solution: Referring to Fig. 4.5(a) and Eqs. (4.7) and (4.8), the voltage that appears across the void is given as

$$V_1 = \frac{V d_1}{\left(d_1 + \frac{\varepsilon_0}{\varepsilon_1} d_2\right)}$$

where,

$$d_1 = 1 \text{ mm}$$

$$d_2 = 9 \text{ mm}$$

$$\varepsilon_0 = 8.89 \times 10^{-12} \text{ F/m}$$

$$\varepsilon_1 = \varepsilon_r \varepsilon_0 = 4.0 \, \varepsilon_0$$

$$V_1 = \frac{V \times 1}{\left(1 + \frac{9}{4}\right)}$$

$$= \left(\frac{4V}{13}\right)$$

The voltage at which the air void of 1 mm thickness breaks down is 3 kV/mm × 1 mm = 3 kV

$$\therefore V_1 = \frac{13V}{4} = \frac{13 \times 3}{4} = \frac{39}{4}$$

$$= 9.75 \text{ kV (peak)}$$

The internal discharges appear in the sinusoidal voltage 80 sin ωt kV when the voltage reaches a value of 9.75 kV (see Fig. 4.6 for the discharge pattern).

Example 4.3: A coaxial cylindrical capacitor is to be designed with an effective length of 20 cm. The capacitor is expected to have a capacitance of 1000 pF and to operate at 15 kV, 500 kHz. Select a suitable insulating material and give the dimensions of the electrodes.

Solution: The capacitance of the coaxial cylindrical capacitor is

$$\frac{2\pi \, \varepsilon_0 \, \varepsilon_r \, l}{\ln \dfrac{d_2}{d_1}} \tag{1}$$

where l = length in metres, d_1 and d_2 are the diameters of the inner and outer electrodes, and ε_r = dielectric constant. The dielectric material that can be selected is either polyethylene or P.T.F.E.

Choosing high density polyethylene, its dielectric constant ε_r = 2.3, and its breakdown stress is taken as 500 V/mil or 200 kV/cm. Allowing a factor of safety of 4, the maximum stress E_{max} = 50 kV/cm. E_{max} occurs near the inner electrode and is given by

$$E_{max} = \frac{V}{r_1 \ln \dfrac{r_2}{r_1}} \tag{2}$$

from equation (1),

$$\ln \frac{d_2}{d_1} = \ln \frac{r_2}{r_1} = \frac{2\pi\varepsilon_0\varepsilon_r l}{\text{capacitance}}$$

$$= \frac{2\pi\dfrac{10^{-9}}{36\pi} \times 2.3 \times 0.2}{100 \times 10^{-12}}$$

$$= 0.02556$$

$$\therefore \frac{r_2}{r_1} = 1.026$$

From equation (2),

$$r_1 = \frac{V}{E_{max}} \ln \frac{1}{\dfrac{r_2}{r_1}}$$

$$= \frac{15}{50 \times 0.02556}$$

$$= 11.74 \text{ cm}$$

$$\therefore \quad r_2 = 1.026 \times 11.74$$

$$= 12.05 \text{ cm}$$

The thickness of the insulation is 3.1 mm (refer to Tables 4.8 and 4.9 for the properties of the material).

REFERENCES

1. O'Dwyer, J.J., *Theory of Dielectric Breakdown in Solids*, Clarendon Press, Oxford (1964).
2. Whitehead, S., *Dielectric Breakdown of Solids*, Oxford University Press, Oxford (1951).
3. Von Hippel, A., *Dielectric Materials and Applications*, John Wiley, New York (1964).
4. Mason, J.H., "Electrical insulation", *Electrical Energy*, Vol. 1, 68-75 (1956).
5. Taylor, H.E., *Modern Dielectric Materials* (Ed. J.B. Birks), Chap. 9, Haywood, London (1960).
6. Clark, F.M., *Insulating Materials for Design and Engineering Practice*, John Wiley, New York (1962).
7. Bradley, A., *Electrical Insulation*, Peter Peregrinus, London (1984).

5

Applications of Insulating Materials

5.1 INTRODUCTION

There is no piece of electrical equipment that does not depend on electrical insulation in one form or other to maintain the flow of electric current in desired paths or circuits. If due to some reasons the current deviates from the desired path, the potential will drop. An example of this is a short circuit and this should always be avoided. This is done by proper choice and application of insulation wherever there is a potential difference between neighbouring conducting bodies that carry current. There are four principal areas where insulation must be applied. They are

(a) between coils and earth (phase-to-earth),
(b) between coils of different phases (phase-to-phase),
(e) between turns in a coil (inter-turn), and
(d) between the coils of the same phase (inter-coil).

As we know, there are three broad categories of insulating materials, gases, liquids and solids. The insulating materials are classified mainly based on the thermal endurance. The insulation is primarily meant to resist electrical stresses. In addition, it should also be able to withstand certain other stresses which the insulation encounters during manufacture, storage and operation. The performance of the insulation depends on its operating temperature. The higher the temperature, the higher will be the rate of its chemical deterioration, and hence the lower will be its useful life. If a reasonably long life of an insulation is expected, its operating temperature must be maintained low. Therefore, it is necessary to determine the limits of temperature for the insulation, which will ensure safe operation over its expected life. Thus the insulating materials are grouped into different classes O, A, B, and C with temperature limits of 90°C, 105°C and 130°C for the first three classes and no specific limit fixed for class C. Classes O and A cover the various organic materials without and with impregnation respectively, while classes B and C cover inorganic materials, respectively with and without a binder. With the advent of newer insulating materials, namely, the plastics and silicones, during the middle of this century, a need was felt to reorganize the classification of the insulating materials. This led IEC (International Electrotechnical Commission) to come up with the new categories:

Class Y (formerly O): 90°C: Paper, cotton, silk, natural rubber, polyvinyl chloride, etc. without impregnation.

Class A: 105°C: Same as class Y but impregnated, plus nylon.

Class E: 120°C: Polythylene terephthalate (terylene fibre, melinex film), cellulose triacetate, polyurethanes, polyvinyl acetate enamel.

Class B: 130°C: Mica, fibreglass (alkali free alumino borosilicate), bitumenized asbestos, bakelite, polyester enamel.

Class F: 155°C: As class B but with alkyd and epoxy based resins.

Class H: 180°C: As class B with silicone resin binder, silicone rubber, aromatic polyamide (nomex paper and fibre), polymide film (enamel, varnish and film) and estermide enamel.

Class C: Above 180°C: As class B but with suitable non-organic binders; teflon (polytetraflouroethylene).

The temperatures mentioned above cannot be regarded as the limiting operating temperatures but only as an index to compare the various insulating materials. All the national standards permit the equipment to work up to these temperatures, but in practice, certain differentials are allowed because of the over loads, other manufacturing advantages and economics.

In this chapter we deal with the insulation systems in electrical and electronic equipments. First we deal with insulation in power apparatus under which insulation in rotating electrical machines, transformers and switchgear are discussed followed by the insulation in capacitors and cables. The insulation of electronic equipment is discussed later.

5.2 APPLICATIONS IN POWER TRANSFORMERS

Transformers are the first to encounter lightning and other high voltage surges. The transformer insulation has to withstand very high impulse voltages many times the power frequency operating voltages. The transformer insulation is broadly divided into

(a) conductor or turn-to-turn insulation,

(b) coil-to-coil insulation,

(c) low voltage coil-to-earth insulation,

(d) high voltage coil-to-low voltage coil insulation, and

(e) high voltage coil-to-ground insulation.

The low voltage coil-to-ground and the high voltage coil-to-low voltage coil insulations normally consist of solid tubes combined with liquid or gas filled spaces. The liquid or gas in the spaces help to remove the heat from the core and coil structure and also help to improve the insulation strengths. The inter-turn insulation is directly applied on the conductor as organic enamel in smaller rating transformers. In the large transformers paper or glass tape is wrapped on the rectangular conductors. In the case of layer to layer, coil-to-coil and coil-to-ground insulations, Kraft paper is used in smaller transformers, whereas thick radial spacers made of pressboard, glass fabric, or porcelain are used in the case of higher rating transformers.

Of all the materials, oil impregnated paper, and pressboard are extensively used in liquid filled transformers. The lack of thermal stability at higher temperatures limits the use of this type of insulation to be used continuously up to 105°C. Paper and its products absorb moisture very rapidly from the atmosphere, and hence this type of insulation should be kept free of moisture during its life in a transformer.

Transformer oil provides the required dielectric strength and insulation and also cools the transformer by circulating itself through the core and the coil structure. The transformer oil, therefore, should be in the liquid state over the complete operating range of temperatures between $-40°C$ and $+50°C$. The oil gets oxidized when exposed to oxygen at high temperatures, and the oxidation results in the formation of peroxides, water, organic acids and sludge. These products cause chemical deterioration of the paper insulation and the metal parts of the transformer. Sludge being heavy, reduces the heat transfer capabilities of the oil, and also forms as a heat insulating layer on the coil structure, the core and the tank walls. In present-day transformers the effects of oxidation are minimized by designing them such that access to oxygen itself is limited. This is done by the use of (a) sealed transformers, (b) by filling the air space with nitrogen gas, and (c) providing oxygen absorbers like activated clay or alumina.

When an arc discharge occurs inside a transformer, the oil decomposition occurs. The decomposition products consist of hydrogen and gaseous hydrocarbons which may lead to explosion. And hence, oil insulated transformers are seldom used inside buildings or other hazardous locations like mines. Under such conditions dry type and askarel or sulphur hexafluoride (SF_6) gas filled transformers are used. Askarel is a fireproof liquid and is the generic name for a number of synthetic chlorinated aromatic hydrocarbons. These are more stable to oxidation and do not form acids or sludge. Under arcing they are very stable and do not give rise to inflammable gases. However they give out hydrochloric acid which is toxic and which attacks the paper insulation. This is removed by using tin or tetraphenyl. However, if the arc is very heavy, the hydrochloric acid cannot be absorbed completely. For these reasons SF_6 gas insulated transformers are popular. Also, askarel cannot be used in high voltage transformers, because the impulse strength of askarel impregnated paper is very low compared to that of oil impregnated paper. Moreover, its dielectric strength deteriorates rapidly at high voltages and at high frequencies liberating hydrochloric acid.

Even today there is no perfect all purpose transformer fluid. In recent years, progress has been made with the use of fluorocarbon liquids and SF_6 gas. However, these liquids have not become very popular because of their high cost.

5.3 APPLICATIONS IN ROTATING MACHINES

Rotating machines are normally divided into two categories: those with voltage ratings less than 6,600 V are called low voltage machines, and the others are high voltage machines. Because of the difficulty of insulating high voltages, machines above 22 kV rating are not built except under special conditions. Classes Y and C insulation find no application in rotating machines. Class E which was widely used in low voltage machines for over 20 years is now being replaced by class F which is meant for the high voltage machines. Also, Class F is being increasingly used in place of class B. Thus class F appears to be the insulation of the future. Considerable progress has been made in recent years, in reducing the size of the machines for a given rating by use of class H materials, particularly, for small machines. However, the cost of class H materials (silicones, teflon) is very high, and hence they are used only under special conditions like severe over loads in traction motors and mill motors. The various materials used in modern rotating machines are tabulated in

Table 5.1 Typical Modern Insulating Materials for Rotating Machines

Component		Low voltage machines			High voltage machines	
		Class E	Class B	Class F	Class B	Class F
Turn-to-turn insulation		Polyvinyl acetal for both wire and strip conductors	Polyester enamel (wire) or phenolic bonded fibreglass (strip)	Estermide enamel (wire) or alkyd bonded fibreglass (strip)	Phenolic bonded fibre-glass (strip)	Alkyd bonded fibre-glass (strip)
Coil-to-coil and phase-to-phase insulation	Inside the slots	Bakelized fabric strips	Bakelized fabric strips	Epoxy fibreglass strips	Shellac or bitumen bonded mica foil or tape on straight portions of the coils	Epoxy impregnated mica paper foil or tape on straight portions of the coil.
	On over-hangs	Melinex film bonded to press paper	Alkyd bonded mica glass sheet	Nomex sheet	Alkyd varnished glass tape on coil ends and alkyd bonded mica sheet between layers	Epoxy varnished glass tape on coil ends and alkyd bonded mica glass sheet between layers
Phase (or coil) to earth insulation		Melinex film bonded to press paper	Mica alkyd bonded to glass cloth	Nomex sheet	No extra insulation because the phase-to-phase insulation itself is sufficient	
Slot closure (wedge)		Bakelized fabric strip	Bakelized fabric strip	Epoxy fibreglass strip	Bakelized fabric strip	Epoxy fibreglass strip
Insulation for leads		Alyd varnished terylene or glass tape or sleeving			Alkyd varnished glass tape	
Varnish for impregnation treatment		Alkyd phenolic	Alkyd phenolic	Estermide or epoxy	Alkyd-phenolic	Epoxy

Table 5.1. This is only a typical listing and may vary depending on the choice of the design engineer.

Mica has been used in the electrical industry since its inception. Normally, mica is available in the form of very thin splittings. Hence it is bound to a supporting sheet of electrical grade paper or glass cloth with a suitable binding agent. The resulting mica sheets are known as micanite. Since mica splittings of fairly large surface area were not available, methods were evolved to make mica paper using mica of any size. The mica paper so obtained is not sufficiently strong or self-supporting. Hence, it has to be given a backing of glass cloth or other binding material such as epoxy resin. Epoxy resin bounded mica paper is extensively used in both low and high voltage machines. For non-epoxy system a varnish impregnation is essential to fill the voids and also to act as a barrier against moisture and chemicals present in the atmosphere. For this purpose the varnish should have the property of forming an unbroken tightly adhesive and yet reasonably flexible film. The solvent in the varnish must not attack any of the insulating materials used, and the resin should have a long-term compatibility with these materials.

The maintenance of good mechanical properties is also equally important for the reliable operation of machines. The insulation should withstand the expansion and contraction during temperature cycles in large machines. These effects become very severe at the high temperatures observed in power generators of a very large size. Maintenance of good mechanical properties and thermal endurance are very essential in low voltage machines also.

5.4 APPLICATIONS IN CIRCUIT BREAKERS

A circuit breaker is a switch which automatically interrupts the circuit when a critical current or voltage rating is exceeded. a.c. currents are considerably easier to interrupt than d.c. currents. a.c. current interruption generally requires first to substitute an arc for part of the metallic circuit and then its deionization when the current goes through zero, so that the arc will not reestablish again.

Circuit breakers are also divided into two categories, namely, the low voltage and high voltage types. Low voltage breakers use synthetic resin mouldings to carry the metallic parts. For higher temperatures ceramic parts are used. When the arc is likely to come into contact with moulded parts, melanine or some special kind of alkyd resins are used because of their greater arc resistance. The high voltage circuit breakers are further classified into air circuit breakers and oil circuit breakers. Many insulating fluids are suitable for arc extinction and the choice of the fluid depends on the rating and type of the circuit breaker. The insulating fluids commonly used are atmospheric air, compressed air, high vacuum, SF_6 and oil. In some ancillary equipment used with circuit breakers, the fluid serves the purpose of providing only insulation. Many insulants are available for this purpose.

The oils used in circuit breakers normally has the same characteristics as transformer oil. In circuit breakers oil serves an additional purpose of interrupting the arc. Since the gases (mainly hydrogen) help to extinguish the arc, a liquid which generates the maximum amount of the gas for one unit of arc energy is preferred. Transformer oil possesses these characteristics. Many other oils have been tried but with no success. Askarels produce large quantities of toxic and corrosive products.

The circuit breaker bushings of lower voltage ratings may consist of solid cylinders of porcelain and shellac or resin treated paper wrapped on the current carrying electrode. High voltage bushings of voltages of 66 kV and above are filled with oil. The constructional details vary widely. In certain designs, the system of coaxial porcelain or treated paper cylinders are used with space between them filled with oil. In the condenser type bushings, paper is wound on the electrode and metal foils are wrapped on it at intervals throughout the diameter such that the capacitance between successive foils is constant. This ensures uniform voltage distribution, and hence higher dielectric strength.

The different types of insulating materials used in the construction of high voltage switchgear are classified in Table 5.2. This includes some of the modern insulating materials for future applications. Of these, a few are widely used as major insulants. They are, porcelain, insulating oil, synthetic resin bonded paper laminates, epoxy resins, and SF_6 gas.

5.5 APPLICATIONS IN CABLES

In the recent years natural rubber has been completely replaced by synthetic rubbers and plastics as cable insulation. The physical properties required for wire and cable insulation depend on the type of application. It should have good elongation and tensile strength and toughness, so that it will withstand handling during installation and service. It should also have low dielectric constant and power factor but high dielectric strength and insulation resistance. Also, during operation, because of over loading, the insulation may be exposed to high temperatures for long periods of time. This necessitates the insulation to have excellent resistance to ageing at high temperatures. The insulation should also be able to withstand long exposure to sunlight and various chemicals. High voltage cables also give rise to ozone and the insulation will deteriorate in its presence. This is particularly severe for the insulation near the conductors. Cables are also laid in rivers and under the sea. For these applications.it should have very low water absorption. When cables have to operate at low temperature the insulation should not become stiff and brittle. The partial discharges in the cable insulation should also be kept as low as possible.

Table 5.2 Insulating in High Voltage Switchgear

Materials	Applications
Epoxy resins	Low pressure castings for bushings, switchgear orifices, bus-bars, instrument transformers. Fluidized bed dip coating for bus-bar insulation and dough moulding for bus-bar barriers and secondary terminals.
Epoxy resin bonded glassfibre	For components such as arc control devices, circuit breaker operating rod and high pressure feed pipes for air blast circuit breakers.

(Contd.)

Materials	Applications
Polyester resins	Insulating lever for circuit breaker and phase barrier plate in switch board.
Porcelain	Insulators and bushings of power transformers, circuit breakers and instrument transformers.
Vulcanized Fibre	Arc chamber segments.
Synthetic resin bonded paper	Bushings, arc chambers, etc.
Nylon	Injunction mouldings for arc control devices in circuit breakers.
Silicone rubber	Filling for moulded joint boxes in air insulated circuit breakers.
Butyl rubber	Presssure moulding of current transfcrmers.
Chloro-sulphonated polyethylene	Cable insulation for use in air or oil.

The main types of insulants used in the cable industries are paper, rubber, plastics and compressed gas. Paper insulated lead sheathed cables are still used because of their reliability, high dielectric strength, low dielectric loss, and long life. The most commonly used insulating materials for low and medium voltage (up to 3.3 kV) cables is polyvinylychloride (P.V.C.) Polyethylene and cross-linked polyethylene are also used. P.V.C. is not suitable for high voltage applications because of its high dielectric constant and high loss. It cannot be operated continuously at higher voltages, although it can be used up to 85°C continuous at low voltages. On the other hand, polyethylene has low dielectric constant and low loss but high dielectric strength. The best material for high voltage and high temperature operation is teflon (P.T.F.E.) which can be used up to 250°C. Silicone rubber has a high degree of heat resistance for continuous operation up to 150°C. It gives rise to very little carbon formation when destroyed by fire, and as such it continues to function even after the fire. Hence it is used for aircraft cables where contamination with aircraft fuel can occur at very high temperatures.

In paper insulated cables the paper is impregnated with oil by the process of mass impregnation using free flowing liquid. Table 5.3 gives the various insulating materials used in cables and the maximum cable operating voltages and temperatures.

Table 5.3 Cable Insulations

Insulation	Maximum cable operating voltage a.c. (kV)	Range of operating temperature (°C)
(a) *Impregnated Paper*		
Solid type	95.0	−10 to 85
Oil-filled	400.0	−20 to 70
Gas-filled	400.0	−20 to 70
Varnished cloth	28.0	−10 to 80

(Contd.)

Insulation	Maximum cable operating voltage a.c. (kV)	Range of operating temperature (°C)
(b) *Rubber*		
Natural	3.0	– 40 to 70
Synthetic-latex	0.6	– 40 to 75
Synthetic-neoprene	0.6	– 30 to 90
Synthetic-silicone	5.0	– 40 to 150
Synthetic-butyl	28.0	– 40 to 80
(c) *Plastics*		
P.V.C.	0.6	– 30 to 105
Polyethylene	15.0	– 60 to 80
Teflon	5.0	– 54 to 250
Fluorothenes	5.0	– 54 to 150

5.6 APPLICATIONS IN POWER CAPACITORS

It is well-recognized that power capacitors are indispensable for power system administration and are used for voltage regulation of power transmission systems and for the improvement of power factor of power distribution networks.

In most of the industrial applications, the power requirements are reactive in nature and a lagging current is drawn from the power lines. This requires additional generating capacity. This can be compensated by using capacitors which take a leading current in a.c. circuits. Hence the greatest use of power capacitors is with the power frequency systems. Capacitors are made in simple units with voltage ratings for 220 to 13800 V with kVAR ratings varying from 0.5 to 25 kVAR. Power capacitors are normally made using impregnated paper dielectric. Power capacitors are also used for high frequency applications such as power factor correction in high frequency heaters and induction furnaces. At high frequencies the dielectric losses increase very rapidly, and the capacitors have to be cooled externally using water cooling. Capacitors are also used in d.c. applications such as impulse voltage generators, energy storage, welding and high intensity flash x-ray and light photography.

Generally, power capacitors are made of several layers of insulation paper of adequate thickness and aluminium foil of 6 microns thickness as electrodes interleaved and wound. Single units are connected in parallel depending on the rating of the capacitor unit to be manufactured. These are placed in containers hermetically sealed, thoroughly dried, and then impregnated with insulating oil. The impregnating oils used are either mineral oil or chlorinated diphenyl oil. Capacitors made with mineral oil are quite expensive, and hence capacitors made with chlorinated diphenyl are preferred for power factor correction applications because of their low cost and non-inflammability.

Properties required for the insulation paper for capacitor applications are high dielectric strength, low dielectric loss, high dielectric constant, uniform thickness, and minimum conducting particles. The recent discovery of polypropylene film has considerable power dielectric loss and higher operating voltage. However, paper is still widely used partly, mainly due to the reason that paper after impregnation offers many desirable properties required for use at high voltages in addition to economy.

The impregnant for power capacitors should provide high dielectric strength, dielectric constant equal to that of paper, high permeability to paper, and sufficient viscosity to enter the voids in paper. Its flash and solidifying points should be above 120°C and below −10°C respectively. The impregnants used are broadly classified into mineral oil and synthetic oil (askarels). The dielectric properties of the tissue paper and the capacitor impregnants are given in Tables 5.4 and 5.5.

Table 5.4 Characteristics of Tissue Paper

(a) Fibre Composition		
	Unbalanced sulphate	100% ASTM
	Ash	less than 0.3%
	Moisture	4 to 8%
(b) Water Extract		
	Conductivity	less than 4 μ Siemens/cm
	pH value	6.0-7.5
	Chloride content	less than 5 mg/kg
	Standard thickness	10, 12, 13, 14, 15, and 18 micron
	Conducting particles	nil
	Density	1 or 1.2 g/cm^3
	Dielectric strength	30–40 V (d.c.)/micron

Table 5.5 Properties of Impregnating Oils

Property	Mineral oil	Trichlorodiphenyl (Askarel)
Specific gravity	0.90	1.378
Boiling point (°C)	270-300	325-360
Flash point (°C)	152	None
Solidifying point (°C)	−40	−19
Dielectric constant	2.25	5-8
Minimum resistivity	10^{12} ohms at 100°C	10^{12} ohms at 100°C
tan δ at 50°C	5×10^{-4}	2×10^{-3}

The electrode material extensively used in aluminium foil of 6 microns thickness because of its high tensile strength, low specific resistance, high melting point, low specific gravity, low cost and easy availability.

As already mentioned, the latest trend in capacitor manufacture is to replace paper by polypropylene plastic films. This results in a drastic reduction in size. Its use results in cheaper capacitors for high voltage ratings because of its high working stress. As regards impregnants, askarels are harmful to the environment and hence are being banned. The latest trend is to develop other types of materials. With this in view, research is being directed towards the use of vegetable oils like castor oil.

5.7 APPLICATIONS IN ELECTRONIC EQUIPMENT

The progress in electronic industry depends on the availability of dielectric materials. Modern electronic components and instruments are very complex, and their performance will depend on the nature and reliability of dielectric materials used in their construction. The electronic equipments have to operate with d.c. and a.c. voltages, under varying humidity and temperature conditions. No single material can meet these requirements, and different materials are used under different conditions.

The general properties of insulating materials for electronic use are the same as those in electrical industries, but the relative importance is different. Properties such as electrical and mechanical strength, and thermal and electrochemical endurance are important. In addition, size, reliability, and failure rates are important factors in selecting materials for electronic industries. There is an ever increasing demand from the electronics industries to make components which are smaller, more reliable, more stable, and capable of operating under adverse electrical environmental conditions.

Dielectric materials are divided into two different groups as far as applications in the electronic industry are concerned based on the frequency range over which they operate, namely polar dielectrics and the nonpolar dielectrics. The polar dielectrics are normally used for d.c. and power frequency (50 Hz) applications, while the nonpolar dielectrics are used at very high frequencies. Typical non-polar or low loss materials are polystyrene, polyethylene, teflon, quartz, etc. and typical polar substances are phenolic plastics, nylon, plasticised cellulose acetate, castor oil, etc.

5.7.1 Materials for Low Frequency Applications (Polar Materials)

Large quantities of wires and cables are used for connecting various components in electronic equipments. All these, except the high frequency cables, utilize polar dielectrics. Early polar dielectrics were natural rubber and textile fibre. These were highly susceptible to moisture absorption. They were replaced later on, by synthetic resins of thermoplastic type such as polyvinylchloride. The presence of plasticizer limits the use of these wires to high temperature work. Therefore, nylon jacketed P.V.C. wires were considered. But the susceptibility of nylon at high temperatures and the resultant brittleness made this unsuitable for high temperature applications. This problem was overcome by the use of teflon for the insulation of cables. Teflon insulated cables are very compact, possess very high resistance, and can be operated up to 250°C. The only disadvantage is their very high cost.

5.7.2 Materials for High Frequency Applications

As already mentioned, the examples of materials of this group are teflon, polyethylene and polystyrene which possess the required ideal properties for high frequency applications, such as low dielectric constant, high dielectric strength, and low dielectric loss. The properties of all these materials were already discussed in the previous chapter.

5.7.3 Materials for Resistors

Resistors are of two types, namely fixed and variable. The necessity to select proper insulating materials is more important in the case of fixed resistors to achieve stability and compactness. A resistor will have two insulating components, the core and the encapsulant. The core is normally made out of high grade non-porous electro-ceramic material. The core material may have to withstand about 1000°C, when a carbon film is deposited on it. Epoxy resin is a very popular encapsulant. It has high insulation resistance, is impervious to moisture, and has good heat resistance and thermal conductivity. For wire wound resistors, again high grade ceramic is used for the core, and vitreous enamel is used as a protective coating. The power handling capacity of this type of capacitors is high, and the protective covering has to withstand much higher temperatures.

5.7.4 Materials of Electronic Capacitors

Capacitors were also made of both fixed and variable types. Electrolytic capacitors are used for d.c. applications, while capacitors made of paper, plastic film, mica, ceramic, and semiconductor materials are used for a.c. applications.

As already discussed, paper capacitors are widely used for power and high voltage applications. In the field of electronics and communication engineering other types of capacitors are popular. Polystyrene film, polyester, and polyprophylene are the popular dielectrics for these applications. Polystyrene film capacitors are extremely stable in capacitance value and can be used up to several mega-cycles. The main disadvantage is their poor heat resistance. They can be used only in the temperature range between – 40°C and + 70°C. Except the limitations on the operating temperature range, polystyrene film capacitors are very popular and have replaced mica capacitors for many applications.

Polyester capacitors, on the other hand, have a wide operating temperature range, low water absorption, and low dielectric losses and are fast replacing paper capacitors. Polyester film capacitors coated with epoxy resin are extensively used in radio frequency circuits for by-pass, inter-stage coupling, etc. They can be used up to very high frequencies. Metallized polyester film, developed in recent years, is a big boon to the capacitor industry. Use of this film drastically reduces the size of the capacitors.

Polypropylene capacitors are also becoming popular. They have similar properties as polyster capacitors, but are much cheaper in price. They can be used as low frequency a.c. capacitors, but they have to compete in price with paper impregnated capacitors.

Polyethylene film has excellent electrical properties but poor mechanical properties and hence cannot be drawn into a very thin film. Therefore, it cannot be used for low voltage capacitors, but can be employed in high voltage capacitors.

Teflon also cannot be easily drawn into thin film, and hence it has not become very popular as a capacitor dielectric. Another factor is its high cost.

Mica capacitors are made using mica as the dielectric. Mica has got all the good properties of a dielectric. The encapsulant is again an epoxy resin to protect the capacitors from the environment. To limit the loss of properties of the mica element, to fully realize the stability and the operating temperature, it is housed in a metal can filled with an inert atmosphere, and the leads are brought through ceramic seals. Ceramic capacitors are made of ceramic, and are encapsulated by a dip coating in phenolic resin; if higher operating temperatures are expected, an epoxy resin coating is given.

For electrolytic capacitors, the dielectric is the oxide layer formed on etched aluminium or tantalum foil. The foils are separated by electrolytic grade Kraft paper. The dielectric layer is very thin; hence the insulation resistance is low, and the leakage current is high. The stability of the oxide layer and hence the stability of the capacitance value depends on the type and purity of the electrolyte. Tantalum capacitors give stable capacitance with wider operating temperatures and lower leakage currents.

5.7.5 Materials for Professional Grade Electronic Components

Resistors, capacitors and other electronic components used in defence equipment, computers, space applications, etc. should be highly reliable and must be capable of stable operation under severe environmental conditions. These are called professional grade components, and they should be superior in characteristics regarding reliability, stability, close tolerance, wider range of operating temperatures, superior electrical and electronic properties, and capacity to operate under severe environmental conditions. The materials to achieve these superior characteristics should be chosen very carefully.

Dielectric materials used in professional grade components should serve the dual purpose of providing electrical insulation and protection from adverse environmental conditions. We have already elaborated on the materials with superior insulation properties used in resistors and capacitors. The protection against atmospheric influences has to be almost perfect, and this is achieved by placing the capacitors in a nonferrous metal can and bringing out the leads through glass to metal seals. The next method in order of preference is encasing in metal cans and using ceramic to metal seals. Further down in priority would be to use metal can with ceramic bushings, or an epoxy free moulded can with epoxy end sealing, or a ceramic can with epoxy end sealing, etc.

5.7.6 Materials for Electromechanical Components

There are a large number of electromechanical components used in electronic industry such as relays, connectors, valve bases, terminals, terminal boards, bushes, plugs and sockets, etc. These are made from pressed parts or moulded parts.

One of the most commonly used material for pressed parts is industrial grade phenolic resin laminate. There are various grades in these laminates, and choice can be made based on the insulation properties and resistance to water absorption. This is commonly used for entertainment grade components used in radio and television sets. Glass epoxy laminates are used for printed circuit boards and some terminal boards. Phenolic resin bonded cellulose acetate paper laminates and polyester filled with traces of glass fibre and laminated with epoxy resin are becoming increasingly popular now-a-days. The latter have excellent electrical properties, low water absorption, and good punchability. Insulating plastic parts subjected to wear and those that come into contact with metal parts are made from polypropylene sheets.

Moulded insulating parts for electromechanical components are made from a number of plastic materials. When higher operating temperatures are not required, high impact polystyrene is used. For operation with higher temperatures, polycarbonate moulding is employed. When higher mechanical strength is also desired along with higher operating temperatures, nylon moulding is used. When higher mechanical strength, high insulation resistance, high operating temperatures, and dimensional stability are required, mica filled bakelite or glass-filled diallylphthalate are used, and parts are compression moulded. Diallyphthalate is used for professional grade components such as, valve bases, printed circuit board connectors, multipin connectors, etc. For professional grade components, where vibration and shock are not present, steatite is also used.

REFERENCES

1. Clark, F.M. *Insulating Materials for Design and Engineering Practice,* John Wiley, New York (1962).
2. Black, R.M. and Reynolds, E.H., *Proc. I.E.E.* **112**(6), 1226 (1965).
3. Birks, J.B., *Modern Insulating Materials,* Haywood, London (1960).
4. Koritsky, Y.U., *Electrical Engineering Materials,* Mir Publishers, Moscow (1970).
5. Mason, R.W. and Gorton. G., *Insulation for Small Transformers,* ERA Publication, Leatherhead, Surrey, England (1959).
6. Insulation Encyclopedia, *Journal of Insulation,* June-July (1970).
7. Insulation Directory, p. 163-207 (1969).
8. Palmer, S. and Sharpley, W.A., *Proc. I.E.E.,* **116,** 2029 (1969).
9. Micafil News, MNV 45/3e, August (1968).
10. Sillars, R.W., *Electrical Insulating Materials and their Applications,* Peter Peregrinus, London (1973).
11. Schewarz, K.K., *Proc. I.E.E.,* **116** (10), 1735 (1969).
12. Guide for the Determination of Thermal Endurance Properties of Electrical Insulating Materials, I.E.C. *Publication,* 216-1, 1974.
13. Ringer W., "Some Problems of High Voltage Instrument Transformers", *Bulletin No. 22, Swiss Association of Electrical Engineers* (1963).
14. Greenway, R. and Ryan, H.M., *I.E.E. Conference on Metalclad Switchgears,* Pub. No. 83, Part 1, London (1972).
15. Indian Standard Specifications *IS: 6380-1971, IS: 692-1973, IS: 6474-1971, IS: 1554-1964,* and *IS: 7098-1973.*
16. Von Hippel, A., Dielectric Materials and Applications, M.I.T. Press, Boston, Mass. (1962).

6

Generation of High Voltages and Currents

In the fields of electrical engineering and applied physics, high voltages (d.c., a.c., and impulse) are required for several applications. For example, electron microscopes and x-ray units require high d.c. voltages of the order of 100 kV or more. Electrostatic precipitators, particle accelerators in nuclear physics, etc. require high voltages (d.c.) of several kilovolts and even megavolts. High a.c. voltages of one million volts or even more are required for testing power apparatus rated for extra high transmission voltages (400 kV system and above). High impulse voltages are required for testing purposes to simulate overvoltages that occur in power systems due to lightning or switching surges. For electrical engineers, the main concern of high voltages is for the insulation testing of various components in power systems for different types of voltages, namely, power frequency a.c., high frequency, switching or lightning impulses. Hence, generation of high voltages in laboratories for testing purposes is essential and is discussed in this chapter.

Different forms of high voltages mentioned above are classified as

    (*i*) high d.c. voltages

    (*ii*) high a.c. voltages of power frequency.

    (*iii*) high a.c. voltages of high frequency.

    (*iv*) high transient or impulse voltages of very short duration such as lightning overvoltages, and

    (*v*) transient voltages of longer duration such as switching surges.

Normally, in high voltage testing, the current under conditions of failure is limited to a small value (less than an ampere in the case of d.c. or a.c. voltages and few amperes in the case of impulse or transient voltages). But in certain cases, like the testing of surge diverters or the short circuit testing of switchgear, high current testing with several hundreds of amperes is of importance. Tests on surge diverters require high surge currents of the order of several kiloamperes. Therefore, test facilities require high voltage and high current generators. High impulse current generation is also required along with voltage generation for testing purposes.

6.1 GENERATION OF HIGH d.c. VOLTAGES

Generation of high d.c. voltages is mainly required in research work in the areas of pure and applied physics. Sometimes, high direct voltages are needed in insulation tests on cables and capacitors. Impulse generator charging units also require high d.c.

voltages of about 100 to 200 kV. Normally, for the generation of d.c. voltages of up to 100 kV, electronic valve rectifiers are used and the output currents are about 100 mA. The rectifier valve require special construction for cathode and filaments since a high electrostatic field of several kV/cm exists between the anode and the cathode in the non-conduction period. The a.c. supply to the rectifier tubes may be of power frequency or may be of audio frequency from an oscillator. The latter is used when a ripple of very small magnitude is required without the use of costly filters to smoothen the ripple.

6.1.1 Half and Full Wave Rectifier Circuits

Rectifier circuits for producing high d.c. voltages from a.c. sources may be (a) half wave, (b) full wave, or (c) voltage doubler type rectifiers. The rectifier may be an electron tube or a solid state device. Now-a-days single electron tubes are available for peak inverse voltages up to 250 kV, and semiconductor or solid state diodes up to 20 kV. For higher voltages, several units are to be used in series. When a number of units are used in series, transient voltage distribution along each unit becomes non-uniform and special care should be taken to make the distribution uniform.

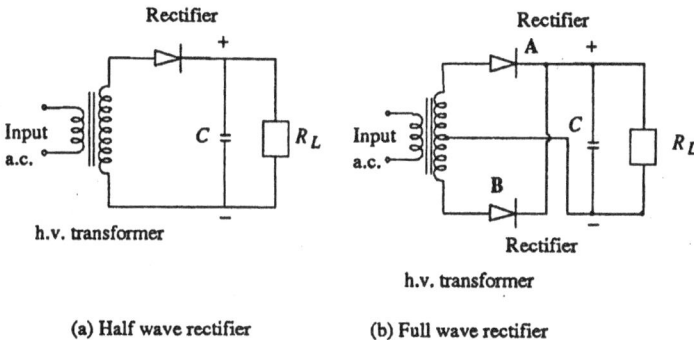

(a) Half wave rectifier   (b) Full wave rectifier

Fig. 6.1   Full and half wave rectifiers

Commonly used half wave and full wave rectifiers are shown in Fig. 6.1. In the half wave rectifier (Fig. 6.1a) the capacitor is charged to V_{max}, the maximum a.c. voltage of the secondary of the high voltage transformer in the conducting half cycle. In the other half cycle, the capacitor is discharged into the load. The value of the capacitor C is chosen such that the time constant CR_L is at least 10 times that of the period of the a.c. supply. The rectifier valve must have a peak inverse rating of at least $2\,V_{max}$. To limit the charging current, an additional resistance R is provided in series with the secondary of the transformer (not shown in the figure).

A full wave rectifier circuit is shown in Fig. 6.1b. In the positive half cycle, the rectifier A conducts and charges the capacitor C, while in the negative half cycle the rectifier B conducts and charges the capacitor. The source transformer requires a centre tapped secondary with a rating of $2V$.

For applications at high voltages of 50 kV and above, the rectifier valves used are of special construction. Apart from the filament, the cathode and the anode, they contain a protective shield or grid around the filament and the cathode. The anode will be usually a circular plate. Since the electrostatic field gradients are quite large, the heater and the cathode experience large electrostatic forces during the non-conduction periods. To protect the various elements from these forces, the anode is firmly fixed to the valve cover on one side. On the other side, where the cathode and filament are located, a steel mesh structure or a protective grid kept at the cathode potential surrounds them so that the mechanical forces between the anode and the cathode are reflected on the grid structure only.

In modern high voltage laboratories and testing installations, semiconductor rectifier stacks are commonly used for producing d.c. voltages. Semiconductor diodes are not true valves since they have finite but very small conduction in the backward direction. The more commonly preferred diodes for high voltage rectifiers are silicon diodes with peak inverse voltage (P.I.V.) of 1 kV to 2 kV. However, for laboratory applications the current requirement is small (a few milliamperes, and less than one ampere) and as such a selenium element stack with P.I.V. of up to 500 kV may be employed without the use of any voltage grading capacitors.

Both full wave and half wave rectifiers produce d.c. voltages less than the a.c. maximum voltage. Also, ripple or the voltage fluctuation will be present, and this has to be kept within a reasonable limit by means of filters.

Ripple Voltage with Half Wave and Full Wave Rectifiers

When a full wave or a half wave rectifier is used along with the smoothing condenser C, the voltage on no load will be the maximum a.c. voltage. But when on load, the condenser gets charged from the supply voltage and discharges into load resistance R_L whenever the supply voltage waveform varies from peak value to zero value. These waveforms are shown in Fig. 6.2. When loaded, a fluctuation in the output d.c. voltage δV appears, and is called a ripple. The ripple voltage δV is larger for a half wave rectifier than that for a full wave rectifier, since the discharge period in the case of half wave rectifier is larger as shown in Fig. 6.2. The ripple δV depends on (a) the supply voltage frequency f, (b) the time constant CR_L, and (c) the reactance of the supply transformer X_L. For half wave rectifiers, the ripple frequency is equal to the supply frequency and for full wave rectifiers, it is twice that value. The ripple voltage is to be kept as low as possible with the proper choice of the filter condenser and the transformer reactance for a given load R_L.

6.1.2 Voltage Doubler Circuits

Both full wave and half wave rectifier circuits produce a d.c. voltage less than the a.c. maximum voltage. When higher d.c. voltages are needed, a voltage doubler or cascaded rectifier doubler circuits are used. The schematic diagram of voltage doublers are given in Figs. 6.3a and b.

In voltage doubler circuit shown in Fig. 6.3a, the condenser C_1 is charged through rectifier **R** to a voltage of $+V_{max}$ with polarity as shown in the figure during the negative half cycle. As the voltage of the transformer rises to positive V_{max} during the next half cycle, the potential of the other terminal of C_1 rises to a voltage of $+2V_{max}$.

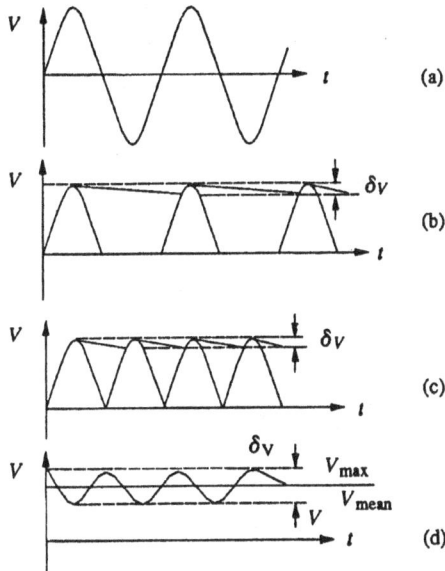

Fig. 6.2 Input and output waveforms of half and full wave rectifiers

(a) Input sine wave
(b) Output with half wave rectifier and condenser filter
(c) Output with full wave rectifier and condenser filter
(d) V_{max}, V_{mean} and ripple voltage and δV with condenser filter of a full wave rectifier

Thus, the condenser C_2 in turn is charged through R_2 to $2V_{max}$. Normally the d.c. output voltage on load will be less than $2V_{max}$, depending on the time constant $C_2 R_L$ and the forward charging time constants. The ripple voltage of these circuits will be about 2% for $R_L/r \leq 10$ and $X/r \leq 0.25$, where X and r are the reactance and resistance of the input transformer. The rectifiers are rated to a peak inverse voltage of $2V_{max}$, and the condensers C_1 and C_2 must also have the same rating. If the load current is large, the ripple also is more.

Cascaded voltage doublers are used when larger output voltages are needed without changing the input transformer voltage level. A typical voltage doubler is shown in Fig. 6.3b and its input and output waveforms are shown in Fig. 6.3(c). The rectifiers R_1 and R_2 with transformer T_1 and condensers C_1 and C_2 produce an output voltage of $2V$ in the same way as described above. This circuit is duplicated and connected in series or cascade to obtain a further voltage doubling to $4V$. T is an isolating transformer to give an insulation for $2V_{max}$ since the transformer T_2 is at a potential of $2V_{max}$ above the ground. The voltage distribution along the rectifier string R_1, R_2, R_3 and R_4 is made uniform by having condensers C_1, C_2, C_3 and C_4 of equal values. The arrangement may be extended to give $6V$, $8V$, and so on by repeating further stages with suitable isolating transformers. In all the voltage doubler circuits, if valves are used, the filament transformers have to be suitably designed and insulated, as all the cathodes will not be at the same potential from ground. The arrangement becomes cumbersome if more than $4V$ is needed with cascaded steps.

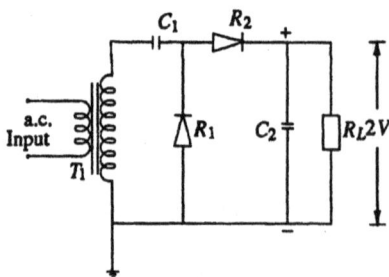

Fig. 6.3a   Simple voltage doubler

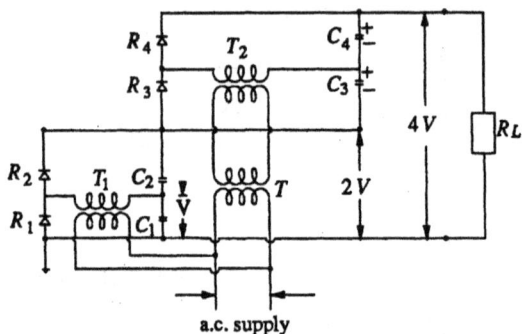

Fig. 6.3b   Cascaded voltage doubler

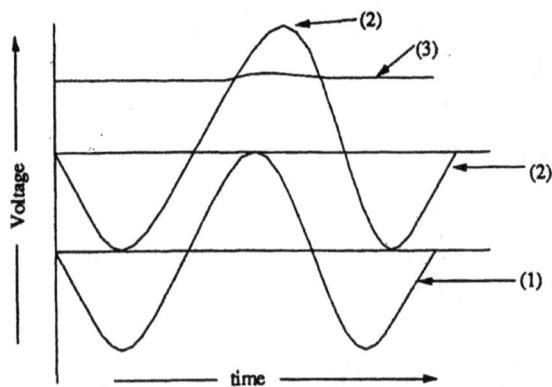

Fig. 6.3 (c)   Waveforms of a.c. voltage and the d.c. output voltage on
no-load of the voltage doubler shown in Fig. 6.3 (b)

(1) a.c input voltage waveform, (2) a.c. output voltage waveform without
condenser filter, (3) a.c. output voltage waveform with condenser filter

Fig. 6.3   Voltage doubler circuits

T_1, T_2 – h.v. transformers; R_1, R_2, R_3, R_4 – rectifiers;
C_1, C_2, C_3 – condensers; R_L – Load resistance; T – isolating transformer

6.1.3 Voltage Multiplier Circuits

Cascaded voltage multiplier circuits for higher voltages are cumbersome and require too many supply and isolating transformers. It is possible to generate very high d.c. voltages from single supply transformers by extending the simple voltage doubler circuits. This is simple and compact when the load current requirement is less than one milliampere, such as for cathode ray tubes, etc. Valve type pulse generators may be used instead of conventional a.c. supply and the circuit becomes compact. A typical circuit of this form is shown in Fig. 6.4a.

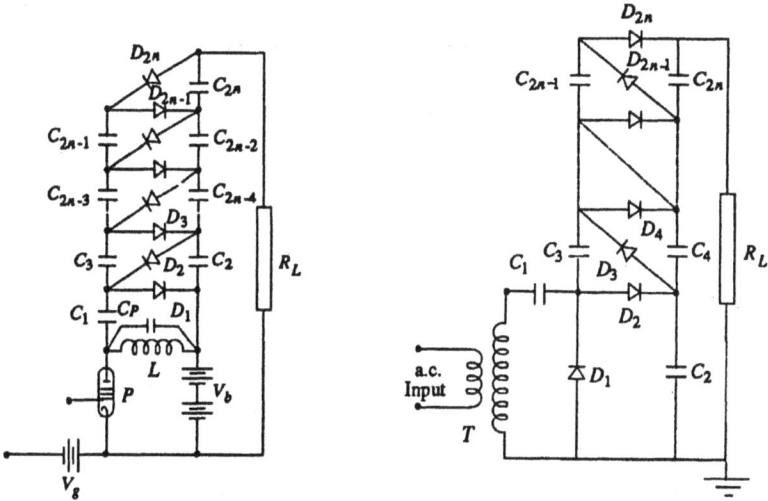

Fig. 6.4a Cascaded rectifier unit with pulse generator

Fig. 6.4b Cockcroft-Walton voltage multiplier circuit

P — Pulse generator

V_b — D.C. supply to pulse generator

V_g — Bias voltage

The pulses generated in the anode circuit of the valve P are rectified and the voltage is cascaded to give an output of $2nV_{max}$ across the load R_L. A trigger voltage pulse of triangular waveform (ramp) is given to make the valve switched on and off. Thus, a voltage across the coil L is produced and is equal to $V_{max} = I \sqrt{L/C_P}$, where C_P is the stray capacitance across the coil of inductance L. A d.c. power supply of about 500 V applied to the pulse generator, is sufficient to generate a high voltage d.c. of 50 to 100 kV with suitable number of stages. The pulse frequency is high (about 500 to 1000 Hz) and the ripple is quite low (<1%). The voltage drop on load is about 5% for load currents of about 150 μ A. The voltage drops rapidly at high load currents.

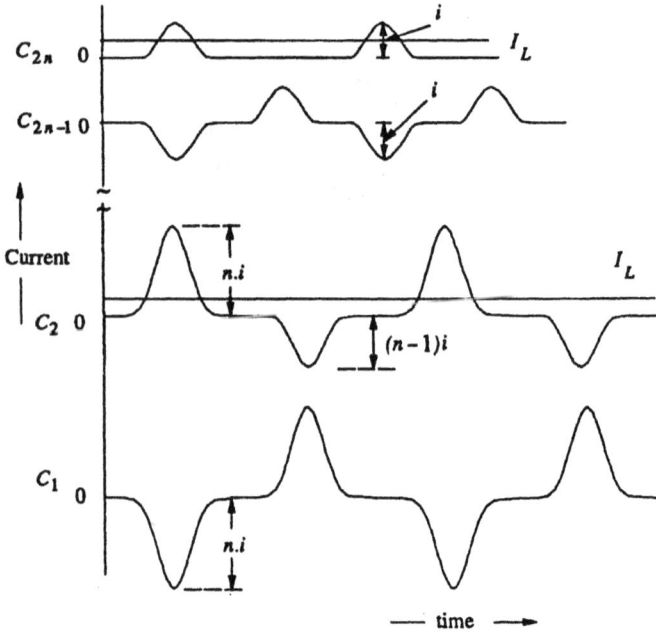

Fig. 6.4c   Schematic current waveforms across the first and the last capacitors of the cascaded voltage multiplier circuit shown in Fig. 6.4(b)

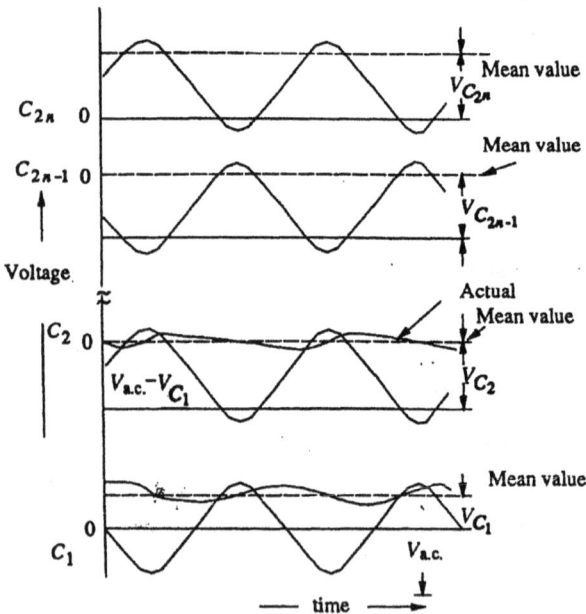

Fig. 6.4d   Voltage waveforms across the first and the last capacitors of the cascaded voltage multiplier circuit shown in Fig. 6.4(b)

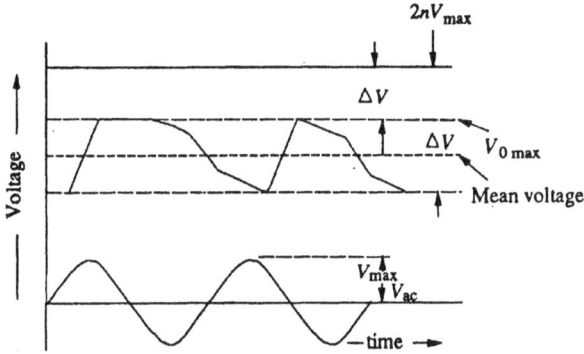

Fig. 6.4e   Ripple voltage δV and the voltage drop ΔV in a cascaded voltage multiplier circuit shown in Fig. 6.4(b)

Voltage multiplier circuit using the Cockcroft-Walton principle is shown in Fig. 6.4b. The first stage, i.e. D_1, D_2, C_1, C_2, and the transformer T are identical as in the voltage doubler shown in Fig. 6.3. For higher output voltage of $4, 6, \ldots 2n$ of the input voltage V, the circuit is repeated with cascade or series connection. Thus, the condenser C_4 is charged to $4V_{max}$ and C_{2n} to $2nV_{max}$ above the earth potential. But the volt across any individual condenser or rectifier is only $2V_{max}$.

The rectifiers $D_1, D_3, \ldots D_{2n-1}$ shown in Fig. 6.4b operate and conduct during the positive half cycles while the rectifiers $D_2, D_4 \ldots D_{2n}$ conduct during the negative half cycles. Typical current and voltage waveforms of such a circuit are shown in Figs. 6.4c and 6.4d respectively. The voltage on C_2 is the sum of the input a.c. voltage, V_{ac} and the voltage across condenser C_1, V_{c_1}) as shown in Fig 6.4. The mean voltage on C_2 is less than the positive peak charging voltage $(V_{ac} + V_{c_1})$. The voltages across other condensers C_2 to C_{2n} can be derived in the same manner, (i.e.) from the difference between voltage across the previous condenser and the charging voltage. Finally the voltage after $2n$ stages will be V_{ac} $(n_1 + n_2 + \ldots)$, where $n_1, n_2, \ldots$ are factors when ripple and regulation are considered in the next rectifier. The ripple voltage δV and the voltage drop ΔV in a cascaded voltage multiplier unit are shown in Fig. 6.4e.

Ripple in Cascaded Voltage Multiplier Circuits

With load, the output voltage of the cascaded rectifiers is less than $2nV_{max}$, where n is the number of stages. The ripple and the voltage regulation of the rectifier circuit may be estimated as follows.

Let f = supply frequency,

q = charge transferred in each cycle,

I_1 = load current from the rectifier,

t_1 = conduction period of the rectifiers

t_2 = non-conduction period of rectifiers, and

δV = ripple voltage.

Referring to Fig. 6.3a, when load current I_1 is supplied from condenser C_2 to load R_L during the non-conducting period, the charge transferred per cycle from the condenser C_2 to the load during the non-conduction period t_2 is q, and is related as follows.

$$I_1 = \frac{dq}{dt} \approx \frac{q}{t_2}$$

Since $t_1 \ll t_2$ and $t_1 + t_2 = \frac{1}{f}$ (i.e. the period of these a.c. supply voltage),

$$t_2 = \frac{1}{f}$$

Also,

$$q = C_2 \, \delta V$$

Hence,

$$\delta V = \text{the ripple} = \frac{I_1}{f C_2}$$

At the same time a charge q is transferred from C_1 to C_2 during each cycle of $x = \frac{I_1}{f C_2}$.

Hence, regulation = mean voltage drop from $2V_{max}$

$$= \frac{I_1}{f}\left[\frac{1}{C_1} + \frac{1}{2C_2}\right] \tag{6.1}$$

Therefore, the mean output voltage = $2V_{max} - \frac{I_1}{f}\left(\frac{1_1}{C_1} + \frac{1}{2C_2}\right)$.

For the cascade circuit, on no load, the voltages between stages are raised by $2V_{max}$ giving an output voltage of $2nV_{max}$ for n stages.

Referring to Fig. 6.4b, to find an expression for the total ripple voltage, let it be assumed that all capacitances C_1, C_2, ..., C_{2n} be equal to C. Let q be the charge transferred from C_{2n} to the load per cycle.

Then the ripple at the condenser C_{2n} will be $\frac{I_1}{fC}$.

Simultaneously, C_{2n-2} transfers as charge q to the load and to C_{2n-1}.

Hence, the ripple at the condenser C_{2n-2} is $\frac{2I_1}{fC}$.

Similarly, C_{2n-4} transfers a charge q to the load, to C_{2n-3}, and to C_{2n-2}.

Therefore, the ripple at condenser C_{2n-4} is $\frac{3I_1}{fC}$.

Proceeding in the same way, the ripple at C_2 will be $\frac{nI_1}{fC}$.

Hence, for n stages the total ripple will be

$$\delta V_{total} = \frac{I_1}{fC}[1 + 2 + 3 \ldots + n] = \frac{I_1'}{fC} \frac{n(n+1)}{2}. \tag{6.2}$$

It can be seen from the above expression that the lowest capacitances (C_2, C_4, etc.) contribute most for the ripple. If these capacitances are increased proportionately, i.e. C_1 and C_2 are made equal to nC, C_3 and C_4 are made equal to $(n-1)C$ and so on, the total ripple will be only $\frac{nI_1}{fC}$.

Regulation or Voltage Drop on Load

In addition to the ripple δV there is a voltage drop ΔV, which is the difference between the theoretical no load and the on load voltage.

To find ΔV, the voltage drop under load condition, it is assumed that all the capacitors C_1, C_2 ...C_{2n} have equal value, C. From the analysis of the ripple voltage it may be seen that all the capacitors are not charged to $2V_{max}$. The condenser C_2 is charged to $[2V_{max} - 2nI_1/fC]$ volts. Similarly, condenser C_3 is charged to a voltage of $[2V_{max} - 2nI_1/fC] - (2n - 1)I_1/fC]$ volts. This is because, condenser C_3 during the conduction period transfers a charge of $(2n - 1)I_1/fC$ to the load and to the next stage.

Thus, the voltage drops at various condenser stages. Taking them in pairs, for a stack of $2n$ stages, the drop will be

$$\Delta V_{2n} = nI_1/fC \text{ (due to the ripple from } n \text{ stages.)}$$
$$= I_1/fC \, (2 \, n - n).$$
$$\Delta V_{2n-2} = (n-1) I_1/fC + 2n \, I_1/fC \text{ (due to ripple and voltage drop due to charge transfer to the next stage)}$$
$$= (I_1/fC) \, [2n + (n-1)]$$
$$= I_1/fC \, [2n + 2(n-1) - (n-1)]$$
$$\Delta V_4 = I_1/fC \, [2n + 2(n-1) + \ldots + 2]$$
$$= I_1/fC \, [2n + 2(n-1) + \ldots + 2\cdot2 - 2]$$
$$\Delta V_2 = I_1/fC \, [2n + 2(n-1) + \ldots + 2\cdot2 - 1]$$
$$= I_1/fC \, [2 \, n + 2 \, (n-1) + \ldots + 2\cdot2 + 2\cdot1 - 1].$$

So that,

$$\Delta V = (I_2/fC) \sum_{r=1}^{n} \Delta V_{2r} = (I_1/fC) \left[\sum_1^n n(2n) - \sum_1^n n \right]$$

$$= (I_1/fC) \sum_1^n n(2n-1)$$

$$= (I_1/fC) \, [2.n.(n+1)(2n+1)/6 - n(n+1)/2]$$
$$= (I_1/fC) \, [(2/3).n^3 + n^2/2 - n/6] \tag{6.3}$$

Here also, it is seen that most of the voltage drop is due to the lowest stage condensers C_1, C_2, etc. Hence, it is advantageous to increase their values proportional to the number of the stage from the top.

For large values of n (≥ 5), $\dfrac{n^2}{2}$ and $\dfrac{n}{6}$ terms in Eq. (6.3) will becomes small compared to $\dfrac{2}{3} n^3$ and may be neglected; then the optimum number of stages for the minimum voltage drop may be expressed as

$$n_{optimum} = \sqrt{\frac{V_{max} fC}{I}} \qquad (6.4)$$

where I is the load current.

Thus, for a multiplier or a cascaded circuit with $f = 50$ Hz, $C = 0.1\ \mu F$, $V_{max} = 100$ kV and $I = 5$ mA, the number of stages $n \approx 10$.

The regulation can be improved by increasing f, but an upper limit is set by the high voltage appearing across the inductances and high capacitor currents which are considerable. At present, the Cockcroft-Walton type voltage multipliers are available using selenium rectifiers and a.c. supply frequencies of 500 to 1000 Hz for output voltages of more than one million volts and load currents of 30 mA.

Cascaded Modular Voltage Multipliers or "Deltatron" Circuits for Very High Voltages

A combination of Cockcroft-Walton type voltage multiplier with cascaded transformer d.c. rectifier is developed recently for very high voltages but limited output currents having high stability, small ripple factor and fast regulation. One such unit is recently patented by "ENGE" in U.S.A., called "ENGETRON" or "DELTATRON". The schematic diagram of a typical Deltatron unit is shown in Fig. 6.5.

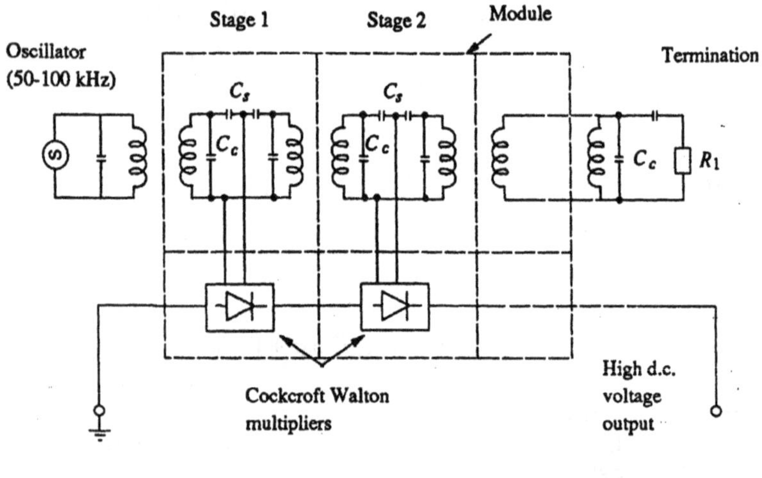

Fig. 6.5 Deltatron unit for generation of very high d.c. voltages

Basically this circuit consists of Cockcroft-Walton multiplier units fed from a supply transformer unit (stage 1). These supply transformers are air-cored to have low inductance and are connected in series through capacitors C. In addition, the windings of the transformers are shunted by a capacitor C to compensate for the magnetising current. The entire unit is terminated by a load resistor R_1. All the Cockcroft-Walton multipliers are connected in series and the entire unit is enclosed in a cylindrical vessel insulated by SF_6 gas. Each stage of the unit is typically rated for 10 to 50 kV, and about 20 to 25 stages are used in a unit. The whole assembly is usually of smaller size and weight than a cascaded rectifier unit. The supply frequency to the transformers is from a high frequency oscillator (50 to 100 kHz) and as such the capacitors used are of smaller value. The voltage regulation system is controlled by a parallel R-C divider which in turn controls the supply oscillator. Regulation due to load variations or power source voltage variations is very fast (response time < 1 ms). The disadvantage of this circuit is that the polarity of the unit cannot be reversed easily. Typical units of this type may have a rating of 1 MV, 2 mA with each module or stage rated for 50 kV with ripple content less than 1%.

6.1.4 Electrostatic Machines: Basic Principle

In electromagnetic machines, current carrying conductors are moved in a magnetic field, so that the mechanical energy is converted into electrical energy. In electrostatic machines charged bodies are moved in an electric field against an electrostatic field in order that mechanical energy is converted into electrical energy. Thus, if an insulated belt with a charge density δ moves in an electric field "$E(x)$" between two electrodes with separation 's' then

   (i) the charge on the strip of belt at a distance dx is $dq = \delta\, b.dx$ where b is the width of the belt, and
   (ii) the force on the belt, F is

$$F = \int_{o}^{s} E(x) \cdot dq = \int_{o}^{s} \delta \cdot b \cdot E(x)dx$$

If the belt moves with a velocity, v, then the mechanical power P, required to move the belt is $P = F \cdot v = \delta \cdot b \cdot v \int_{o}^{s} E(x)\, dx$

The current, I, in the system is given as

$$I = dq/dt = \delta\, b \cdot dx/dt = \delta \cdot b \cdot v$$

and the potential difference, V, between the electrodes is

$$V = \int_{o}^{s} E(x)\, dx$$

Thus, in an electrostatic machine, the mechanical power required to move the belt at a velocity v, i.e. $P = F.v$ is converted into the electrical power, $P = V.I$, assuming that there are no losses in the system. The Van de Graaff generator is one such electrostatic machine which generates very high voltages, with small output current.

Van de Graaff Generators

The schematic diagram of a Van de Graaff generator is shown in Fig. 6.6. The generator is usually enclosed in an earthed metallic cylindrical vessel and is operated under pressure or in vacuum. Charge is sprayed on to an insulating moving belt from corona points at a potential of 10 to 100 kV above earth and is removed and collected from the belt connected to the inside of an insulated metal electrode through which the belt moves. The belt is driven by an electric motor at a speed of 1000 to 2000 metres per minute. The potential of the high voltage electrode above the earth at any instant is $V = Q/C$, where Q is the charge stored and C is the capacitance of the high voltage electrode to earth. The potential of the high voltage electrode rises at a rate

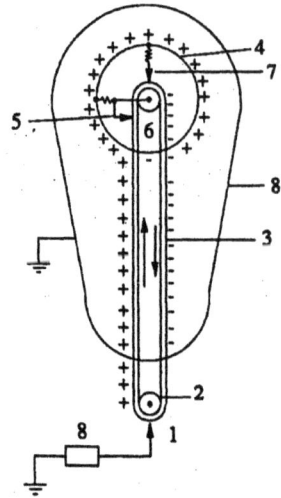

Fig. 6.6 Van de Graaff generator

1. Lower spray point
2. Motor driven pulley
3. Insulated belt
4. High voltage terminal
5. Collector
6. Upper pulley insulated from terminal
7. Upper spray point
8. Earthed enclosure

$$\frac{dV}{dt} = \frac{1}{C}\frac{dQ}{dt} = \frac{I}{C}$$ where I is the net charging current.

A steady potential will be attained by the high voltage electrode when the leakage currents and the load current are equal to the charging current. The shape of the high voltage electrode is so made with re-entrant edges as to avoid high surface field gradients, corona and other local discharges. The shape of the electrode is nearly spherical.

The charging of the belt is done by the lower spray points which are sharp needles and connected to a d.c. source of about 10 to 100 kV, so that the corona is maintained between the moving belt and the needles. The charge from the corona points is collected by the collecting needles from the belt and is transferred on to the high voltage electrode as the belt enters into the high voltage electrode. The belt returns with the charge dropped, and fresh charge is sprayed on to it as it passes through the lower corona point. Usually in order to make the charging more effective and to utilize the return path of the belt for charging purposes, a self-inducing arrangement or a second corona point system excited by a rectifier inside the high voltage terminal is employed. To obtain a self-charging system, the upper pulley is connected to the collector needle and is therefore maintained at a potential higher than that of the high voltage terminal. Thus a second row of corona points connected to the inside of the high voltage terminal and directed towards the pulley above its point of entry into the terminal gives a corona discharge to the belt. This neutralizes any charge on the belt and leaves an excess of opposite polarity to the terminal to travel down with the belt to the bottom charging point. Thus, for a given belt speed the rate of charging is doubled.

The charging current for unit surface area of the belt is given by $I = bv \, \delta$, where b is the breadth of the belt in metres, v is the velocity of the belt in m/sec, and δ is the surface charge density in coulombs/m^2. It is found that δ is $\leq 1.4 \times 10^{-5}$ C/m^2 to have a safe electric field intensity normal to the surface. With $b = 3$ m and $v = 3$ m/sec, the charging current will be approximately 125 μA. The generator is normally worked in a high pressure gaseous medium, the pressure ranging from 5 to 15 atm. The gas may be nitrogen, air, air-freon (CCl_2F_2) mixture, or sulphur hexafluoride (SF_6).

Van de Graaff generators are useful for very high voltage and low current applications. The output voltage is easily controlled by controlling the corona source voltage and the rate of charging. The voltage can be stabilized to 0.01%. These are extremely flexible and precise machines for voltage control.

Electrostatic Generators

Van de Graaff generators are essentially high voltage but low power devices, and their power rating seldom exceeds few tens of kilowatts. As such electrostatic machines which effectively convert mechanical energy into electrical energy using variable capacitor principle were developed. These are essentially duals of electromagnetic machines and are constant voltage variable capacitance machines. An electrostatic generator consists of a stator with interleaved rotor vanes forming a variable capacitor and operates in vacuum.

The current through a variable capacitor is given by $I = \dfrac{dV}{Cdt} + V\dfrac{dC}{dt}$ where C is a capacitor charged to a potential V.

The power input into the circuit at any instant is

$$P = VI = CV\frac{dV}{dt} + V^2\frac{dC}{dt} \tag{6.5}$$

If $\dfrac{dC}{dt}$ is negative, mechanical energy is converted into electrical energy.

With the capacitor charged with a d.c. voltage V, $\dfrac{dC}{dt} = 0$ and the power output will be

$P = V^2 \dfrac{dC}{dt}$.

A schematic diagram of a synchronous electrostatic generator with interleaved stator and rotor plates is shown in Fig. 6.7. The rotor is insulated from the ground, and is maintained at a potential of $+V$. The rotor to stator capacitance varies from C_0 to C_m, and the stator is connected to a common point between two rectifiers across the d.c. output which is $-E$ volts. When the capacitance of the rotor is maximum (C_m), the rectifier **B** does not conduct and the stator is at ground potential. The potential E is

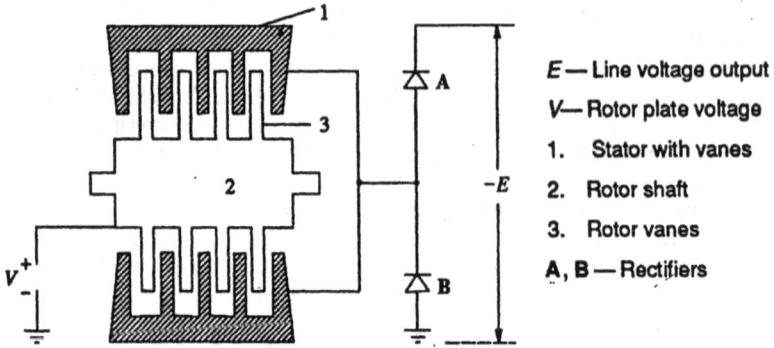

Fig. 6.7 Electrostatic generator

E — Line voltage output
V — Rotor plate voltage
1. Stator with vanes
2. Rotor shaft
3. Rotor vanes
A, B — Rectifiers

applied across the rectifier A and V is applied across C_m. As the rotor rotates, the capacitance C decreases and the voltage across C increases.

Thus, the stator becomes more negative with respect to ground. When the stator reaches the line potential $-E$ the rectifier A conducts, and further movement of the rotor causes the current to flow from the generator. Rectifier B will now have E across it and the charge left in the generator will be $Q_0 = C_0 (V + E) + E(C_s + C_r)$, where C_s is the stator capacitance to earth, C_r is the capacitance of rectifier B to earth, and C_0 is the minimum capacitance value of C (stator to rotor capacitance).

A generator of this type with an output voltage of one MV and a field gradient of 1 MV/cm in high vacuum and having 16 rotor poles, 50 rotor plates of 4 feet maximum and 2 feet minimum diameter, and a speed of 4000 rpm would develop 7 MW of power.

6.1.5 Regulation of d.c. Voltages

The output voltage of a d.c. source, whether it is a rectifier or any other machine, changes with the load current as well as with the input voltage variations. In order to maintain a constant voltage at the load terminals, it is essential to have a regulator circuit which corrects the variation in voltage. It is essential to keep the change in voltage between ± 0.1% and ± 0.001% depending on the applications.

A d.c. voltage regulator consists of detecting elements which sense the change in voltage from the desired value and controlling elements actuated by the detector in

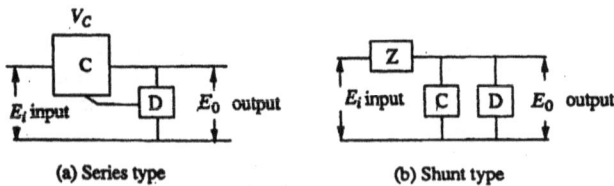

(a) Series type   (b) Shunt type

Fig. 6.8 Schematic diagram of voltage stabilizers

such a manner as to correct the changes. The regulators are generally of two types (*a*) series type, and (*b*) shunt or parallel type. Schematic diagrams of these regulators are given in Fig. 6.8.

If ΔE_0 is the change in E_0 as a result of a change of ΔE_i in E_i, then the stabilization ratio S is defined as

$$S = [\Delta E_i / E_i] / [\Delta E_0 / E_0]$$

$$= \frac{\Delta E_i}{\Delta E_0} \frac{E_0}{E_i}. \tag{6.6}$$

A second parameter R_0 is introduced to define completely the functional performance of a regulator. R_0 is the effective internal resistance of the regulator as seen from the output terminals.

$$R_0 = \frac{\Delta E_0}{\Delta I_0}$$

where ΔE_0 is the change in the output voltage caused due to a change in load current of ΔI_0, with the input voltage remaining constant.

The regulation R is defined as,

$$R = [\Delta E_0 / E_0] / [\Delta I_0 / I_0] = \frac{R_0}{R_L} \tag{6.7}$$

where R_L is the load resistance.

Typical series type regulator and degenerative feedback system of voltage regulation are shown in Figs. 6.9a and b respectively.

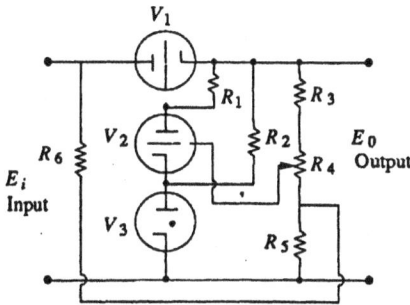

Fig. 6.9a   Series d.c. voltage regulator

V_1— Series regulating tube

V_2— Sensing and amplifying tube

V_3— Reference tube

R_1— Grid control resistance for tube 1

R_3, R_4, R_5 — Output potential divider

R_6— Input fluctuation correction resistance

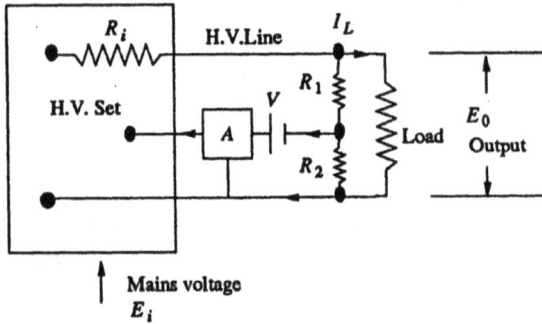

Fig. 6.9b Degenerative feedback system of voltage regulation

R_1, R_2 — Potential divider
A — d.c. amplifier
R_i — Internal resistance of control source, the h.v. set

Figure 6.9a shows the simplified diagram of a series type of voltage regulator. Here the gas tube V_3 provides the reference voltage which keeps the potential of the cathode of the tube V_2 constant. When the output voltage E_0 changes, the current through the divider R_2, R_4 and R_5 changes, and hence the bias voltage of the tube V_2 with respect to its cathode changes in the opposite manner (reduces when the output voltage increases and vice versa). Thus, the grid potential of V_1 is altered in such a manner as to oppose the variation in output voltage E_0; that is, the voltage drop across the tube V_1 reduces, if the output voltage decreases and vice versa. Resistance R_6 together with R_5 provides a voltage input to the d.c. amplifier (the detector) in proportion to fluctuations in the input voltage in such a way as to reduce the effect of these variations in the output voltage.

A convenient method of regulating high d.c. voltages is the degenerative feedback system, shown in Fig. 6.9b. The fluctuations in the output voltage are measured by a detecting device, amplified by a d.c. amplifier and fed back into the high voltage set so as to correct for the original fluctuations. The detecting device in the above circuit is the potential divider of ratio $\beta = R_1/(R_1 + R_2)$. The amplifier is directly coupled and the difference in potential between the tapping on the divider and the amplifier grid voltage is made up by the battery voltage V.

The output voltage E_0 is given as follows:

$$E_0 = NE_i - I_l R_i - Ae \tag{6.7a}$$

where N = ratio between the output voltage of the h.v. set and the main supply voltage, E_i,

I_l = load current,

R_i = internal resistance of the h.v. set,

A = gain of the feedback amplifier,

e = amplifier input voltage.

Since e = $\beta E_0 - V$, the output voltage becomes

$$E_0 = NE_i - I_l R_i - (\beta E_v - V) A = NE_i - I_l R_i - A\beta E_0 + AV$$

or, $$E_0(1 + A\beta) = NE_i + AV - I_l R_i$$

or, $$E_0 = \frac{NE_i}{(1+A\beta)} + \frac{AV}{(1+A\beta)} - \frac{I_l R_i}{(1+A\beta)}.$$

For large values of A

$$E_0 = \frac{AV}{(1+A\beta)} \approx \frac{V}{\beta}.$$

The stabilization ratio = $[\Delta E_i / E_i]/[\Delta E_0/E_0] = 1 + A \beta$.

In practice, values of $A\beta$ of the order of 1000 are common. It can be seen that for given value of A, stabilization ratio can be increased by making β as large as possible.

Stabilities to about 2 parts in 10^4 in high voltage are easily obtained by this method if load changes are kept small.

6.2 GENERATION OF HIGH ALTERNATING VOLTAGES

When test voltage requirements are less than about 300 kV, a single transformer can be used for test purposes. The impedance of the transformer should be generally less than 5% and must be capable of giving the short circuit current for one minute or more depending on the design. In addition to the normal windings, namely, the low and high

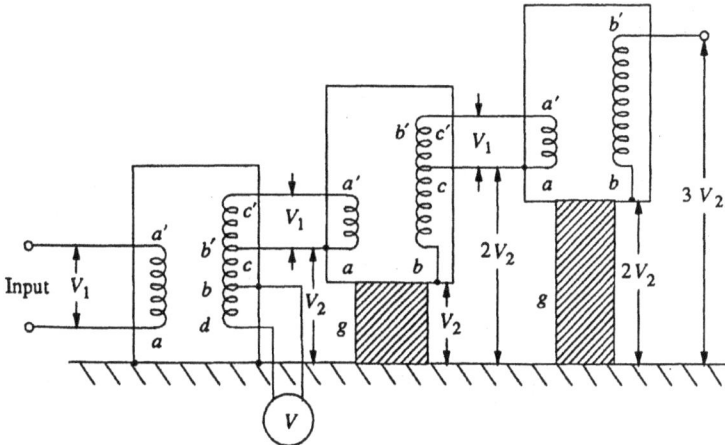

Fig. 6.10 Cascade transformer connection (schematic)

V_1 — Input voltage
V_2 — Output voltage
aa' — L.V. primary winding
bb' — H.V. secondary winding
cc' — Excitation winding
bd — Meter winding (200 to 500 V)

voltage windings, a third winding known as meter winding is provided to measure the output voltage. For higher voltage requirements, a single unit construction becomes difficult and costly due to insulation problems. Moreover, transportation and erection of large transformers become difficult. These drawbacks are overcome by series connection or cascading of the several identical units of transformers, wherein the high voltage windings of all the units effectively come in series.

Schematic diagrams of the cascade transformer units are shown in Figs. 6.10 and 6.11.

6.2.1 Cascade Transformers

Figure 6.10 shows the cascade transformer units in which the first transformer is at the ground potential along with its tank. The second transformer is kept on insulators and maintained at a potential of V_2, the output voltage of the first unit above the ground. The high voltage winding of the first unit is connected to the tank of the second unit. The low voltage winding of this unit is supplied from the excitation winding of the first transformer, which is in series with the high voltage winding of the first transformer at its high voltage end. The rating of the excitation winding is almost identical to that of the primary or the low voltage winding. The high voltage

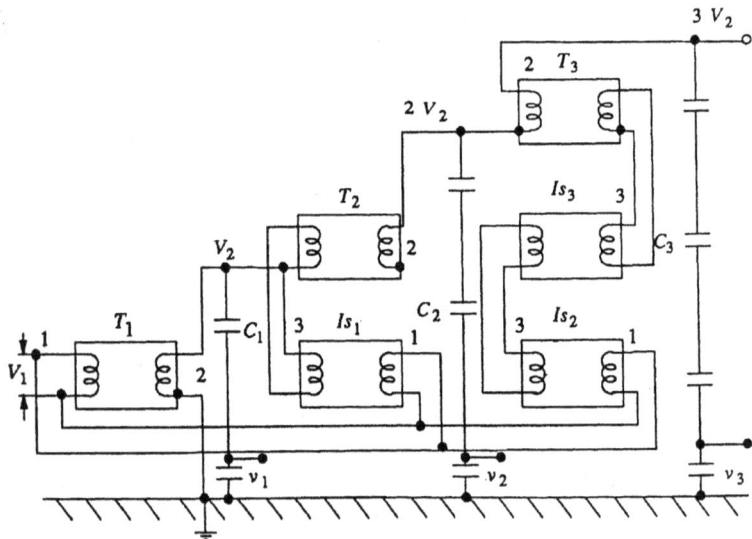

Fig. 6.11   Cascade transformer unit with isolating transformers for excitation

T_1, T_2, T_3 — Cascade transformer units
Is_1, Is_2, Is_3 — Isolation transformer units
C_1, C_2, C_3 — Capacitance voltage dividers for h.v. measurement after 1st, 2nd and 3rd stages
V_1, V_2, V_3 — For metering after 1st, 2nd and 3rd stages
  1. Primary (l.v. winding), 2.h.v. winding, 3. Excitation winding.

connection from the first transformer winding and the excitation winding terminal are taken through a bushing to the second transformer. In a similar manner, the third transformer is kept on insulators above the ground at a potential of $2V_2$ and is supplied likewise from the second transformer. The number of stages in this type of arrangement are usually two to four, but very often, three stages are adopted to facilitate a three-phase operation so that $\sqrt{3}V_2$ can be obtained between the lines.

Supply to the units can be obtained from a motor-generator set or through an induction regulator for variation of the output voltage. The rating of the primary or the low voltage winding is usually 230 or 400 V for small units up to 100 kVA. For larger outputs the rating of the low voltage winding may be 3.3kV, 6.6 kV or 11 kV.

In Fig. 6.11, a second scheme for providing the excitation to the second and the third stages is shown. Isolating transformers I_{s1}, I_{s2}, and I_{s3} are 1 : 1 ratio transformers insulated to their respective tank potentials and are meant for supplying the excitation for the second and the third stages at their tank potentials. Power supply to the isolating transformers is also fed from the same a.c. input. This scheme is expensive and requires more space. The advantage of this scheme is that the natural cooling is sufficient and the transformers are light and compact. Transportation and assembly is easy. Also the construction is identical for isolating transformers and the high voltage cascade units. Three phase connection in delta or star is possible for three units. Testing transformers of ratings up to 10 MVA in cascade connection to give high voltages up to 2.25 MV are available for both indoor and outdoor applications.

Testing of an H.V. apparatus or insulation always involves supplying of capacitive loads with very low power dissipation. Thus if C is the capacitance of the test object, V is the rms value of the nominal output voltage of the transformer at an angular frequency ω, then the nominal rating of the transformer in kVA will be $P = K \cdot V^2 \omega C$, where K (> 1.0) is a factor to account for any extra capacitance in the test circuit like that of the measuring capacitance divider etc. K may have values of the order of 2 or more for very high voltages (> 1MV). Typical capacitance values for high capacitance test objects like power transformers, cables etc. are as follows:

Power transformers (rating < 1MVA)	1000 pF
Power transformers (rating > 1MVA)	1000-10,000 pF
High Voltage power cables (with solid insulation)	250-300 pF/m
High Voltage power cables (with gas insulation)	50-80 pF/m
Metal Clad sub-station with gas insulation (G.I.S.)	100-10,000 pF

The charging currents for the test apparatus may range from 10 mA at 100 kV to a few milliamperes in the megavolt range. As such the transformers should have only a short time rating (10 to 15 min) for high power ratings, as compared to those with nominal power ratings.

Large testing transformers rated for more than 1 MVA at 1 MV are now-a-days designed for outdoor use only. The design is of the type mentioned in the second scheme and ensures that the units are enclosed by large-size metal rings to prevent corona, and are terminated with near spherical polycone electrodes. Modern test transformers are built to withstand transients during the flashover of the test object. However, particular care need to be taken to see that steady state voltage distribution within the cascade units is uniform. Sometimes reactor compensation may have to be provided for excessive load currents of capacitive nature. Cascade transformers are very expensive apparatus and are difficult to repair. Therefore it is necessary to limit the high short-circuit currents by using limiting reactors in the input stage.

Power Supply for a.c. Test Circuits

Large cascade transformers units are supplied power through a separate motor-generator set or by means of voltage regulators. Supply through a voltage regulator will be cheaper, and will be more flexible in the sense that the units in the cascade set can be operated in cascade, or in parallel, or as three-phase units. It is also necessary that the impedance of the voltage regulating transformer is low in all voltage positions, from the minimum to the rate value.

6.2.2 Resonant Transformers

The equivalent circuit of a high voltage testing transformer consists of the leakage reactances of the windings, the winding resistances, the magnetizing reactance, and the shunt capacitance across the output terminal due to the bushing of the high voltage terminal and also that of the test object. This is shown in Fig. 6.12 (a) with its equivalent circuit in Fig. 6.12(b). It may be seen that it is possible to have series resonance at power frequency ω, if $(L_1 + L_2) = 1/\omega C$. With this condition, the current in the test object is very large and is limited only by the resistance of the circuit. The waveform of the voltage across the test object will be purely sinusoidal. The magnitude of the voltage across the capacitance C of the test object will be

$$V_C = \left| \frac{-jVX_C}{R+j(X_L-X_C)} \right| = \frac{V}{R}X_C = \frac{V}{\omega CR} \tag{6.8}$$

where R is the total series resistance of the circuit.

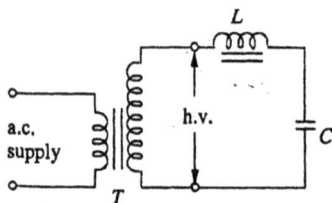

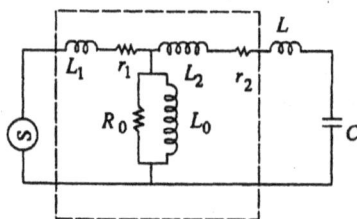

(a) (b)

Fig. 6.12a Transformer   **Fig. 6.12b** Equivalent circuit

T — Testing transformer	L_1, L_2 — Leakage inductances of the
L — Choke	transformer
C — Capacitance of h.v. terminal	r_1, r_2 — Resistances of the windings
and test object	R_0 — Resistance due to core loss
L_0 — Magnetizing inductance	

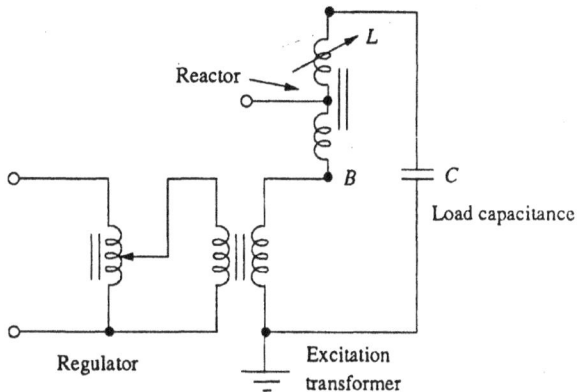

Fig. 6.12c   Series resonant a.c. test system

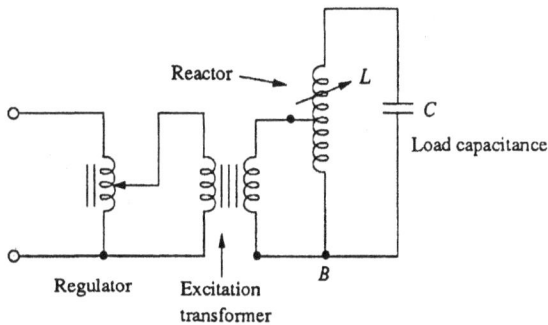

Fig. 6.12d   Parallel resonant a.c. test system

Ratings: Regulator : 10 – 100 kVA
Excitation transformer : 10 – 100 kVA with an output voltage of about 10 kV.
Reactor voltage – each unit up to 300 kV.

Fig. 6.12   Resonant transformer and equivalent circuit

The factor $X_C/R = 1/\omega CR$ is the Q factor of the circuit and gives the magnitude of the voltage multiplication across the test object under resonance conditions. Therefore, the input voltage required for excitation is reduced by a factor $1/Q$, and the output kVA required is also reduced by a factor $1/Q$. The secondary power factor of the circuit is unity.

This principle is utilized in testing at very high voltages and on occasions requiring large current outputs such as cable testing, dielectric loss measurements, partial discharge measurements, etc. A transformer with 50 to 100 kV voltage rating and a relatively large current rating is connected together with an additional choke, if necessary. The test condition is set such that $\omega(L_e + L) = 1/\omega C$ where L_e is the total equivalent leakage inductance of the transformer including its regulating transformer. The chief advantages of this principle are:

(a) it gives an output of pure sine wave,

(b) power requirements are less (5 to 10% of total kVA required),

(c) no high-power arcing and heavy current surges occur if the test object fails, as resonance ceases at the failure of the test object,

(d) cascading is also possible for very high voltages,

(e) simple and compact test arrangement, and

(f) no repeated flashovers occur in case of partial failures of the test object and insulation recovery. It can be shown that the supply source takes Q number of cycles at least to charge the test specimen to the full voltage.

The disadvantages are the requirements of additional variable chokes capable of withstanding the full test voltage and the full current rating.

A simplified diagram of the series resonance test system is given in Fig. 6.12c and that of the parallel resonant test system in 6.12d. A voltage regulator of either the auto-transformer type or the induction regulator type is connected to the supply mains and the secondary winding of the exciter transformer is connected across the H.V. reactor, L, and the capacitive load C. The inductance of the reactor L is varied by varying its air gap and operating range is set in the ratio 10 : 1. Capacitance C comprises of the capacitance of the test object, capacitance of the measuring voltage divider, capacitance of the high voltage bushing etc. The Q-factor obtained in these circuits will be typically of the order of 50. In the parallel resonant mode the high voltage reactor is connected as an auto-transformer and the circuit is connected as a parallel resonant circuit. The advantage of the parallel resonant circuit is that more stable output voltage can be obtained along with a high rate of rise of test voltage, independent of the degree of tuning and the Q-factor. Single unit resonant test systems are built for output voltages up to 500 kV, while cascaded units for outputs up to 3000 kV, 50/60 Hz are available.

6.2.3 Generation of High Frequency a.c. High Voltages

High frequency high voltages are required for rectifier d.c. power supplies as discussed in Sec. 6.1. Also, for testing electrical apparatus for switching surges, high frequency high voltage damped oscillations are needed which need high voltage high frequency transformers. The advantages of these high frequency transformers are:

(i) the absence of iron core in transformers and hence saving in cost and size,

(ii) pure sine wave output,

(iii) slow build-up of voltage over a few cycles and hence no damage due to switching surges, and

(iv) uniform distribution of voltage across the winding coils due to subdivision of coil stack into a number of units.

The commonly used high frequency resonant transformer is the Tesla coil, which is a doubly tuned resonant circuit shown schematically in Fig. 6.13a. The primary voltage rating is 10 kV and the secondary may be rated to as high as 500 to 1000 kV. The primary is fed from a d.c. or a.c. supply through the condenser C_1. A spark gap G connected across the primary is triggered at the desired voltage V_1 which induces

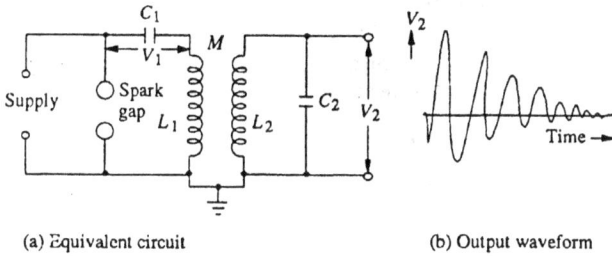

(a) Equivalent circuit   (b) Output waveform

Fig. 6.13 Tesla coil equivalent circuit and its output waveform

a high self-excitation in the secondary. The primary and the secondary windings (L_1 and L_2) are wound on an insulated former with no core (air-cored) and are immersed in oil. The windings are tuned to a frequency of 10 to 100 kHz by means of the condensers C_1 and C_2.

The output voltage V_2 is a function of the parameters L_1, L_2, C_1, C_2, and the mutual inductance M. Usually, the winding resistances will be small and contribute only for damping of the oscillations.

The analysis of the output waveform can be done in a simple manner neglecting the winding resistances. Let the condenser C_1 be charged to a voltage V_1 when the spark gap is triggered. Let a current i_1 flow through the primary winding L_1 and produce a current i_2 through L_2 and C_2.

Then,

$$V_1 = \frac{1}{C_1}\int_0^t i_1\, dt + L_1 \frac{di_1}{dt} + M \frac{di_2}{dt}$$

(6.9)

and,

$$0 = \frac{1}{C_2}\int_0^t i_2\, dt + L_2 \frac{di_2}{dt} + M \frac{di_1}{dt}$$

The Laplace transformed equations for the above are,

$$\frac{V_1}{s} = \left[L_1 s + \frac{1}{C_1 s}\right] I_1 + M s I_2$$

(6.10)

and,

$$0 = [Ms] I_1 + \left[L_2 s + \frac{1}{[C_2 s]}\right] I_2$$

where I_1 and I_2 are the Laplace transformed values, of i_1 and i_2.

The output voltage V_2 across the condenser C_2 is

$$V_2 = \frac{1}{C_2}\int_0^t i_2 dt; \quad \text{or its transformed equation is}$$

$$V_2(s) = \frac{I_2}{C_2 s} \qquad (6.11)$$

where $V_2(s)$ is the Laplace transform of V_2.

The solution for V_2 from the above equations will be

$$V_2 = \frac{MV_1}{\sigma L_1 L_2 C_1} \frac{1}{\gamma_2^2 - \gamma_1^2} [\cos \gamma_1 t - \cos \gamma_2 t] \qquad (6.12)$$

where,

$$\sigma^2 = 1 - \frac{M^2}{L_1 L_2} = 1 - K^2$$

K = coefficient of coupling between the windings L_1 and L_2

$$\gamma_{1,2} = \frac{\omega_1^2 + \omega_2^2}{2} \pm \sqrt{\left(\frac{\omega_1^2 + \omega_2^2}{2}\right)^2 - \omega_1^2 \omega_2^2 (1 - K^2)}$$

$$\omega_1 = \frac{1}{\sqrt{L_1 C_1}} \quad \text{and} \quad \omega_2 = \frac{1}{\sqrt{L_2 C_2}}$$

The output waveform is shown in Fig. 6.13b.

The peak amplitude of the secondary voltage V_2 can be expressed as,

$$V_{2max} = V_1 e \sqrt{\frac{L_2}{L_1}} \qquad (6.13)$$

where,

$$e = \frac{2\sqrt{(1 - \sigma)}}{\sqrt{(1 + a)^2 - 4\sigma a}}$$

$$a = \frac{L_2 C_2}{L_1 c_1} = \frac{W_1^2}{W_2^2}$$

A more simplified analysis for the Tesla coil may be presented by considering that the energy stored in the primary circuit in the capacitance C_1 is transferred to C_2 via the magnetic coupling. If W_1 is the energy stored in C_1 and W_2 is the energy transferred to C_2 and if the efficiency of the transformer is η, then

$$W_1 = \frac{1}{2} \eta C_1 V_1^2 = (\tfrac{1}{2} C_2 V_2^2) \qquad (6.14)$$

from which

$$V_2 = V_1 \sqrt{\eta \frac{C_1}{C_2}} \qquad (6.14a)$$

It can be shown that if the coefficient of coupling K is large the oscillation frequency is less, and for large values of the winding resistances and K, the waveform may become a unidirectional impulse. This is shown in the next sections while dealing with the generation of switching surges.

6.3 GENERATION OF IMPULSE VOLTAGES

6.3.1 Standard Impulse Waveshapes

Transient overvoltages due to lightning and switching surges cause steep build-up of voltage on transmission lines and other electrical apparatus. Experimental investigations showed that these waves have a rise time of 0.5 to 10 μ s and decay time to 50% of the peak value of the order of 30 to 200 μ s. The waveshapes are arbitrary, but mostly unidirectional. It is shown that lightning overvoltage wave can be represented as double exponential waves defined by the equation

$$V = V_0 [\exp(-\alpha t) - \exp(-\beta t)] \tag{6.15}$$

where α and β are constants of microsecond values.

The above equation represents a unidirectional wave which usually has a rapid rise to the peak value and slowly falls to zero value. The general waveshape is given in Fig. 6.14. Impulse waves are specified by defining their rise or front time, fall or tail time to 50% peak value, and the value of the peak voltage. Thus 1.2/50 μs, 1000 kV wave represents an impulse voltage wave with a front time of 1.2 μs, fall time to 50% peak value of 50 μs, and a peak value of 1000 kV. When impulse waveshapes are recorded, the initial portion of the wave will not be clearly defined or sometimes will be missing. Moreover, due to disturbances it may contain superimposed oscillations in the rising portion. Hence, the front and tail times have to be defined.

Referring to the waveshape in Fig. 6.14, the peak value A is fixed and referred to as 100% value. The points corresponding to 10% and 90% of the peak values are located in the front portion (points C and D). The line joining these points is extended to cut the time axis at O_1. O_1 is taken as the virtual origin. 1.25 times the interval between times t_1 and t_2 corresponding to points C and D (projections on the time axis)

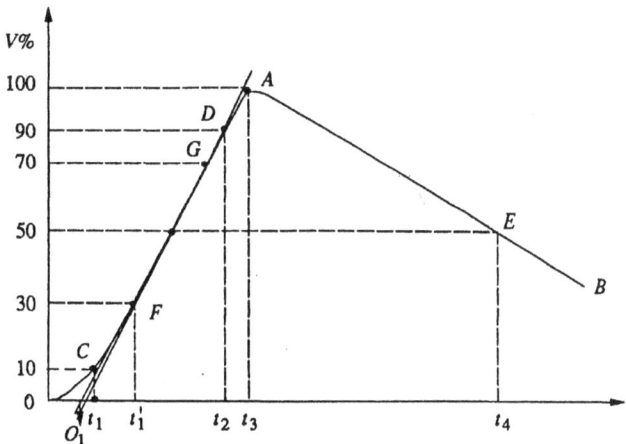

Fig. 6.14   Impulse waveform and its definitions

is defined as the front time, i.e. $1.25\,(O_1t_2 - O_1t_1)$. The point E is located on the wave tail corresponding to 50% of the peak value, and its projection on the time axis is t_4. Ot_4 is defined as the fall or tail time. In case the point C is not clear or missing from the waveshape record, the point corresponding to 30% peak value F is taken and its projection t'_1 is located on time axis. The wavefront time in that case will be defined as $1.67\,(O_1t_3 - O_1t'_1)$. The tolerances that can be allowed in the front and tail times are respectively ± 30% and ± 20%. Indian standard specifications define 1.2/50 μs wave to be the standard impulse. The tolerance allowed in the peak value is ± 3%.

6.3.2 Theoretical Representation of Impulse Waves

The impulse waves are generally represented by the Eq. (6.15) given earlier. V_0 in the equation represents a factor that depends on the peak value. For impulse wave of 1.2/50 μs, $\alpha = -0.0146$, $\beta = -2.467$, and $V_0 = 1.04$ when time t is expressed in microseconds, α and β control the front and tail times of the wave respectively.

6.3.3 Circuits for Producing Impulse Waves

A double exponential waveform of the type mentioned in Eq. (6.15) may be produced in the laboratory with a combination of a series R-L-C circuit under over damped conditions or by the combination of two R-C circuits. Different equivalent circuits that produce impulse waves are given in Figs. 6.15a to d. Out of these circuits, the ones shown in Figs. 6.15a to d are commonly used. Circuit shown in Fig. 6.15a is limited to model generators only, and commercial generators employ circuits shown in Figs. 6.15b to 6.15d.

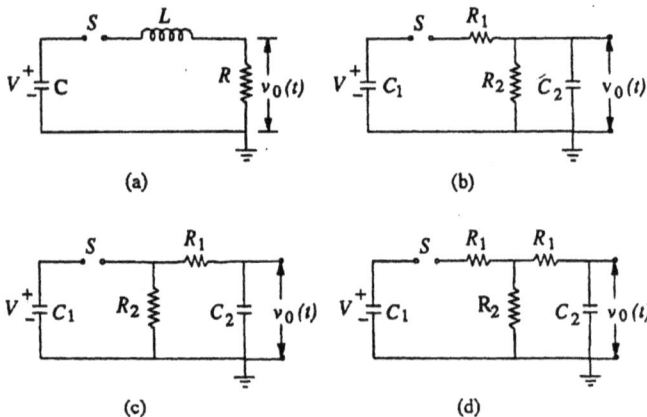

Fig. 6.15 Circuits for producing impulse waves

A capacitor (C_1 or C) previously charged to a particular d.c. voltage is suddenly discharged into the waveshaping network (LR, $R_1R_2C_2$ or other combination) by closing the switch S. The discharge voltage $V_0\,(t)$ shown in Fig. 6.15 gives rise to the desired double exponential waveshape.

Analysis of Impulse Generator Circuit of Series R-L-C Type

Referring to Fig. 6.15 the current through the load resistance R is given by

$$V = \frac{1}{C}\int_0^t i\, dt + Ri + L\frac{di}{dt} \tag{6.16}$$

with initial condition at $t = 0$ being $i(0) = 0$ and the net charge in the circuit $i = dq/dt$ = 0. Writing the above equation as a Laplace transform equation,

$$V/s = \left(\frac{1}{Cs} + R + Ls\right)I(s)$$

or,

$$I(s) = \frac{V}{L}\left[\frac{1}{s^2 + \dfrac{Rs}{L} + \dfrac{1}{LC}}\right]$$

The voltage across the resistor R (which is the output voltage) is, $v_0(s) = I(s)\,R$; hence,

$$v_0(s) = V\frac{R}{L}\frac{1}{s^2 + \dfrac{Rs}{L} + \dfrac{1}{LC}}$$

For an overdamped condition, $R/2L \geq 1/\sqrt{LC}$

Hence, the roots of the equation $s^2 + \dfrac{Rs}{L} + \dfrac{1}{LC}$ are

$$\alpha = s_1 = -\frac{R}{2L} + \sqrt{\left(\frac{R}{2L}\right)^2 - \frac{1}{LC}}$$

$$\beta = s_2 = -\frac{R}{2L} - \sqrt{\left(\frac{R}{2L}\right)^2 - \frac{1}{LC}}$$

The solution of the equation for $v_0(t)$ is

$$v_0(t) = \frac{V\left(\dfrac{R}{2L}\right)}{\left[\dfrac{R^2}{4L^2} - \dfrac{1}{LC}\right]^{\frac{1}{2}}}[\exp(-\alpha t) - \exp(-\beta t)] \tag{6.17}$$

$$= V_0[\exp(-\alpha t) - \exp(-\beta t)] \tag{6.17a}$$

where,

$$V_0 = \frac{V\left(\dfrac{R}{2L}\right)}{\left[\dfrac{R^2}{4L^2} - \dfrac{1}{LC}\right]^{\frac{1}{2}}} = \frac{V}{\left[1 - \dfrac{4L}{CR^2}\right]^{\frac{1}{2}}} \tag{6.17b}$$

The sum of the roots $(\alpha + \beta) = -\dfrac{R}{2L}$

and, the product of the roots $\alpha\beta = \dfrac{1}{LC}$ (6.17c)

The wave front and the wave tail times are controlled by changing the values of R and L simultaneously with a given generator capacitance C; choosing a suitable value for L. β or the wave front time is determined and α or the wave tail time is controlled by the value of R in the circuit. The advantage of this circuit is its simplicity. But the waveshape control is not flexible and independent. Another disadvantage is that the basic circuit is altered when a test object which will be mainly capacitive in nature, is connected across the output. Hence, the waveshape gets changed with the change of test object.

Analysis of the Other Impulse Generator Circuits

The most commonly used configurations for impulse generators are the circuits shown in Figs. 6.15b and c. The advantages of these circuits are that the wave front and wave tail times are independently controlled by changing either R_1 or R_2 separately. Secondly, the test objects which are mainly capacitive in nature form part of C_2.

For the configuration shown in Fig. 6.15b, the output voltage across C_2 is given by, $v_0(t) = \dfrac{1}{C_2}\displaystyle\int_0^t i_2\, dt.$

Performing Laplace transformation, $\dfrac{1}{C_2 s} I_2(s) = v_0(s)$

where i_2 is the current through C_2.
Taking the current through C_1 as i_1 and its transformed value as $I_1(s)$,

$$I_2(s) = \left[\frac{R_2}{R_2 + \dfrac{1}{C_2 s}}\right] I_1(s)$$

$$I_1(s) = \frac{V}{s} \cfrac{1}{\dfrac{1}{C_1 s} + R_1 + \cfrac{R_2 \cdot \dfrac{1}{C_2 s}}{R_2 + \dfrac{1}{C_2 s}}}$$

where, $\dfrac{R_2 \cdot \dfrac{1}{C_2 s}}{R_2 + \dfrac{1}{C_2 s}}$ represents the impedance of the parallel combination of R_2 and C_2.

Substitution of $I_1(s)$ gives

$$v_0(s) = \frac{1}{C_2 s} \frac{R_2}{R_2 + \dfrac{1}{C_2 s}} \frac{V}{s} \cfrac{1}{\dfrac{1}{C_1 s} + R_1 + \dfrac{R_2(1/C_2 s)}{R_2 + (1/C_2 s)}}$$

After simplification and rearrangement,

$$v_0(s) = \frac{V}{R_1 C_2} \left[\frac{1}{s^2 + \left(\frac{1}{C_1 R_1} + \frac{1}{C_2 R_2} + \frac{1}{C_2 R_1} \right) s + \frac{1}{C_1 C_2 R_1 R_2}} \right] \qquad (6.18)$$

Hence, the roots of the equation

$$s^2 + \left[\frac{1}{C_1 R_1} + \frac{1}{C_2 R_2} + \frac{1}{C_2 R_1} \right] s + \frac{1}{C_1 C_2 R_1 R_2}$$

are found from the relations,

$$\alpha + \beta = -\left[\frac{1}{C_1 R_1} + \frac{1}{C_2 R_2} + \frac{1}{C_2 R_1} \right] \qquad (6.19)$$

$$\alpha\beta = \frac{1}{C_1 C_2 R_1 R_2}$$

Taking inverse transform of $v_0(s)$ gives

$$v_0(t) = \frac{V}{R_1 C_2 (\alpha - \beta)} [\exp(-\alpha t) - \exp(-\beta t)] \qquad (6.20)$$

Usually, $\frac{1}{C_1 R_1}$ and $\frac{1}{C_2 R_2}$ will be much smaller compared to $\frac{1}{R_1 C_2}$.

Hence, the roots may be approximated as

$$\alpha \approx \frac{1}{R_1 C_2} \text{ and, } \beta = \frac{1}{R_2 C_1} \qquad (6.21)$$

Following a similar analysis, it may be shown that the output waveform for the circuit configuration of Fig. 6.15c will be

$$v_0(t) = \frac{V C_1 R_2 \alpha \beta}{(\beta - \alpha)} [\exp(-\alpha t) - \exp(-\beta t)]$$

where α and β are the roots of the Eq. (6.19). The approximate values of α and β given by Eq. (6.21) are valid for this circuit also.

The equivalent circuit given in Fig. 6.15d is a combination of the configurations of Fig. 6.15b and Fig. 6.15c. The resistance R_1 is made into two parts and kept on either side of R_2 to give greater flexibility for the circuits.

Restrictions on the Ratio of the Generator and Load Capacitances, C_1/C_2 on the Circuit Performance

For a given waveshape, the choice of R_1 and R_2 to control the wave front and wave tail times is not entirely independent but depends on the ratio of C_1/C_2. It can be shown mathematically that

$$R_2 = P(y)/C_1 \text{ and } R_1 = Q(y)/C_1$$

where $y = C_1/C_2$ and P and Q are functions of y. In order to get real values for R_1 and R_2 for a given waveshape, a maximum and minimum value of y exists in practice. This is true whether the configuration of Fig. 6.15b or 6.15c is used. For example, with the circuit of Fig. 6.15b, the ratio of C_1/C_2 cannot exceed 3.35 for a 1/5 μ s waveshape. Similarly, for a 1/50 μ s waveshape the ratio C_1/C_2 lies between 6 and 106.5. If the configuration chosen is 6.15c, the minimum value of C_1/C_2 for 1/5 μ s waveshape is about 0.3 and that for the 1/50 μ s waveshape is about 0.01. The reader is referred to *High Voltage Laboratory Techniques* by Craggs and Meek for further discussion on the restrictions imposed on the ratio C_1/C_2.

Effect of Circuit Inductances and Series Resistance on the Impulse Generator Circuits

The equivalent circuits shown in Figs. 6.15a to d, in practice comprise several stray series inductances. Further, the circuits occupy considerable space and will be spread over several metres in a testing laboratory. Each component has some residual inductance and the circuit loop itself contributes for further inductance. The actual value of the inductance may vary from 10 μ H to several hundreds of microhenries. The effect of the inductance is to cause oscillations in the wave front and in the wave tail portions. Inductances of several components and the loop inductance are shown in Fig. 6.16a. Figure 6.16b gives a simplified circuit for considering the effect of inductance. The effect of the variation of inductance on the waveshape is shown in Fig. 6.16c. If the series resistance R_1 is increased, the wave front oscillations are damped, but the peak value of the voltage is also reduced. Sometimes, in order to control the front time a small inductance is added.

Impulse Generators for Testing Objects having Large Capacitance

When test objects with large capacitances are to be tested ($C > 5$ nF), it is difficult to generate standard impulses with front time within the specified tolerance of ± 30% and the specified less than 5% tolerance in the overshoot. This is mainly because of the effect of the inductance of the impulse generator and the front resistors. Normally the inductance of the impulse generator will be about 3 to 5 μH per stage and that of the leads about 1 μH/m. Also the front resistor which is usually of bifilar type, has inductance of about 2 μH/unit. An overshoot in the voltage wave of more than 5% will occur if $R/(\sqrt{L/c}) \leq 1.38$, where R is the front resistance, L is the generator inductance and C is the equivalent capacitance of the generator given by $C = C_1 \cdot C_2/(C_1 + C_2)$.

Impulse Generators for Test Objects with Inductance

Often impulse generators are required to test equipment with large inductance, such as power transformers and reactors. Usually, generating the impulse voltage wave of proper time-to-front and obtaining a good voltage efficiency are easy, but obtaining the required time-to-tail as per the standards will be very difficult. The equivalent circuit for medium and low inductive loads will be as shown in Fig. 6.16(d). For the calculation of time-to-tail the circuit can still be approximated as a series C-R-L circuit. As the value $R/(\sqrt{L/C})$ decreases, the overshoot and the swing of the wave to the opposite polarity increases thereby deviating from the standard wave shape. Therefore, it is necessary to keep the value of the effective resistance R in the circuit large. One method of doing this is to connect a large resistance, R_2 in parallel with the test object or to connect the untested winding of the transformer (load) with a suitable resistance. Another method that can be used is to increase the generator capacitance with which the time-to-tail also increases, but without altering the time-to-front and the overshoot. Figure 6.16(d) gives the circuit arrangement for inductive loads and 6.16(e) gives the requirement of energy and capacitance of the impulse voltage generator. Figure B of 6.16(e) gives the generator capacitance required to give the time-to-tail values in the range of 40 to 60 μ s at different inductive loads.

Waveshape Control

Generally, for a given impulse generator of Fig. 6.15b or c the generator capacitance C_1 and load capacitance C_2 will be fixed depending on the design of the generator and the test object. Hence, the desired waveshape is obtained by controlling R_1 and R_2. The following approximate analysis is used to calculate the wave front and wave tail times.

The resistance R_2 will be large. Hence, the simplified circuit shown in Fig. 6.16b is used for wave front time calculation. Taking the circuit inductance to be negligible during charging, C_1 charges the load capacitance C_2 through R_1. Then the time taken for charging is approximately three times the time constant of the circuit and is given by

$$t_1 = 3.0\,R_1\,\frac{C_1 C_2}{C_1 + C_2} = 3\,R_1\,C_e \qquad (6.22)$$

where $C_e = \dfrac{C_1 C_2}{C_1 + C_2}$. If R_1 is given in ohms and C_e in microfarads, t_1 is obtained in microseconds.

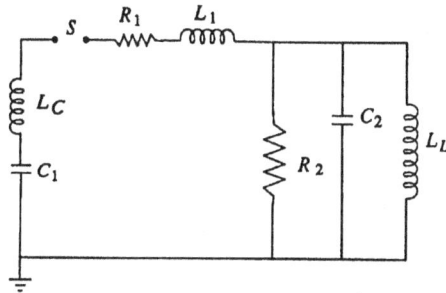

Fig. 6.16 (a) Series inductances in impulse generator circuit

L_C — Inductance of the generator capacitance C_1 and lead capacitances

L_1 — Inductance of the series resistance and the circuit loop inductance

L_L — Test object inductance

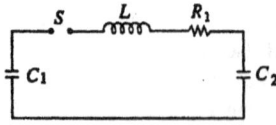

Fig. 6.16 (b) Simplified circuit for calculation of wave front time
$L = L_C + L_1 +$ any other added inductance

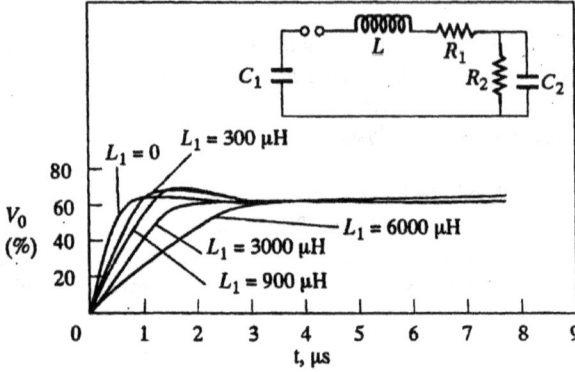

Fig. 6.16 (c) Effect of series inductance on wave front time. v_0 is the percentage of charging voltage V, to which C_1 is charged

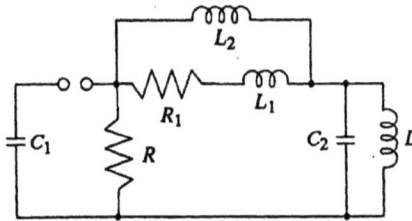

Medium inductance load $(L = 0.4$ to 15 mH$)$

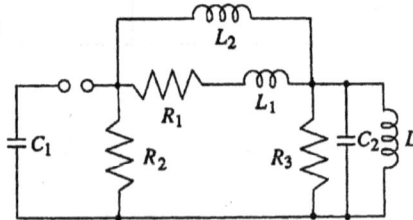

Low inductance load $(L = 0.4$ to 4mH$)$

Fig. 6.16 (d) Effect of inductive loads on impulse voltage generator circuits

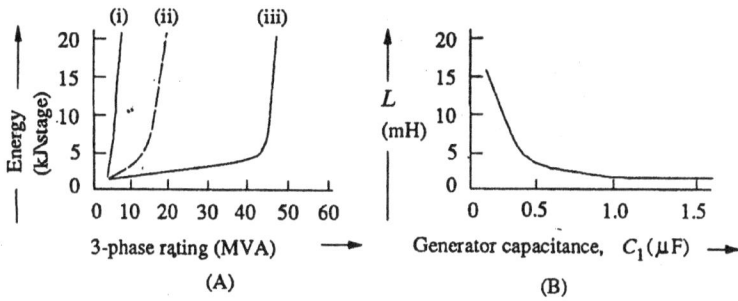

(A)

. Energy requirement per 200 kV stage of an impulse voltage generator as a function of the MVA rating of 3-phase reactor transformer. Curves (i) for 11 kV, (ii) for 22 kV and (iii) for 33 kV winding ratings.

(B)

Generator capacitance (C_1) required for different inductive loads (L) to give the standard tail time of an impulse voltage wave.

Fig. 6.16   (e) Requirements of an impulse voltage generator energy and capacitance for the testing of the transformer (reactor) winding using standard impulse voltages

Fig. 6.16   Series inductance in impulse generator circuits and its effect on waveshape

For discharging or tail time, the capacitances C_1 and C_2 may be considered to be in parallel and discharging occurs through R_1 and R_2. Hence, the time for 50% discharge is approximately given by

$$t_2 = 0.7 (R_1 + R_2) (C_1 + C_2)$$ (6.23)

These formulae for t_1 and t_2 hold good for the equivalent circuits shown in Fig. 6.15b and c. For the circuit given in Fig. 6.15d, R is to be taken as $2R_1$. With the approximate formulae, the wave front and wave tail times can be estimated to within $\pm 20\%$ for the standard impulse waves.

6.3.4   Multistage Impulse Generators—Marx Circuit

In the above discussion, the generator capacitance C_1 is to be first charged and then discharged into the wave shaping circuits. A single capacitor C_1 may be used for voltages up to 200 kV. Beyond this voltage, a single capacitor and its charging unit may be too costly, and the size becomes very large. The cost and size of the impulse generator increases at a rate of the square or cube of the voltage rating. Hence, for producing very high voltages, a bank of capacitors are charged in parallel and then discharged in series. The arrangement for charging the capacitors in parallel and then connecting them in series for discharging was originally proposed by Marx. Now-a-days modified Marx circuits are used for the multistage impulse generators.

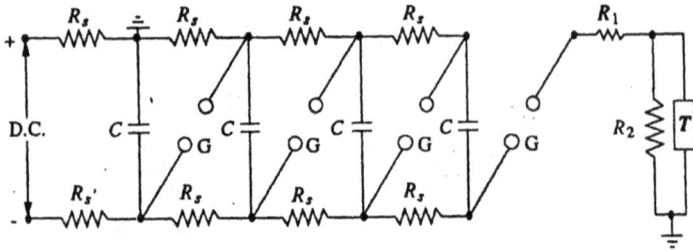

Fig. 6.17a Schematic diagram of Marx circuit arrangement for multistage impulse generator

C — Capacitance of the generator
R_s — Charging resistors
G — Spark gap
R_1, R_2 — Wave shaping resistors
T — Test object

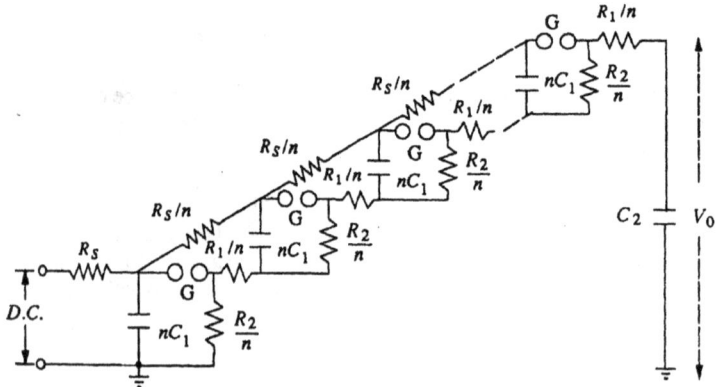

Fig. 6.17b Multistage impulse generator incorporating the series and wave tail resistances within the generator

The schematic diagram of Marx circuit and its modification are shown in Figs. 6.17a and 6.17b, respectively. Usually the charging resistance R_s is chosen to limit the charging current to about 50 to 100 mA, and the generator capacitance C is chosen such that the product CR_s is about 10 s to 1 min. The gap spacing is chosen such that the breakdown voltage of the gap G is greater than the charging voltage V. Thus, all the capacitances are charged to the voltage V in about 1 minute. When the impulse generator is to be discharged, the gaps G are made to spark over simultaneously by some external means. Thus, all the capacitors C get connected in series and discharge into the load capacitance or the test object. The discharge time constant CR_1/n (for n stages) will be very very small (microseconds), compared to the charging time constant CR_s which will be few seconds. Hence, no discharge takes place through the charging resistors R_s. In the Marx circuit is of Fig. 6.17a the impulse wave shaping circuit is connected externally to the capacitor unit. In Fig. 6.17b, the modified Marx circuit is shown, wherein the resistances R_1 and R_2 are incorporated inside the unit. R_1 is divided into n parts equal to R_1/n and put in series with the gap G. R_2 is also divided into n parts and arranged across each capacitor unit after the gap G. This arrangement saves space, and also the cost is reduced. But, in case the waveshape is to be varied widely, the variation becomes difficult. The additional advantages gained by distributing R_1 and R_2 inside the unit are that the control resistors are smaller in size and the efficiency (V_0/nV) is high.

Impulse generators are nominally rated by the total voltage (nominal), the number of stages, and the gross energy stored. The nominal output voltage is the number of stages multiplied by the charging voltage. The nominal energy stored is given by $\frac{1}{2} C_1 V^2$ where $C_1 = C/n$ (the discharge capacitance) and V is the nominal maximum voltage (n times charging voltage). A 16-stage impulse generator having a stage capacitance of 0.280 μ F and a maximum charging voltage of 300 kV will have an energy rating of 192 kW sec. The height of the generator will be about 15 m and will occupy a floor area of about 3.25 × 3.00 m. The waveform of either polarity can be obtained by suitably changing the charging unit polarity.

6.3.5 Components of a Multistage Impulse Generator

A multistage impulse generator requires several components parts for flexibility and for the production of the required waveshape. These may be grouped as follows:

(i) d.c. Charging Set

The charging unit should be capable of giving a variable d.c. voltage of either polarity to charge the generator capacitors to the required value.

(ii) Charging Resistors

These will be non-inductive high value resistors of about 10 to 100 kilo-ohms. Each resistor will be designed to have a maximum voltage between 50 and 100 kV.

(iii) Generator Capacitors and Spark Gaps

These are arranged vertically one over the other with all the spark gaps aligned. The capacitors are designed for several charging and discharging operations. On dead short circuit, the capacitors will be capable of giving 10 kA of current. The spark gaps will be usually spheres or hemispheres of 10 to 25 cm diameter. Sometimes spherical ended cylinders with a central support may also be used.

(iv) Wave-shaping Resistors and Capacitors

Resistors will be non-inductive wound type and should be capable of discharging impulse currents of 1000 A or more. Each resistor will be designed for a maximum voltage of 50 to 100 kV. The resistances are bifilar wound or non-inductive thin flat insulating sheets. In some cases, they are wound on thin cylindrical formers and are completely enclosed. The load capacitor may be of compressed gas or oil filled with a capacitance of 1 to 10 μF.

Modern impulse generators have their wave-shaping resistors included internally with a flexibility to add additional resistors outside, when the generator capacitance is changed (with series parallel connection to get the desired energy rating at a given test voltage). Such generators optimize the set of resistors. A commercial impulse voltage generator uses six sets of resistors ranging from 1.0 ohm to about 160 ohms with different combinations (with a maximum of two resistors at a time) such that a resistance value varying from 0.7 ohm to 235 ohms per stage is obtained, covering a very large range of energy and test voltages. The resistors used are usually resin cast with voltage and energy ratings of 200 to 250 kV and 2.0 to 5.0 kWsec. The entire range of lightning and switching impulse voltages can be covered using these resistors either in series or in parallel combination.

(v) Triggering System

This consists of trigger spark gaps to cause spark breakdown of the gaps (see Sec. 6.5).

(vi) Voltage Dividers

Voltage dividers of either damped capacitor or resistor type and an oscilloscope with recording arrangement are provided for measurement of the voltages across the test object. Sometimes a sphere gap is also provided for calibration purposes (see Chapter 7 for details).

6.3.6 Generation of Switching Surges

Now-a-days in extra high voltage transmission lines and power systems, switching surge is an important factor that affects the design of insulation. All transmission lines rated for 220 kV and above, incorporate switching surge sparkover voltage for their insulation levels. A switching surge is a short duration transient voltage produced in the system due to a sudden opening or closing of a switch or circuit breaker or due to an arcing at a fault in the system. The waveform is not unique. The transient voltage may be an oscillatory wave or a damped oscillatory wave of frequency ranging from few hundred hertz to few kilo hertz. It may also be considered as a slow rising impulse having a wave front time of 0.1 to 10 ms, and a tail time of one to several ms. Thus, switching surges contain larger energy than the lightning impulse voltages.

Several circuits have been adopted for producing switching surges. They are grouped as (i) impulse generator circuit modified to give longer duration waveshapes, (ii) power transformers or testing transformers excited by d.c. voltages giving oscillatory waves and these include Tesla coils.

Standard switching impulse voltage is defined, both by the Indian Standards and the IEC, as 250/2500 μs wave, with the same tolerances for time-to-front and time-to-tail as those for the lightning impulse voltage wave i.e. time-to-front of (250 ± 50) μs and time-to-half value of (2500 ± 500) μs. Other switching impulse voltage waves commonly used for testing the lightning arresters are 250/1500 μs with a tolerance of ± 500 μs in time-to-half value.

Figure 6.18 shows the impulse generator circuits modified to give switching surges. The arrangement is the same as that of an impulse generator. The values of R_1 and R_2 for producing waveshapes of long duration, such as 100/1000 μs or 400/4000 μs, will range from 1 to 5 kilo-ohms and 5 to 20 kilo-ohms respectively. Thus, R_1 is about 20 to 30% of R_2. The efficiency of the generator gets considerably reduced to about 50% or even less. Moreover, the values of the charging resistors R_1 are to be increased to very high values as these will come in parallel with R_2 in the discharge circuit.

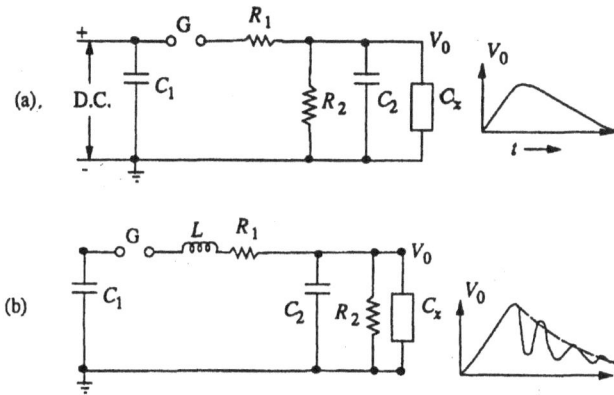

Fig. 6.18 Circuits for producing switching surge voltages. Also shown are the output waveshapes across the load C_x

The circuit given in Fig. 6.18b produces unidirectional damped oscillations. With the use of an inductor L, the value of R_1 is considerably reduced, and the efficiency of the generator increases. The damped oscillations may have a frequency of 1 to 10 kHz depending on the circuit parameters. Usually, the maximum value of the switching surge obtained is 250 to 300 kV with an impulse generator having a nominal rating of 1000 kV and 25 kW sec. Bellaschi et al.[12] used only an inductor L of low resistance to produce switching impulse up to 500 kV. A sphere gap was included in parallel with the test object for voltage measurement and also for producing chopped waves.

Switching surges of very high peaks and long duration can be obtained by using the circuit shown in Fig. 6.19. An impulse generator condenser C_1 charged to a low voltage d.c. (20 to 25 kV) is discharged into the low voltage winding of a power or testing transformer. The high voltage winding is connected in parallel to a load capacitance C_2, a potential divider R_2, a sphere gap S, and the test object. Through an autotransformer action, switching surge of proper waveshape can be generated across

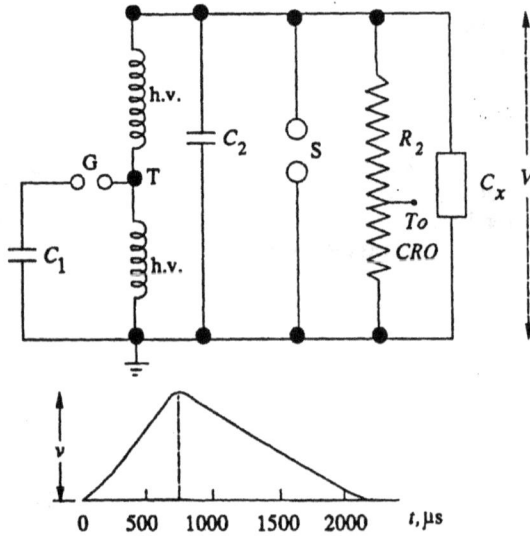

Fig. 6.19 Circuit for producing switching surges using a transformer

C_1 —	Surge generator	G —	Triggering spark gap
T —	Power transformer	S —	Sphere gap
C_2 —	Load capacitance	R_2 —	Potential divider
C_x —	Test object	h.v.,l.v. —	Transformer high and low voltage windings

the test object. The efficiency obtained by this method is high but the transformer should be capable of withstanding very high voltages.

Multi Test Sets for High Voltage Testing

In many small laboratories like in the teaching institutions, small industries and utility organizations, the requirements of high voltages may be less than about 200 kV, 50 Hz, a.c., 400 kV d.c. and 400 kV standard lightning and switching impulse voltages. The power requirements will be around 5 kW or kVA and the energy requirement will be less than 1.5 kJ. For such applications, flexible and universally interchangeable modular systems of the above voltage and energy ratings are available under different trade names. These systems mainly consist of

(i) a.c. Testing Transformers :
With continuous power ratings of 3 to 5 kVA with a short time rating about 150%. The unit can be one single transformer of up to 100 kV(rms), or 2 or 3 units connected in cascade with voltage ratings up to 300 kV(rms).

(ii) d.c Units :

A.C. transformer with the addition of a rectifier unit and a filter capacitor, with ripple factor at rated current less than 5% and a voltage drop or regulation less than 10%, for a single stage output of about 100 kV (half-wave rectifier) constitute a d.c. set. D.C. sets are available as multi-stage voltage doubler units with one pulse output, or as a quadampler unit of up to 400 kV rating with the same specifications. In either case, the power ratings will be about 3 to 5 kW continuous. The rectifier stacks used are the selenium diode type.

(iii) Impulse Voltage Units :

Marx circuit of 2 to 4 stages can be assumed using the transformer and d.c. rectifier unit described earlier for an output voltage of about 400 kV(peak) using a one stage rectifier unit. The necessary wave front and wave tail resistors and load capacitances are normally provided. The units are assembled with modular components mounted on suitable insulating columns. The units normally have voltage efficiency of about 90%.

All the basic units are clearly and compactly arranged. By having increased number of units the system can be expanded to obtain higher and desired type of voltage. Control cubicles/boxes are provided for the control and measurement of voltages. The units can be mounted on wheels or located permanently in a test hall of size 4m × 3m × 3m.

Multi-test sets are currently being manufactured and assembled in India by some leading manufacturers of high voltage test equipments.

6.4 GENERATION OF IMPULSE CURRENTS

Lightening discharges involve both high voltage impulses and high current impulses on transmission lines. Protective gear like surge diverters have to discharge the lightning currents without damage. Therefore, generation of impulse current waveforms of high magnitude ($\approx$ 100 kA peak) find application in testing work as well as in basic research on non-linear resistors, electric arc studies, and studies relating to electric plasmas in high current discharges.

6.4.1 Definition of Impulse Current Waveforms

The waveshapes used in testing surge diverters are 4/10 and 8/20 μ s, the figures respectively representing the nominal wave front and wave tail times (see Fig. 6.14). The tolerances allowed on these times are ± 10% only. Apart from the standard impulse current waves, rectangular waves of long duration are also used for testing. The waveshape should be nominally rectangular in shape. The rectangular waves generally have durations of the order of 0.5 to 5 ms, with rise and fall times of the waves being less than ±10% of their total duration. The tolerance allowed on the peak value is +20% and –0% (the peak value may be more than the specified value but not less). The duration of the wave is defined as the total time of the wave during which the current is at least 10% of its peak value.

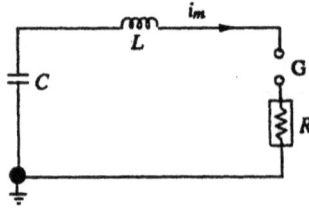

(a) Basic circuit of an impulse current generator

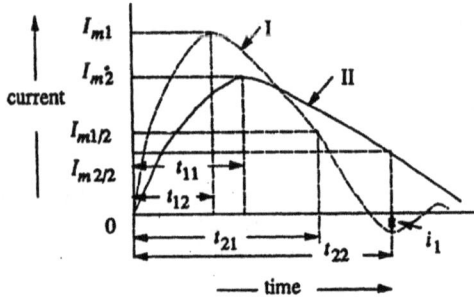

t_1 and t_{12} = time-to-front of waves I and II
t_{21} and t_{22} = time-to-tail of waves I and II

I — damped oscillatory wave
II — overdamped wave
i_1 — overshoot

(b) Types of impulse current waveforms

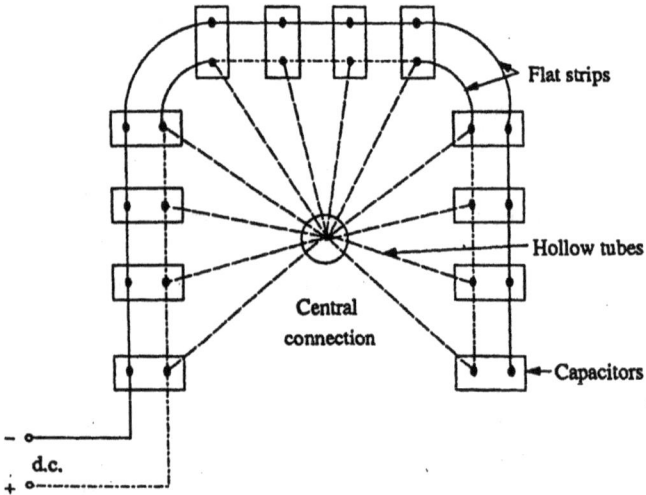

(c) Arrangement of capacitors for high impulse current generation

.**Fig. 6.20** impulse current generator circuit and its waveform

6.4.2 Circuit for Producing Impulse Current Waves

For producing impulse currents of large value, a bank of capacitors connected in parallel are charged to a specified value and are discharged through a series $R\text{-}L$ circuit as shown in Fig. 6.20. C represents a bank of capacitors connected in parallel which are charged from a d.c. source to a voltage up to 200 kV. R represents the dynamic resistance of the test object and the resistance of the circuit and the shunt. L is an air cored high current inductor, usually a spiral tube of a few turns.

If the capacitor is charged to a voltage V and discharged when the spark gap is triggered, the current i_m will be given by the equation

$$V = R\, i_m + L\frac{di_m}{dt} + \frac{1}{C}\int_0^t i_m\, dt \qquad (6.24)$$

The circuit is usually underdamped, so that

$$\frac{R}{2} < \sqrt{L/C}$$

Hence, i_m is given by

$$i_m = \frac{V}{\omega L}[\exp(-\alpha t)]\sin(\omega t) \qquad (6.25)$$

where

$$\alpha = \frac{R}{2L} \text{ and } \omega = \sqrt{\frac{1}{LC} - \frac{R^2}{4L^2}} \qquad (6.25a)$$

The time taken for the current i_m to rise from zero to the first peak value is

$$t_1 = t_f = \frac{1}{\omega}\sin^{-1}\frac{\omega}{\sqrt{LC}} = \frac{1}{\omega}\tan^{-1}\frac{\omega}{\alpha} \qquad (6.26)$$

The duration for one half cycle of the damped oscillatory wave t_2 is,

$$t_2 = \frac{\pi}{\sqrt{\dfrac{1}{LC} - \dfrac{R^2}{4L^2}}} \qquad (6.27)$$

It can be shown that the maximum value of i_m is normally independent of the value of V and C for a given energy $W = \frac{1}{2}CV^2$, and the effective inductance L. It is also clear from Eq. (6.25) that a low inductance is needed in order to get high current magnitudes for a given charging voltage V.

The present practice as per IEC 60.2 is to express the characteristic time t_2 as the time for half value of the peak current, similar to the definition given for standard impulse voltage waves. With this definition, the values of α and ω for 8/20 μs impulse wave will be $\alpha = 0.0535 \times 10^6$ and $\omega = 0.113 \times 10^6$, when R, L, C are expressed in ohms, henries and farads respectively. The product LC will be equal to 65 and the peak value of i_m is given by $(VC)/14$. Here, the charging voltage is in kV and i_m is in kA.

6.4.3 Generation of High Impulse Currents

For producing large values of impulse currents, a number of capacitors are charged in parallel and discharged in parallel into the circuit. The arrangement of capacitors is shown in Fig. 6.20c. In order to minimize the effective inductance, the capacitors are subdivided into smaller units. If there are n_1 groups of capacitors, each consisting of n_2 units and if L_0 is the inductance of the common discharge path, L_1 is that of each group and L_2 is that of each unit, then the effective inductance L is given by

$$L = L_0 + \frac{L_1}{n} + \frac{L_2}{n_1 n_2}$$

Also, the arrangement of capacitors into a horse-shoe shaped layout minimizes the effective load inductance.

The essential parts of an impulse current generator are:
   (i) a d.c. charging unit giving a variable voltage to the capacitor bank,
   (ii) capacitors of high value (0.5 to 5 μF) each with very low self-inductance, capable of giving high short circuit currents,
   (iii) an additional air cored inductor of high current value,
   (iv) proper shunts and oscillograph for measurement purposes, and
   (v) a triggering unit and spark gap for the initiation of the current generator.

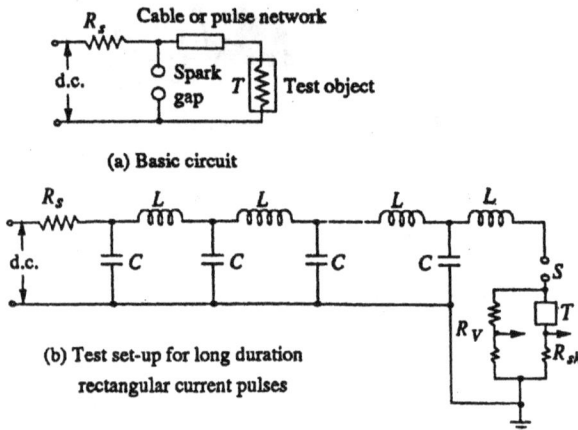

(a) Basic circuit

(b) Test set-up for long duration rectangular current pulses

Fig. 6.21   Basic circuit and schematic set-up for producing rectangular current pulses

R_s — Charging resistor
S — Trigger spark gap
T — Test object
L-C — Pulse forming network
R_v — Potential divider for voltage measurement
R_{sh} — Current shunt for current measurement

6.4.4 Generation of Rectangular Current Pulses

Generation of rectangular current pulses of high magnitudes (few hundred amperes and duration up to 5 ms) can be done by discharging a pulse network or cable previously charged. The basic circuit for producing rectangular pulses is given in Fig. 6.21. The length of a cable or an equivalent pulse forming network is charged to a specified d.c. voltage. When the spark gap is short-circuited, the cable or pulse network discharges through the test object.

To produce a rectangular pulse, a coaxial cable of surge impedance $Z_0 = \sqrt{L_0/C_0}$ (where L_0 is the inductance and C_0 is the capacitance per unit length) is used. If the cable is charged to a voltage V and discharged through the test object of resistance R, the current pulse I is given by $I = V/(Z_0 + R)$. A pulse voltage $RV/(R + Z_0)$ is developed across the test object R, and the pulse current is sustained by a voltage wave $(V-IR)$. For $R = Z_0$, the reflected wave from the open end of the cable terminates the pulse current into the test object, and the pulse voltage becomes equal to $V/2$.

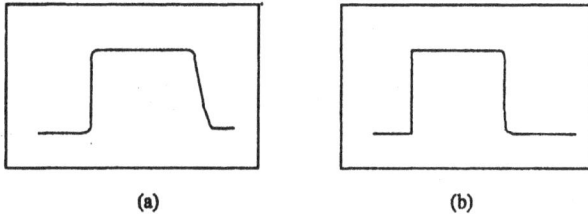

(a) (b)

(a) Pulse waveform using pulse forming network (3.5 μ s)
(b) Pulse waveform using a coaxial cable (3.6 μ s)

Fig. 6.22 Current waveforms produced by rectangular current generators

In practice, it is difficult to get a coaxial cable of sufficient capacitance and length. Often artificial transmission lines with lumped L and C as shown in Fig. 6.21b are used. Usually, 6 to 9 L-C sections will be sufficient to give good rectangular waves. The duration of the pulse time is seconds (t) is given by $t = 2 (n - 1) \sqrt{LC}$, where n is the number of sections used, C is the capacitance per stage or section, and L is the inductance per stage or section.

The current waveforms produced by an artificial line or pulse network and a coaxial cable are shown in Figs. 6.22a and b.

6.5 TRIPPING AND CONTROL OF IMPULSE GENERATORS

In large impulse generators, the spark gaps are generally sphere gaps or gaps formed by hemispherical electrodes. The gaps are arranged such that sparking of one gap results in automatic sparking of other gaps as overvoltage is impressed on the other. In order to have consistency in sparking, irradiation from an ultra-violet lamp is provided from the bottom to all the gaps.

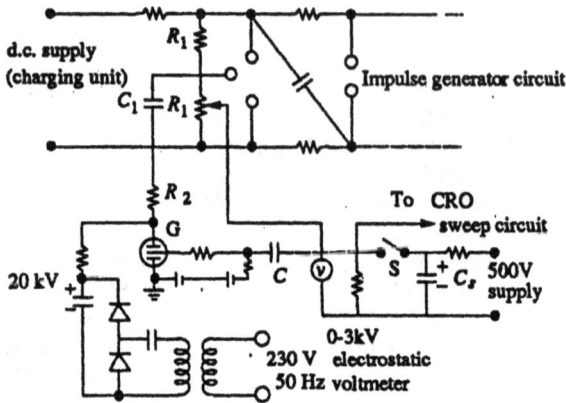

Fig. 6.23 Tripping of an impulse generator with a three electrode gap

To trip the generator at a predetermined time, the spark gaps may be mounted on a movable frame, and the gap distance is reduced by moving the movable electrodes closer. This method is difficult and does not assure consistent and controlled tripping.

A simple method of controlled tripping consists of making the first gap a three electrode gap and firing it from a controlled source. Figure 6.23 gives the schematic arrangement of a three electrode gap. The first stage of the impulse generator is fitted with a three electrode gap, and the central electrode is maintained at a potential in between that of the top and the bottom electrodes with the resistors R_1 and R_L. The tripping is initiated by applying a pulse to the thyratron G by closing the switch S. The capacitor C produces an exponentially decaying pulse of positive polarity. The pulse goes and initiates the oscillograph time base. The thyratron conducts on receiving the pulse from the switch S and produces a negative pulse through the capacitance C_1 at the central electrode of the three electrode gap. Hence, the voltage between the central electrode and the top electrode of the three electrode gap goes above its sparking potential and thus the gap conducts. The time lag required for the thyratron firing and breakdown of the three electrode gap ensures that the sweep circuit of the oscillograph

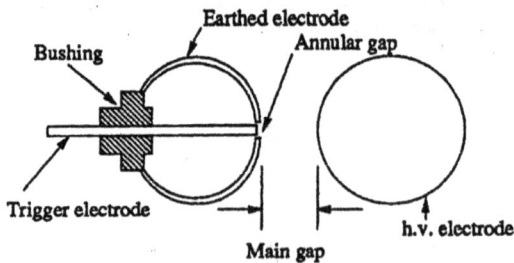

Fig. 6.24 (a) Trigatron gap

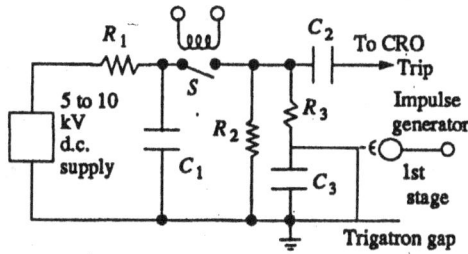

(b) Tripping circuit using a trigatron

Fig. 6.24 Trigatron gap and tripping circuit

begins before the start of the impulse generator voltage. The resistance R_2 ensures decoupling of voltage oscillations produced at the spark gap entering the oscilloscope through the common trip circuit.

The three electrode gap requires larger space and an elaborate construction. Now-a-days a trigatron gap shown in Fig. 6.24 is used, and this requires much smaller voltage for operation compared to the three electrode gap. A trigatron gap consists of a high voltage spherical electrode of suitable size, an earthed main electrode of spherical shape, and a trigger electrode through the main electrode. The trigger electrode is a metal rod with an annular clearance of about 1 mm fitted into the main electrode through a bushing. The trigatron is connected to a pulse circuit as shown in Fig. 6.24b. Tripping of the impulse generator is effected by a trip pulse which produces a spark between the trigger electrode and the earthed sphere. Due to space charge effects and distortion of the field in the main gap, sparkover of the main gap

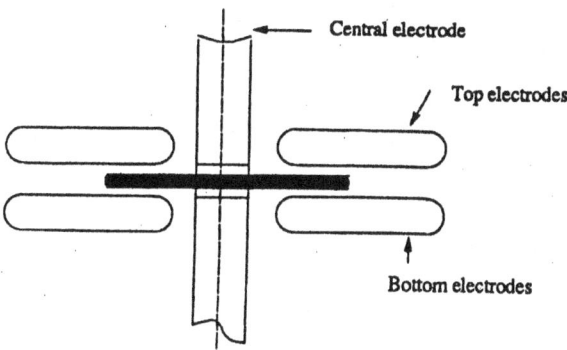

Fig. 6.25 Three electrode gap for high current switching

occurs. The trigatron gap is polarity sensitive and a proper polarity pulse should be applied for correct operation.

Three Electrode Gap for Impulse Current Generator

In the case of impulse current generators using three electrode gaps for tripping and control, a certain special design is needed. The electrodes have to carry high current from the capacitor bank. Secondly, the electrode has to switch large currents in a small duration of time (in about a microsecond). Therefore, the switch should have very low inductance. The erosion rate of the electrodes should be low.

For high current capacitor banks, a number of spark gap switches connected in parallel as shown in Fig. 6.25 are often used to meet the requirement. Recently, trigatron gaps are being replaced by triggered vacuum gaps, the advantage of the latter being fast switching at high currents (> 100 kA) in a few nanoseconds. Triggering of the spark gaps by focused laser beams is also adopted since the performance is better than the conventional triggering methods.

QUESTIONS

Q.6.1 Explain with diagrams, different types of rectifier circuits for producing high d.c. voltages.

Q.6.2 What are the special features of high voltages rectifier valves? How is proper voltage division between the valves ensured, if a number of tubes are used in series ?

Q.6.3 Why is a Cockcroft-Walton circuit preferred for voltage multiplier circuits ? Explain its working with a schematic diagram.

Q.6.4 Give the expression for ripple and regulation in voltage multiplier circuits. How are the ripple and regulation minimized ?

Q.6.5 Describe, with a neat sketch, the working of a Van de Graaff generator. What are the factors that limit the maximum voltage obtained ?

Q.6.6 Explain the different schemes for cascade connection of transformers for producing very high a.c. voltages.

Q.6.7 Why is it preferable to use isolating transformers for excitation with cascade transformer units, if the power requirement is large ?

Q.6.8 What is the principle of operation of a resonant transformer ? How is it advantageous over the cascade connected transformers ?

Q.6.9 What is a Tesla coil ? How are damped high frequency oscillations obtained from a Tesla coil ?

Q.6.10 Define the front and tail times of an impulse wave. What are the tolerances allowed as per the specifications ?

Q.6.11 Give different circuits that produce impulse waves explaining clearly their relative merits and demerits.

Q.6.12 Give the Marx circuit arrangement for multistage impulse generators. How is the basic arrangement modified to accommodate the wave time control resistances ?

Q.6.13 How are the wave front and wave tail times controlled in impulse generator circuits ?

Q.6.14 Explain the different methods of producing switching impulses in test laboratories.

Q.6.15 Explain the effect of series inductance on switching impulse waveshapes produced.

Q.6.16 Describe the circuit arrangement for producing lightning current waveforms in laboratories.

Q.6.17 How is the circuit inductance controlled and minimized in impulse current generators ?

Q.6.18 How are rectangular current pulses generated for testing purposes ? How is their time duration controlled ?

Q.6.19 Explain one method of controlled tripping of impulse generators. Why is controlled tripping necessary ?

Q.6.20 What is a trigatron gap ? Explain its functions and operation.

Q.6.21 An impulse generator has 12 capacitors of $0.12\,\mu F$, and 200 kV rating. The wave front and wave tail resistances are 1.25 kΩ and 4 kΩ respectively. If the load capacitance including that of the test object is 10 000 pF, find the wave front and wave tail times and the peak voltage of impulse wave produced.

Q.6.22 A 8-stage impulse generator has $0.12\,\mu F$ capacitors rated for 167 kV. What is its maximum discharge energy ? If it has to produce a 1/50 μs waveform across a load capacitor of 15,000 pF, find the values of the wave front and wave tail resistances.

Q.6.23 Calculate the peak current and waveshape of the output current of the following generator. Total capacitance of the generator is 53 μF. The charging voltage is 200 kV. The circuit inductance is 1.47 mH, and the dynamic resistance of the test object is 0.051 ohms.

Q.6.24 A single phase testing transformer rated for 2 kV/350 kV, 3500 kVA, 50 Hz on testing yields the following data: (i) No-load voltage on H.V. side = 2% higher than the rated value when the input voltage is 2 kV on the L.V. side. (ii) Short circuit test with H.V. side shorted, rated current was obtained with 10% rated voltage on the input side. Calculate the self-capacitance on the H.V. side and the leakage reactance referred to the H.V. side. Neglect resistance.

Q.6.25 Determine the ripple voltage and regulation of a 10 stage Cockcroft-Walton type d.c. voltage multiplier circuit having a stage capacitance = 0.01 μF, supply voltage = 100 kV at a frequency of 400 Hz and a load current = 10mA.

Q.6.26 A voltage doubler circuit has $C_1 = C_2 = 0.01\,\mu F$ and is supplied from a voltage source of $V = 100 \sin 314t$ kV. If the d.c. output current is to be 4 mA, calculate the output voltage and the ripple.

Q.6.27 The primary and secondary winding inductances of a Tesla coil are 0.093 H and 0.011 H respectively with a mutual inductance between the windings equal to 0.026 H. The capacitance included in the primary and secondary circuits are respectively 1.5 μF and 18 ηF. If the Tesla coil is charged through a 10 kV DC supply, determine the output voltage and its waveform. Neglect the winding resistances.

Q.6.28 An impulse current generator is rated for 60 kW sec. The parameters of the circuit are $C = 53\,\mu F$, $L = 1.47\,\mu H$ and the dynamic resistance = 0.0156 ohm. Determine the peak value of the current and the time-to-front and the time-to-tail of the current waveform.

WORKED EXAMPLES

Example 6.1: A Cockcroft-Walton type voltage multiplier has eight stages with capacitances, all equal to 0.05 μF. The supply transformer secondary voltage is 125 kV at a frequency of 150 Hz. If the load current to be supplied is 5 mA, find (a) the percentage ripple, (b) the regulation, and (c) the optimum number of stages for minimum regulation or voltage drop.

Solution: (a) Calculation of Percentage Ripple

$$\text{The ripple voltage } \delta V = \frac{I}{fC} \frac{(n)\,(n+1)}{2}$$

$I = $ 5 mA, $f = 150$ Hz, $C = 0.05\,\mu F$, and $n = 8$,

$\therefore$
$$\delta V = \frac{5 \times 10^{-3}}{150 \times 0.05 \times 10^{-6}} \times \frac{8 \times 9}{2}$$

$$= 24 \text{ kV}$$

$$\% \text{ ripple} = \frac{\delta V \times 100}{2nV_{max}} = \frac{24 \times 100}{2 \times 125 \times 8}$$

$$= 1.2\%$$

(b) Calculation of Regulation

$$\text{Voltage drop, } \Delta V = \frac{I}{fC}\left(\frac{2}{3}n^3 + \frac{n^2}{2} - \frac{n}{6}\right)$$

$$= \frac{5 \times 10^{-3}}{150 \times 0.05 \times 10^{-6}}\left[\left(\frac{2}{3} \times 8^3\right) + \left(\frac{1}{2} \times 8^2\right) - \frac{8}{6}\right]$$

$$= 248 \text{ kV}$$

$\therefore$
$$\text{regulation}\left(\frac{V}{2nV_{max}}\right) = \frac{248}{2 \times 8 \times 125} = \frac{124}{1000}$$

$$= 12.4\%$$

(c) Calculation of Optimum Number of Stages ($n_{optimum}$)

Since $n > 5$,

$$n_{optimum} = \sqrt{V_{max} fC/I}$$

$$= \sqrt{\frac{125 \times 150 \times 0.05 \times 10^{-6} \times 10^{+3}}{5 \times 10^{-3}}}$$

$$= \sqrt{125 \times 1.5}$$

$$= 13.69$$

$$= 14 \text{ stages}$$

Example 6.2: A 100 kVA, 400 V/250 kV testing transformer has 8% leakage reactance and 2% resistance on 100 kVA base. A cable has to be tested at 500 kV using the above transformer as a resonant transformer at 50 Hz. If the charging current of the cable at 500 kV is 0.4 A, find the series inductance required. Assume 2% resistance for the inductor to be used and the connecting leads. Neglect dielectric loss of the cable. What will be the input voltage to the transformer ?

Solution : The maximum current that can be supplied by the testing transformer is

$$\frac{100 \times 10^3}{250 \times 10^3} 0.4 \text{ A}$$

$X_C =$ Reactance of the cable is

$$\frac{V_C}{I} = \frac{500 \times 10^3}{0.4} = 1250 \text{ k}\Omega$$

$X_L =$ Leakage reactance of the transformer is

$$\frac{\%X}{100} \times \frac{V}{I} = \frac{8}{100} \times \frac{250 \times 10^3}{0.4} = 50 \text{ k}\Omega$$

At resonance, $X_C = X_L$.
Hence, additional reactance needed

$$= 1250 - 50 = 1200 \text{ k}\Omega$$

Inductance of additional reactance (at 50 Hz frequency)

$$\frac{1200 \times 10^3}{2\pi \times 50} = 3820 \text{ H}$$

R = Total resistance in the circuit on 100 kVA base is 2% + 2% = 4%.
Hence, the ohmic value of the resistance

$$= \frac{4}{100} \times \frac{250 \times 10^3}{0.4} = 25 \text{ k}\Omega$$

Therefore, the excitation voltage E_2 on the secondary of the transformer

$$= I \times R$$
$$= 0.4 \times 25 \times 10^3$$
$$= 10 \times 10^3 \text{ V or } 10 \text{ kV}$$

The primary voltage or the supply voltage, E_1

$$= \frac{10 \times 10^3 \times 400}{250 \times 10^3}$$
$$= 16 \text{ V}$$

$$\text{Input kW} = \frac{16}{400} \times 100 = 4.0 \text{ kW}$$

(The magnetizing current and the core losses of the transformer are neglected.)

Example 6.3: An impulse generator has eight stages with each condenser rated for 0.16 μF and 125 kV. The load capacitor available is 1000 pF. Find the series resistance and the damping resistance needed to produce 1.2/50 μs impulse wave. What is the maximum output voltage of the generator, if the charging voltage is 120 kV ?

Solution : Assume the equivalent circuit of the impulse generator to be as shown in Fig. 6.15b.

C_1, the generator capacitance $= \dfrac{0.16}{8} = 0.02 \text{ μF}$

C_2, the load capacitance $\quad = 0.001 \text{ μF}$

t_1, the time to front $\qquad = 1.2 \text{ μs}$

$$= 3.0 \, R_1 \frac{C_1 C_2}{C_1 + C_2}$$

∴

$$R_1 = 1.2 \times 10^{-6} \frac{C_1 + C_2}{C_1 C_2} \times \frac{1}{3}$$

$$= 1.2 \times 10^{-6} \frac{0.021 \times 10^{-6}}{0.02 \times 0.001 \times 10^{-12}} \times \frac{1}{3}$$

$$= 420 \, \Omega$$

t_2, time to tail $= 0.7(R_1 + R_2)(C_1 + C_2)$

$$= 50 \times 10^{-6} \, s$$

or, $0.7(420 + R_2)(0.021 \times 10^{-6}) = 50 \times 10^{-6}$

or, $R_2 = 2981 \, \Omega$

The d.c. charging voltage for eight stages is

$$V = 8 \times 120 = 960 \, kV$$

The maximum output voltage is

$$\frac{V}{R_1 C_2 (\alpha - \beta)} (e^{-\alpha t_1} - e^{-\beta t_1})$$

where $\alpha \approx \dfrac{-1}{R_1 C_2}$, $\beta = \dfrac{1}{R_2 C_1}$ and V is the d.c. charging voltage.

Substituting for R_1, C_1 and R_2, C_2,

$$\alpha = 0.7936 \times 10^{+6}$$

$$\beta = 0.02335 \times 10^{+6}$$

∴ maximum output voltage = 932.6 kV.

Example 6.4: An impulse current generator has a total capacitance of 8 μF. The charging voltage is 25 kV. If the generator has to give an output current of 10kA with 8/20 μs waveform, calculate (a) the circuit inductance and (b) the dynamic resistance in the circuit.

Solution: For an 8/20 μs impulse wave,

$$\alpha = R/2L = 0.0535 \times 10^{+6}$$

and, the product to $LC = 65$. Given $C = 8 \, \mu F$ (L in μH, C in μF, and R in ohms) Therefore, the circuit inductance is

$$\frac{65}{C} = 8.125 \, \mu H$$

The dynamic resistance $2L\alpha = 2 \times \dfrac{65 \times 10^{-6}}{8} \times 0.0535 \times 10^{+6}$

$$= 0.8694 \, ohms$$

Peak current is given by $\dfrac{VC}{14} = 10 \, kA$

(V in kV, C in μF, and I in kA),

∴ charging voltage needed is,

$$V = \frac{14 \times 10}{8} = 15.5 \, kV$$

Example 6.5 (*Alternative Solution*) : Assuming the wave to have a time-to-half value of 20 μs and a time-to-front of 8 μs, the time-to-first half cycle of the damped oscillatory wave will be 20 μs. Then

$$t_1 = t_f = 1/\omega \, [\text{arc tan} \, (\omega/\alpha)] = 8 \, \mu s$$

and $t_2 = \pi/\omega = 20 \, \mu s$

Therefore,
$$\omega = \pi/t_2 = \pi \times 10^6/20 = 0.1571 \times 10^6$$
$$\text{arc tan } (\omega/\alpha) = \omega \cdot t_1 = 1.2566.$$

i.e.
$$\omega/\alpha = 0.8986 \text{ radians}$$

and
$$\alpha = 0.1748 \times 10^6.$$

Then,
$$\sqrt{1/(L\,C) - \alpha^2} = 0.1571 \times 10^6.$$

Substituting the value of α and simplifying,
$$LC = 32.47 \times 10^{12}, \text{ hence } L = 4.06 \text{ }\mu H$$

and
$$R = 2L.\alpha = 1.419 \text{ ohm}$$
$$i_m = V/\omega L \cdot \text{Exp} (-\alpha t) = 10 \text{ kA}$$
$$V = \omega L \times 10 \times \text{Exp} (-\alpha t) = 25.8 \text{ kV}.$$

Example 6.6: A 12-stage impulse generator has 0.126 μF condensers. The wave front and the wave tail resistances connected are 800 ohms and 5000 ohms respectively. If the load condenser is 1000 pF, find the front and tail times of the impulse wave produced.

Solution: The generator capacitance $C_1 = \dfrac{0.126}{12} = 0.0105 \text{ }\mu F$

The load capacitance $C_2 = 0.001 \text{ }\mu F$
Resistances, $R_1 = 800$ ohms and $R_2 = 5000$ ohms

$\therefore$ time to front, $t_1 = 3(R_1) \left(\dfrac{C_1 C_2}{C_1 + C_2} \right)$

$$= 3 \times 800 \times \frac{(0.0105 \times 10^{-6} \times 0.001 \times 10^{-6})}{(0.0105 + 0.001) \times 10^{-6}}$$

$$= 2.19 \text{ }\mu s$$

time to tail, $t_2 = 0.7(R_1 + R_2)(C_1 + C_2)$

$$= 0.7(800 + 5000) \times (0.0105 + 0.001) \times 10^{-6}$$

$$= 46.7 \text{ }\mu s$$

REFERENCES

1. Craggs, J.D. and Meek, J.M., *High Voltage Laboratory Technique*, Butterworths, London (1954).
2. Heller, H., and Veverka, A., *Surge Phenomenon in Electrical Machines*, Illifice and Company, London (1969).
3. Dieter Kind, *An Introduction to High Voltage Experimental Technique*, Wiley Eastern, New Delhi (1979).
4. Gallangher T.J. and Pearman A.J., *High Voltage Measurement, Testing and Design*, John Wiley and Sons, New York (1982).
5. Kuffel E., and Zaengl W., *High Voltage Engineering*, Pergamon Press, Oxford (1984).

6. Begamudre R.D., *E.H.V. a.c. Transmission Engineering*, Wiley Eastern, New Delhi (1986).

7. Niels Hilton Cavillius, *High Voltage Laboratory Planning*, Emil Haefely and Company, Basel, Switzerland (1988).

8. "High Voltage Test Systems: H V Laboratory Equipment", *Bulletins of M/s. Emil Haefely and Company*, Basel, Switzerland.

9. "High Voltage Construction Kits", *Bulletin P2/1e*, MWB (India) Ltd., Bangalore.

10. "EHV Testing Plans", *TUR-WEB Transformatoren Und Rontgen Werk*, Dresden (1964).

11. "High Voltage Technology for Industry and Utilities", *Technical literature of Hippotronics Inc.*, Brewster, New York USA.

12. "Methods of High Voltage Testing", *IS : 2071-1973* and *IS : 4850-1967*.

13. "Testing of Surge Diverters and Lightning Arresters", *IS : 4004-1967*.

14. "High Voltage Testing Techniques", Part-2, *Test Procedures*. IEC Publication Number 60-2-1973.

15. "Standard Techniques for High Voltage Testing", *IEEE Standard no. 4-1978*.

16. Enge H.A., "Cascade Transformer High Voltage Generator", *US Patent No. 3596*. July 1971.

17. Rihond Reid, "High Voltage Resonant Testing", *Proceedings of IEEE PES Winter Conference*, Paper no. C74-038-6, 1974.

18. Kannan S.R., and Narayana Rao Y., "Prediction of Parameters for Impulse Generator for Transformer Testing", *Proceedings of IEE, Vol. 12, no. 5*, pp 535-538, 1973.

19. Glaninger P., "Impulse Testing of Low Inductance Electrical Equipment", *2nd International Symposium on High Voltage Technology*, Zurich, pp 140-144, 1975.

20. Faser K., "Circuit Design of Impulse Generators for Lightning Impulse Voltage Testing of Transformers", *Bulletin SEV/VSE, Vol. 68*, 1977.

7

Measurement of High Voltages and Currents

In industrial testing and research laboratories, it is essential to measure the voltages and currents accurately, ensuring perfect safety to the personnel and equipment. Hence a person handling the equipment as well as the metering devices must be protected against overvoltages and also against any induced voltages due to stray coupling. Therefore, the location and layout of the devices are important. Secondly, linear extrapolation of the devices beyond their ranges are not valid for high voltage meters and measuring instruments, and they have to be calibrated for the full range. Electromagnetic interference is a serious problem in impulse voltage and current measurements, and it has to be avoided or minimized. Therefore, even though the principles of measurements may be same, the devices and instruments for measurement of high voltages and currents differ vastly from the low voltage and low current devices. Different devices used for high voltage measurements may be classified as in Tables 7.1 and 7.2.

7.1 MEASUREMENT OF HIGH DIRECT CURRENT VOLTAGES

Measurement of high d.c. voltages as in low voltage measurements, is generally accomplished by extension of meter range with a large series resistance. The net current in the meter is usually limited to one to ten microamperes for full-scale deflection. For very high voltages (1000 kV or more) problems arise due to large power dissipation, leakage currents and limitation of voltage stress per unit length, change in resistance due to temperature variations, etc. Hence, a resistance potential divider with an electrostatic voltmeter is sometimes better when high precision is needed. But potential dividers also suffer from the disadvantages stated above. Both series resistance meters and potential dividers cause current drain from the source. Generating voltmeters are high impedance devices and do not load the source. They provide complete isolation from the source voltage (high voltage) as they are not directly connected to the high voltage terminal and hence are safer. Spark gaps such as sphere gaps are gas discharge devices and give an accurate measure of the peak voltage. These are quite simple and do not require any specialized construction. But the measurement is affected by the atmospheric conditions like temperature, humidity, etc. and by the vicinity of earthed objects, as the electric field in the gap is affected by the presence of earthed objects. But sphere gap measurement of voltages is independent of the waveform and frequency.

Table 7.1 High voltage Measurement Techniques

Type of voltage	Method or technique
(a) d.c. voltages	(i) Series resistance microammeter
	(ii) Resistance potential divider
	(iii) Generating voltmeters
	(iv) Sphere and other spark gaps
(b) a.c. voltages	(i) Series impedance ammeters
(power frequency)	(ii) Potential dividers (resistance or capacitance type)
	(iii) Potential transformers (electromagnetic or CVT)
	(iv) Electrostatic voltmeters
	(v) Sphere gaps
(c) a.c. high frequency voltages, impulse voltages, and other rapidly changing voltages	(i) Potential dividers with a cathode ray oscillograph (resistive or capacitive dividers)
	(ii) Peak voltmeters
	(iii) Sphere gaps

Table 7.2 High Current Measurement Techniques

Type of current	Device or technique
(a) Direct currents	(i) Resistive shunts with milliammeter
	(ii) Hall effect generators
	(iii) Magnetic links
(b) Alternating currents	(i) Resistive shunts
(Power frequency)	(ii) Electromagnetic current transformers
(c) High frequency a.c.,	(i) Resistive shunts
impulse and rapidly	(ii) Magnetic potentiometers or Rogowski coils
changing currents	(iii) Magnetic links
	(iv) Hall effect generators

7.1.1 High Ohmic Series Resistance with Microammeter

High d.c. voltages are usually measured by connecting a very high resistance (few hundreds of megaohms) in series with a microammeter as shown in Fig. 7.1. Only the current I flowing through the large calibrated resistance R is measured by the moving coil microammeter. The voltage of the source is given by

$$V = IR$$

The voltage drop in the meter is negligible, as the impedance of the meter is only few ohms compared to few hundred mega-ohms of the series resistance R. A protective device like a paper gap, a neon glow tube, or a zener diode with a suitable series resistance is connected across the meter as a protection against high voltages in case the series resistance R fails or flashes over. The ohmic value of the series resistance R is chosen such that a current of one to ten microamperes is allowed for full-scale deflection. The resistance is constructed from a large number of wire wound resistors in series. The voltage drop in each resistor element is chosen to avoid surface flashovers and discharges. A value of less than 5 kV/cm in air or less than 20 kV/cm in good oil is permissible. The resistor chain is provided with corona free terminations. The material for resistive elements is usually a carbon-alloy with temperature coefficient less than $10^{-4}/°C$. Carbon and other metallic film resistors are also used. A resistance chain built with ±1% carbon resistors located in an airtight transformer oil filled P.V.C. tube, for 100 kV operation had very good temperature stability. The limitations in the series resistance design are:

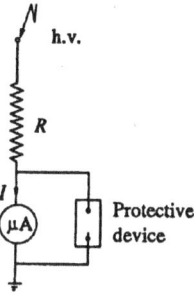

Fig. 7.1 Series resistance micrometer

(*i*) power dissipation and source loading,
(*ii*) temperature effects and long time stability,
(*iii*) voltage dependence of resistive elements, and
(*iv*) sensitivity to mechanical stresses.

Series resistance meters are built for 500 kV d.c. with an accuracy better than 0.2%.

7.1.2 Resistance Potential Dividers for d.c. Voltages

A resistance potential divider with an electrostatic or high impedance voltmeter is shown in Fig. 7.2. The influence of temperature and voltage on the elements is eliminated in the voltage divider arrangement. The high voltage magnitude is given by $[(R_1 + R_2)/R_2]v_2$, where v_2 is the d.c. voltage across the low voltage arm R_2. With sudden changes in voltage, such as switching operations, flashover of the test objects, or source short circuits, flashover or damage may occur to the divider elements due to the stray capacitance across the elements and due to ground capacitances. To avoid these transient voltages, voltage controlling capacitors are connected across the elements. A corona free termination is also necessary to avoid unnecessary discharges at high voltage ends. A series resistor with a parallel capacitor connection for linearization of transient potential distribution is shown in Fig. 7.3. Potential dividers are made with 0.05% accuracy up to 100 kV, with 0.1% accuracy up to 300 kV, and with better than 0.5% accuracy for 500 kV.

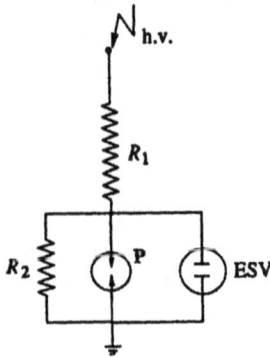

Fig. 7.2 Resistance potential divider with an electrostatic voltmeter

P — Protective device
ESV — Electrostatic volt-meter

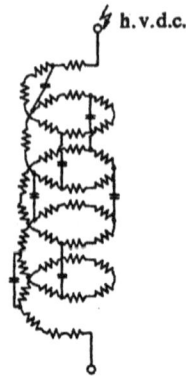

Fig. 7.3 Series resistor with parallel capacitors for potential linearization for transient voltages

7.1.3 Generating Voltmeters

High voltage measuring devices employ generating principle when source loading is prohibited (as with Van de Graaff generators, etc.) or when direct connection to the high voltage source is to be avoided. A generating voltmeter is a variable capacitor electrostatic voltage generator which generates current proportional to the applied external voltage. The device is driven by an external synchronous or constant speed motor and does not absorb power or energy from the voltage measuring source.

Principle of Operation

The charge stored in a capacitor of capacitance C is given by $q = CV$. If the capacitance of the capacitor varies with time when connected to the source of voltage V, the current through the capacitor,

$$i = \frac{dq}{dt} = V\frac{dC}{dt} + C\frac{dV}{dt} \qquad (7.1)$$

For d.c. voltages $dV/dt = 0$. Hence,

$$i = \frac{dq}{dt} = V\frac{dC}{dt} \qquad (7.2)$$

If the capacitance C varies between the limits C_0 and $(C_0 + C_m)$ sinusoidally as

$$C = C_0 + C_m \sin \omega t$$

the current i is

$$i = i_m \cos \omega t$$

where

$$i_m = VC_m\omega$$

(i_m is the peak value of the current). The rms value of the current is given by:

$$i_{r\,ms} = \frac{VC_m\omega}{\sqrt{2}} \tag{7.3}$$

For a constant angular frequency ω, the current is proportional to the applied voltage V. More often, the generated current is rectified and measured by a moving coil meter. Generating voltmeter can be used for a.c. voltage measurements also provided the angular frequency ω is the same or equal to half that of the supply frequency.

A generating voltmeter with a rotating cylinder consists of two excitating field electrodes and a rotating two pole armature driven by a synchronous‚motor at a constant speed n. The a.c. current flowing between the two halves of the armature is rectified by a commutator whose arithmetic mean may be calculated from:

$$i = \frac{n}{30} \Delta C\, V, \qquad \text{where } \Delta C = C_{max} - C_{min}$$

For a symmetric voltage $C_{min} = 0$. When the voltage is not symmetrical, one of the electrodes is grounded and C_{min} has a finite value. The factor of proportionality $\frac{n}{30} \cdot \Delta C$ is determined by calibration.

This device can be used for measuring a.c. voltages provided the speed of the drive-motor is half the frequency of the voltage to be measured. Thus a four-pole synchronous motor with 1500 rpm is suitable for 50 Hz. For peak value measurements, the phase angle of the motor must also be so adjusted that C_{max} and the crest value occur at the same instant.

Generating voltmeters employ rotating sectors or vanes for variation of capacitance. Figure 7.4 gives a schematic diagram of a generating voltmeter. The high voltage source is connected to a disc electrode S_3 which is kept at a fixed distance on the axis of the other low voltage electrodes S_0, S_1, and S_2. The rotor S_0 is driven at a constant speed by a synchronous motor at a suitable speed (1500, 1800, 3000, or 3600 rpm). The rotor vanes of S_0 cause periodic change in capacitance between the insulated disc S_2 and the h.v. electrode S_3. The shape and number of the vanes of S_0 and S_1 are so designed that they produce sinusoidal variation in the capacitance. The generated a.c. current through the resistance R is rectified and read by a moving coil instrument. An amplifier is needed, if the shunt capacitance is large or longer leads are used for connection to rectifier and meter. The instrument is calibrated using a potential divider or sphere gap. The meter scale is linear and its range can be extended

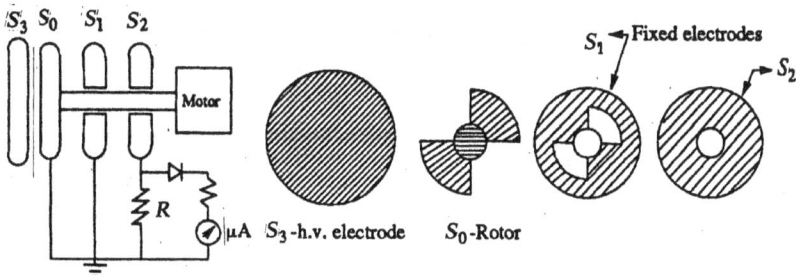

Fig. 7.4   Schematic diagram of a generating voltmeter (rotating vane type)

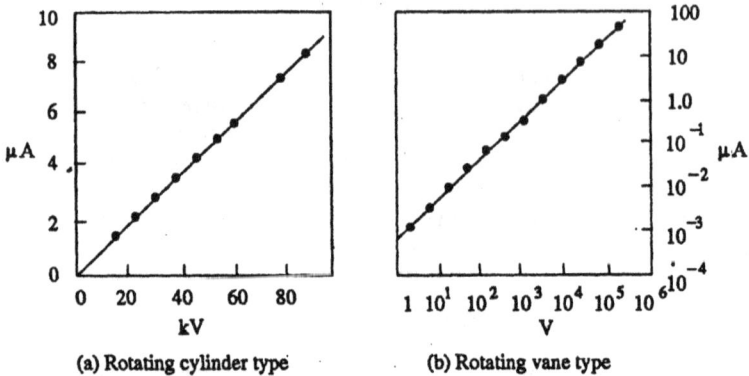

(a) Rotating cylinder type   (b) Rotating vane type

Fig. 7.5 Calibration curves for a generating voltmeter

by extrapolation. Typical calibration curves of a generating voltmeter are given in Figs. 7.5a and b.

Advantages of Generating Voltmeters

- (*i*) No source loading by the meter,
- (*ii*) no direct connection to high voltage electrode,
- (*iii*) scale is linear and extension of range is easy, and
- (*iv*) a very convenient instrument for electrostatic devices such as Van de Graaff generator and particle accelerators.

Limitations of Generating Voltmeters

- (*i*) They require calibration,
- (*ii*) careful construction is needed and is a cumbersome instrument requiring an auxiliary drive, and
- (*iii*) disturbance in position and mounting of the electrodes make the calibration invalid.

7.1.4 Other Methods—Oscillating Spheroid

The period of oscillation of an oscillating spheroid in a uniform electric field is proportional to the applied electric field. This principle is made use of in measuring high d.c. voltages. The period of oscillation of a suspended spheroid between two electrodes with and without an electric field present is measured. If the frequency of the oscillation for small amplitudes is f and f_0 respectively, then the electric field

$$E \propto [f^2 - f_0^2]^{1/2}$$

and hence the applied voltage

$$V \propto [f^2 - f_0^2]^{1/2} \tag{7.4}$$

since $E = V/d$ (d being the gap separation between the electrodes). The proportionality constant can be determined from the dimensions of the spheroid or experimentally.

The uniform electric field is produced by employing two electrodes with a Bruce profile for a spacing of about 50 cm. One of the electrodes is earthed and the other is connected to a high voltage d.c. source. The spheroid is suspended at the centre of the electrodes in the axis of the electric field. The period of oscillation is measured using a telescope and stop watch. Instruments of this type are constructed for voltages up to 200 kV, and the accuracy is estimated to be ±0.1%. In Bruce's design, electrodes of 145 cm diameter with 45 cm spacing were used. An overall accuracy of ±0.03% was claimed up to a maximum voltage of 250 kV. Since this is a very complicated and time consuming method, it is not widely used. The useful range of the spheroidal voltmeter is limited by local discharges.

7.1.5 Measurement of Ripple Voltage in d.c. Systems

It has been discussed in the previous chapter that d.c. rectifier circuits contain ripple, which should be kept low (<< 3%). Ripple voltages are a.c. voltages of non-sinusoidal nature, and as such oscillographic measurement of these voltages is desirable. However, if a resistance potential divider is used along with an oscilloscope, the measurement of small values of the ripple δV will be inaccurate.

A simple method of measuring the ripple voltage is to use a capacitance-resistance (C-R) circuit and measure the varying component of the a.c. voltage by blocking the d.c. component. If V_1 is the d.c. source voltage with ripple (Fig. 7.6a) and V_2 is the voltage across the measuring resistance R, with C acting as the blocking capacitor, then

$$V_2(t) = V_1(t) - V_{d.c.} = \text{ripple voltage}$$

The condition to be satisfied here is $\omega CR >> 1$.

Measurement of Ripple with CRO

The detailed circuit arrangement used for this purpose is shown in Fig. 7.6b. Here, the capacitance 'C' is rated for the peak voltage. It is important that the switch 'S' be closed when the CRO is connected to the source so that the CRO input terminal does not receive any high voltage signal while 'C' is being charged. Further, C should be

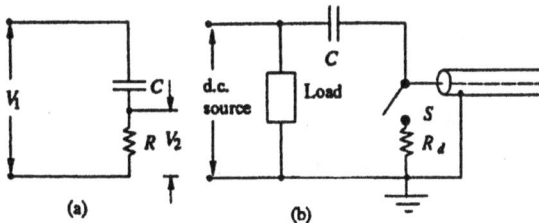

Fig. 7.6 Circuit arrangement for the measurement of ripple voltage

larger than the capacitance of the cable and the input capacitance of the *CRO*, taken together.

7.2 MEASUREMENT OF HIGH a.c. AND IMPULSE VOLTAGES: INTRODUCTION

Measurement of high a.c. voltages employ conventional methods like series impedance voltmeters, potential dividers, potential transformers, or electrostatic voltmeters. But their designs are different from those of low voltage meters, as the insulation design and source loading are the important criteria. When only peak value measurement is needed, peak voltmeters and sphere gaps can be used. Often, sphere gaps are used for calibration purposes. Impulse and high frequency a.c. measurements invariably use potential dividers with a cathode ray oscillograph for recording voltage waveforms. Sphere gaps are used when peak values of the voltage are only needed and also for calibration purposes.

7.2.1 Series Impedance Voltmeters

For power frequency a.c. measurements the series impedance may be a pure resistance or a reactance. Since resistances involve power losses, often a capacitor is preferred as a series reactance. Moreover, for high resistances, the variation of resistance with temperature is a problem, and the residual inductance of the resistance gives rise to an impedance different from its ohmic resistance. High resistance units for high voltages have stray capacitances and hence a unit resistance will have an equivalent circuit as shown in Fig. 7.7. At any frequency ω of the a.c. voltage, the impedance of the resistance R is

$$Z = \frac{R + j\omega L}{(1 - \omega^2 LC) + j\omega CR} \qquad (7.5)$$

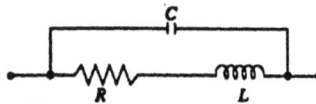

Fig. 7.7 Simplified lumped parameter equivalent circuit of a high ohmic resistance R

L — Residual inductance

C — Residual capacitance

If ωL and ωC are small compared to R,

$$Z \approx R\left[1 + j\left(\frac{\omega L}{R} - \omega CR\right)\right] \qquad (7.6)$$

and the total phase angle is

$$\varphi \approx \left(\frac{\omega L}{R} - \omega CR \right) \tag{7.7}$$

This can be made zero and independent of frequency, if

$$L/C = R^2 \tag{7.8}$$

For extended and large dimensioned resistors, this equivalent circuit is not valid and each elemental resistor has to be approximated with this equivalent circuit. The entire resistor unit then has to be taken as a transmission line equivalent, for calculating the effective resistance. Also, the ground or stray capacitance of each element influences the current flowing in the unit, and the indication of the meter results in an error. The equivalent circuit of a high voltage resistor neglecting inductance and the circuit of compensated series resistor using guard and tuning resistors is shown in Figs. 7.8a and b respectively. Stray ground capacitance effects (refer Fig. 7.8b) can be removed by shielding the resistor R by a second surrounding spiral R_s, which shunts the actual resistor but does not contribute to the current through the instrument. By tuning the resistors R_a, the shielding resistor end potentials may be adjusted with respect to the actual measuring resistor so that the resulting compensation currents between the shield and the measuring resistors provide a minimum phase angle.

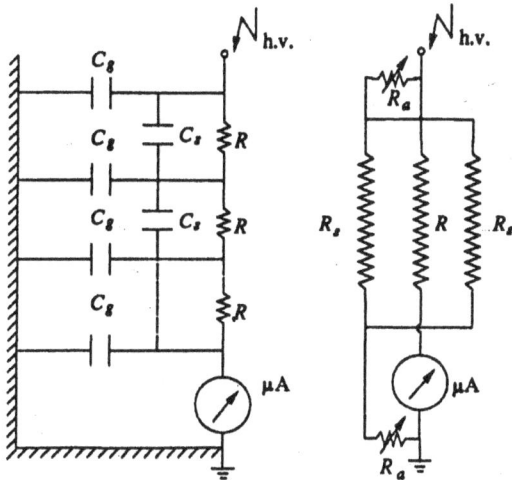

(a) Extended series resistance with inductance neglected

(b) Series resistance with guard and tuning resistances

C_g — Stray capacitance to ground
C_s — Winding capacitance

R — Series resistor
R_s — Guard resistor
R_a — Tuning resistor

Fig. 7.8 Extended series resistance for high a.c. voltage measurements

Series Capacitance Voltmeter

To avoid the drawbacks pointed out earlier, a series capacitor is used instead of a resistor for a.c. high voltage measurements. The schematic diagram is shown in Fig. 7.9. The current I_c through the meter is:

$$I_c = j \omega CV \tag{7.9}$$

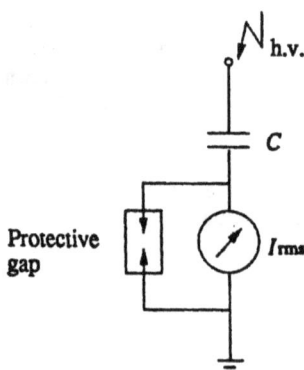

where, C = capacitance of the series capacitor,

ω = angular frequency, and

V = applied a.c. voltage.

If the a.c. voltage contains harmonics, error due to changes in series impedance occurs. The rms value of the voltage V with harmonics is given by

$$V = \sqrt{V_1^2 + V_2^2 + \dots V_n^2} \tag{7.10}$$

where $V_1, V_2, \dots V_n$ represent the rms value of the fundamental, second ... and nth harmonics.

The currents due to these harmonics are

$$
\begin{aligned}
I_1 &= \omega CV_1 \\
I_2 &= 2 \omega CV_2, \dots, \text{ and} \\
I_n &= n \omega CV_n
\end{aligned}
\tag{7.11}
$$

Fig. 7.9 Series capacitance with a milliammeter for measurement of high a.c. voltages

Hence, the resultant rms current is:

$$I = \omega C (V_1^2 + 4V_2^2 + \dots + n^2 V_n^2)^{1/2} \tag{7.12}$$

With a 10% fifth harmonic only, the current is 11.2% higher, and hence the error is 11.2% in the voltage measurement.

This method is not recommended when a.c. voltages are not pure sinusoidal waves but contain considerable harmonics.

Series capacitance voltmeters were used with cascade transformers for measuring rms values up to 1000 kV. The series capacitance was formed as a parallel plate capacitor between the high voltage terminal of the transformer and a ground plate suspended above it. A rectifier ammeter was used as an indicating instrument and was directly calibrated in high voltage rms value. The meter was usually a 0-100 μA moving coil meter and the over all error was about 2%.

7.2.2 Capacitance Potential Dividers and Capacitance Voltage Transformers

The errors due to harmonic voltages can be eliminated by the use of capacitive voltage dividers with an electrostatic voltmeter or a high impedance meter such as a V.T.V.M. If the meter is connected through a long cable, its capacitance has to be taken into account in calibration. Usually, a standard compressed air or gas condenser is used as C_1 (Fig. 7.10), and C_2 may be any large capacitor (mica, paper, or any low loss condenser). C_1 is a three terminal capacitor and is connected to C_2 through a shielded cable, and C_2 is completely shielded in a box to avoid stray capacitances. The applied voltage V_1 is given by

$$V_1 = V_2 \left(\frac{C_1 + C_2 + C_m}{C_1} \right) \quad (7.13)$$

where C_m is the capacitance of the meter and the connecting cable and the leads and V_2 is the meter reading.

Capacitance Voltage Transformer—CVT

Capacitance divider with a suitable matching or isolating potential transformer tuned for resonance condition is often used in power systems for voltage measurements. This is often referred to as CVT. In contrast to simple capacitance divider which requires a high impedance meter like a V.T.V.M. or an electrostatic voltmeter, a CVT can be connected to a low impedance device like a wattmeter pressure coil or a relay coil. CVT can supply a load of a few VA. The schematic diagram of a CVT with its equivalent circuit is given in Fig. 7.11. C_1 is made of a few units of high voltage condensers, and the total capacitance will be around a few thousand picofarads as against a gas filled standard condenser of about 100 pF. A matching transformer is connected between the load or meter M and C_2. The transformer ratio is chosen on economic grounds, and the h.v. winding rating may be 10 to 30 kV with the l.v. winding rated from 100 to 500 V. The value of the tuning choke L is chosen to make the equivalent circuit of the CVT purely resistive or to bring resonance condition. This condition is satisfied when

$$\omega (L + L_T) = \frac{1}{\omega(C_1 + C_2)} \quad (7.14)$$

where, L = inductance of the choke, and

L_T = equivalent inductance of the transformer referred to h.v. side.

The voltage V_2 (meter voltage) will be in phase with the input voltage V_1.

The phasor diagram of CVT under resonant conditions is shown in Fig. 7.11. The meter is taken as a resistive load, and X'_m is neglected. The voltage across the load referred to the divider side will be $V'_2 = (I'_m R'_m)$ and $V_{C_2} = V'_2 + I_m(R_e + X_e)$. It is clear from the phasor diagram that V_1 (input voltage) $= (V_{C_1} + V_{C_2})$ and is in phase

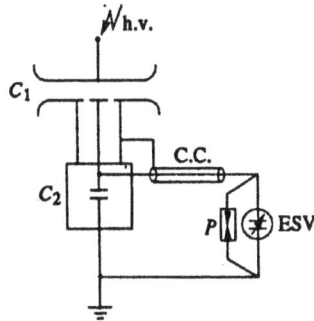

Fig. 7.10 Capacitance potential divider

C_1 — Standard compressed gas h.v. condenser

C_2 — Standard low voltage condenser

ESV — Electrostatic voltmeter

P — Protective gap

C.C. — Connecting cable

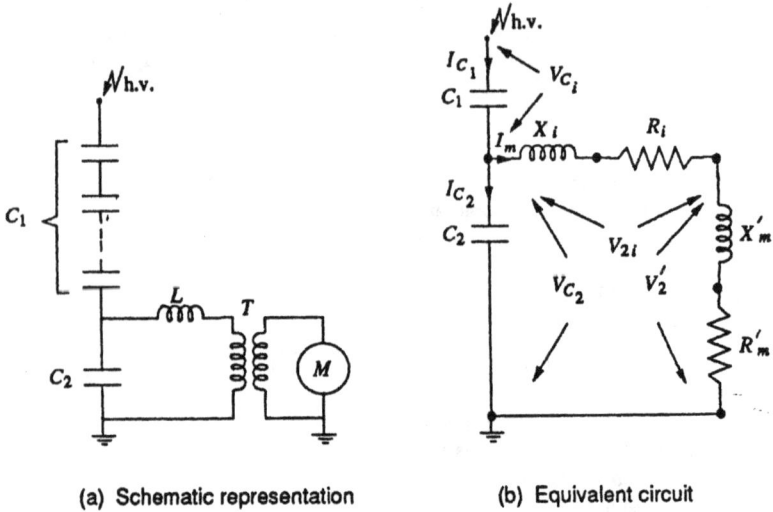

(a) Schematic representation   (b) Equivalent circuit

Fig. 7.11   Capacitive voltage transformer (CVT)

with V_2', the voltage across the meter. R_e and X_e include the potential transformer resistance and leakage reactance. Under this condition, the voltage ratio becomes

$$a = (V_1/V_2) = (V_{C_1} + V_{Ri} + V_2')/V_2' \qquad (7.15)$$

(neglecting the voltage drop $I_m \cdot X_e$ which is very small compared to the voltage V_{C_1}) where V_{Ri} is the voltage drop in the transformer and choke windings.

The advanges of a CVT are:

(i) simple design and easy installation,

(ii) can be used both as a voltage measuring device for meter and relaying purposes and also as a coupling condenser for power line carrier communication and relaying.

(iii) frequency independent voltage distribution along elements as against conventional magnetic potential transformers which require additional insulation design against surges, and

(iv) provides isolation between the high voltage terminal and low voltage metering.

The disadvantages of a CVT are:

(i) the voltage ratio is susceptible to temperature variations, and

(ii) the problem of inducing ferro-resonance in power systems.

Resistance Potential Dividers

Resistance potential dividers suffer from the same disadvantages as series resistance voltmeters for a.c. applications. Moreover, stray capacitances and inductances (Figs 7.7 and 7.8) associated with the resistances make them inaccurate, and compensation has to be provided. Hence, they are not generally used.

7.2.3 Potential Transformers (Magnetic Type)

Magnetic potential transformers are the oldest devices for a.c. measurements. They are simple in construction and can be designed for any voltage. For very high voltages, cascading of the transformers is possible. The voltage ratio is:

$$\frac{V_1}{V_2} = a = \frac{N_1}{N_2} \qquad (7.16)$$

where V_1 and V_2 are the primary and secondary voltages, and N_1 and N_2 are the respective turns in the windings.

These devices suffer from the ratio and phase angle errors caused by the magnetizing and leakage impedances of the transformer windings. The errors are compensated by adjusting the turns ratio with the tappings on the high voltage side under load conditions. Potential transformers (PT) do not permit fast

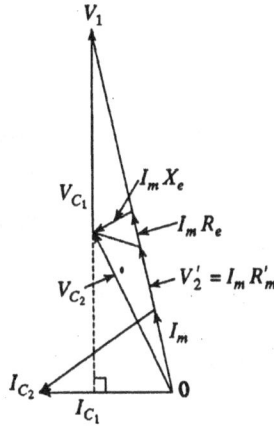

Fig. 7.12 Phasor diagram of a CVT under resonance or tuned condition, Z_m is taken to be equal to resistance R_m

rising transient or high frequency voltages along with the normal supply frequency, but harmonic voltages are usually measured with sufficient accuracy. With high voltage testing transformers, no separate potential transformer is used, but a PT winding is incorporated with the high voltage windings of the testing transformer.

With test objects like insulators, cables, etc. which are capacitive in nature, a voltage rise occurs on load with the testing transformer, and the potential transformer winding gives voltage values less than the actual voltages applied to the test object. If the percentage impedance of the testing transformer is known, the following correction can be applied to the voltage measured by the PT winding of the transformer.

$$V_2 = V_{20}(1 + 0.01\, v_x C/C_N) \qquad (7.17)$$

where, $V_{20} =$ open circuit voltage of the PT winding,

$C_N =$ load capacitance used for testing,

$C =$ test object capacitance ($C \ll C_N$), and

$v_x =$ % reactance drop in the transformer.

7.2.4 Electrostatic Voltmeters

Principle

In electrostatic fields, the attractive force between the electrodes of a parallel plate condenser is given by

$$F = \left|\frac{-\delta W_s}{dS}\right| = \left|\frac{\delta}{\delta S}\left(\tfrac{1}{2}CV^2\right)\right| = \left|\tfrac{1}{2}V^2\frac{\delta C}{\delta S}\right|$$

$$= \tfrac{1}{2}\varepsilon_0 V^2 \frac{A}{s^2} = \tfrac{1}{2}\varepsilon_0 A \left(\frac{V}{s}\right)^2 \qquad (7.18)$$

where,

V = applied voltage between plates,

C = capacitance between the plates,

A = area of cross-section of the plates,

s = separation between the plates,

ε_0 = permittivity of the medium (air or free space), and

W_s = work done in displacing a plate

When one of the electrodes is free to move, the force on the plate can be measured by controlling it by a spring or balancing it with a counter weight. For high voltage measurements, a small displacement of one of the electrodes by a fraction of a millimetre to a few millimetres is usually sufficient for voltage measurements. As the force is proportional to the square of the applied voltage, the measurement can be made for a.c. or d.c. voltages.

Construction

Electrostatic voltmeters are made with parallel plate configuration using guard rings to avoid corona and field fringing at the edges. An absolute voltmeter is made by balancing the plate with a counter weight and is calibrated in terms of a small weight. Usually the electrostatic voltmeters have a small capacitance (5 to 50 pF) and high insulation resistance ($R \geq 10^{13}\ \Omega$). Hence they are considered as devices with high input impedance. The upper frequency limit for a.c. applications is determined from the following considerations:

    (i) natural frequency of the moving system,
    (ii) resonant frequency of the lead and stray inductances with meter capacitance, and
    (iii) the R-C behaviour of the retaining or control spring (due to the frictional resistance and elastance).

An upper frequency limit of about one MHz is achieved in careful designs. The accuracy for a.c. voltage measurements is better than ±0.25%, and for d.c. voltage measurements it may be ±0.1% or less.

The schematic diagram of an absolute electrostatic voltmeter or electrometer is given in Fig. 7.13. It consists of parallel plane disc type electrodes separated by a small distance. The moving electrode is surrounded by a fixed guard ring to make the field uniform in the central region. In order to measure the given voltage with precision, the disc diameter is to be increased, and the gap distance is to be made less. The limitation on the gap distance is the safe working stress (V/s) allowed in air which is normally 5 kV/cm or less. The main difference between several forms of voltmeters lies in the manner in which the restoring force is obtained. For conventional versions of meters, a simple spring control is used, which actuates a pointer to move on the scale of the instruments. In more versatile instruments, only small movements of the moving electrodes is allowed, and the movement is amplified through optical means (lamp and scale arrangement as used with moving coil galvanometers). Two air vane dampers are used to reduce vibrational tendencies in the moving system, and the

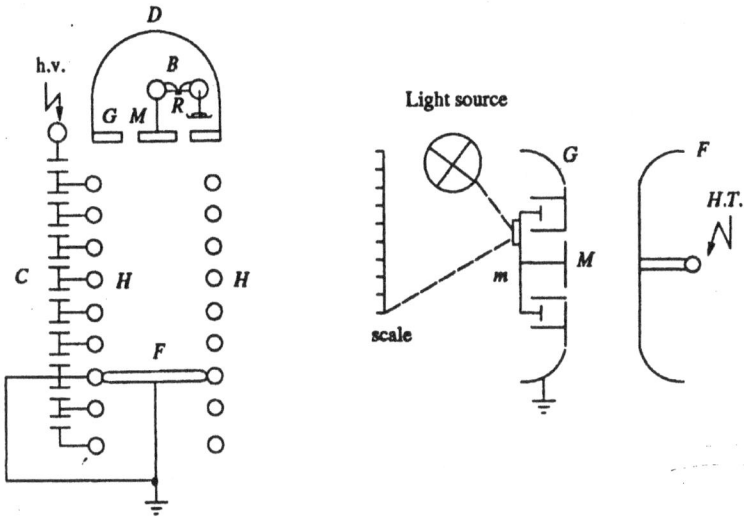

(a) Absolute electrostatic voltmeter

m — mirror
(b) Light beam arrangement

M — Mounting plate
G — Guard plate
F — Fixed plate
H — Guard hoops or rings

B — Balance
C — Capacitance divider
D — Dome
R — Balancing weight

Fig. 7.13   Electrostatic voltmeter

elongation of the spring is kept minimum to avoid field disturbances. The range of the instrument is easily changed by changing the gap separation so that V/s or electric stress is the same for the maximum value in any range. Multi-range instruments are constructed for 600 kV rms and above.

The constructional details of an absolute electrostatic voltmeter is given in Fig. 7.13a. The control torque is provided by a balancing weight. The moving disc M forms the central core of the guard ring G which is of the same diameter as the fixed plate F. The cap D encloses a sensitive balance B, one arm of which carries the suspension of the moving disc. The balance beam carries a mirror which reflects a beam of light. The movement of the disc is thereby magnified. As the spacing between the two electrodes is large, the uniformity of the electric field is maintained by the guard rings H which surround the space between the discs F and M. The guard rings H are maintained at a constant potential in space by a capacitance divider ensuring a uniform special potential distribution.

Some instruments are constructed in an enclosed structure containing compressed air, carbon dioxide, or nitrogen. The gas pressure may be of the order of 15 atm. Working stresses as high as 100 kV/cm may be used in an electrostatic meter in vacuum. With compressed gas or vacuum as medium, the meter is compact and much smaller in size.

7.2.5 Peak Reading a.c. Voltmeters

In some occasions, the peak value of an a.c. waveform is more important. This is necessary to obtain the maximum dielectric strength of insulating solids, etc. When the waveform is not sinusoidal, rms value of the voltage multiplied by $\sqrt{2}$ is not correct. Hence a separate peak value instrument is desirable in high voltage applications.

Series Capacitor Peak Voltmeter

When a capacitor is connected to a sinusoidal voltage source, the charging current i_0

$$= C \int_0^t v dt = j \omega CV$$ where V is the rms value of the voltage and ω is the angular

frequency. If a half wave rectifier is used, the arithmetic mean of the rectifier current is proportional to the peak value of the a.c. voltage. The schematic diagram of the circuit arrangement is shown in Fig. 7.14. The d.c. meter reading is proportional to the peak value of the value V_m or

$$V_m = \frac{I}{2\pi f C}$$

where I is the d.c. current read by the meter and C is the capacitance of the capacitor. This method is known as the Chubb-Frotscue method for peak voltage measurement.

The diode D_1 is used to rectify the a.c. current in one half cycle while D_2 by-passes in the other half cycle. This arrangement is suitable only for positive or negative half

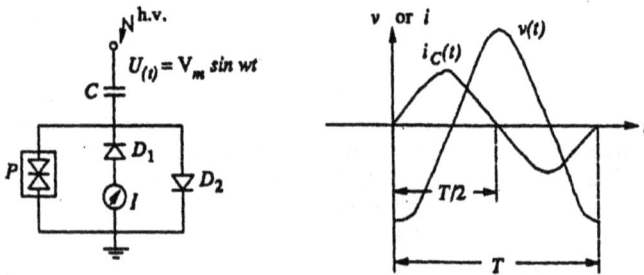

Fig. 7.14 Peak voltmeter with a series capacitor

C — Capacitor	$v(t)$ — Voltage waveform
D_1, D_2 — Diodes	$I_c(t)$ — Capacitor current waveform
P — Protective device	T — Period
I — Indicating meter	
(rectified current indicated)	

cycles and hence is valid only when both half cycles are symmetrical and equal. This method is not suitable when the voltage waveform is not sinusoidal but contains more than one peak or maximum as shown in Fig. 7.14. The charging current through the capacitor changes its polarity within one half cycle itself. The shaded areas in Fig. 7.15 give the reverse current in any one of the half cycles and the current within that period subtracts from the net current. Hence the reading of the meter will be less and is not proportional to V_m as the current flowing during the intervals ($t_1 - t_2$) etc. will not be included in the mean value. The 'second' or the false maxima is easily spotted out by observing the waveform of the charging current on an oscilloscope. Under normal conditions with a.c. testing, such waveforms do not occur and as such do not give rise to errors. But pre-discharge currents within the test circuits cause very short duration voltage drops which may introduce errors. This problem can also be overcome by using a resistance R in series with capacitor C such that $CR \ll 1/\omega$ for 50 Hz application. The error due to the resistance is

$$\frac{\Delta V}{V} = \frac{V - V_m}{V} = \left(1 - \frac{1}{1 + \omega^2 C^2 R^2}\right) \tag{7.19}$$

where,
$$V = \text{actual value, and}$$
$$V_m = \text{measured value}$$

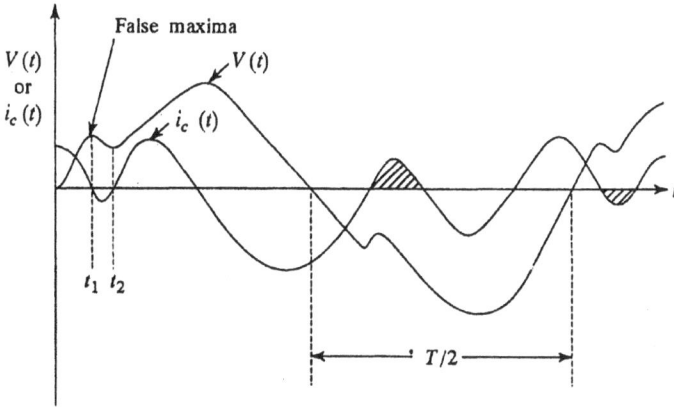

Fig. 7.15 Voltage waveform with harmonic content showing false maxima

In determining the error, the actual value of the angular frequency ω has to be determined.

The different sources that contribute to the error are
  (i) the effective value of the capacitance being different from the measured value of C
  (ii) imperfect rectifiers which allow small reverse currents
  (iii) non-sinusoidal voltage waveforms with more than one peak or maxima per half cycle
  (iv) deviation of the frequency from that of the value used for calibration

As such, this method in its basic form is not suitable for waveforms with more than one peak in each half cycle.

A digital peak reading meter for voltage measurements is shown in Fig. 7.16. Instead of directly measuring the rectified charging current, a proportional analog voltage signal is derived which is then converted into a proportional medium frequency, f_m. The frequency ratio f_m/f is measured with a gate circuit controlled by the a.c. power frequency (f) and a counter that opens for an adjustable number of periods $\Delta t = p/f$. During this interval, the number of impulses counted, n, is

$$n = f_m \cdot \Delta t = p \cdot \frac{f_m}{f} = 2pCV_m AR \qquad (7.20)$$

where p is a constant of the instrument and A represents the conversion factor of the a.c. to d.c. converter. $A = f_m/(R\, i_m)$; i_m is the rectified current through resistance R. An immediate reading of the voltage in kV can be obtained by suitable choice of the parameters R and the number of periods p. The total estimated error in this instrument was less than 0.35%. Conventional instruments of this type are available with less than 2% error.

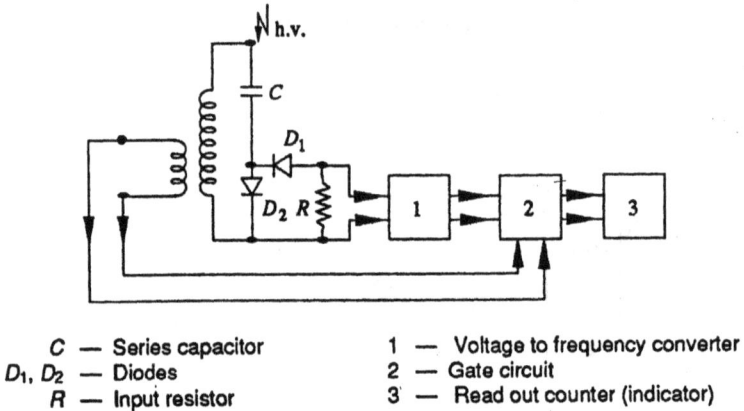

C — Series capacitor	1 — Voltage to frequency converter
D_1, D_2 — Diodes	2 — Gate circuit
R — Input resistor	3 — Read out counter (indicator)

Fig. 7.16   Digital peak voltmeter

Peak Voltmeters with Potential Dividers

Peak voltmeters using capacitance dividers designed by Bowlder *et al.*, are shown in Fig. 7.17a. The voltage across C_2 is made use of in charging the storage capacitor C_s. R_d is a discharge resistor employed to permit variation of V_m whenever V_2 is reduced. C_s is charged to a voltage proportional to the peak value to be measured. The indicating meter is either an electrostatic voltmeter or a high impedance V.T.V.M. The discharge time constant $C_s R_d$ is designed to be about 1 to 10 s. This gives rise to a discharge error which depends on the frequency of the supply voltage. To compensate for the charging and discharging errors due to the resistances, the circuit is modified as shown in Fig. 7.17b. Measurement of the average peak is done by a microameter. Rabus' modification to compensate the charging errors is given in Fig. 7.17c.

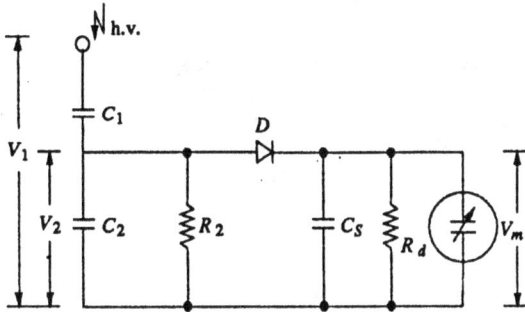

Fig. 7.17a Peak voltmeter with a capacitor potential divider and electrostatic voltmeter

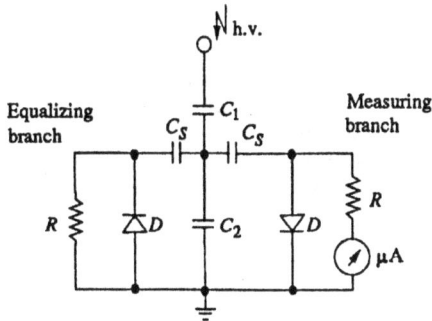

Fig. 7.17b Peak voltmeter as modified by Haefeely (ref. 19)

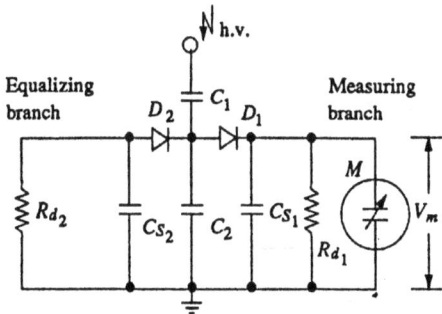

Rabus (ref. 20)

M — Electrostatic voltmeter
or V.T.V.M. of high impedance

C_{S_2} — C_{S_1} + C meter
R_{d_2} — R_{d_1}

Fig. 7.17c Peak voltmeter with equalizing branch as designed by Rabus

7.2.6 Spark Gaps for Measurement of High d.c., a.c. and Impulse Voltages (Peak Values)

A uniform field spark gap will always have a sparkover voltage within a known tolerance under constant atmospheric conditions. Hence a spark gap can be used for measurement of the peak value of the voltage, if the gap distance is known. A sparkover voltage of 30 kV (peak) at 1 cm spacing in air at 20°C and 760 torr pressure occurs for a sphere gap or any uniform field gap. But experience has shown that these measurements are reliable only for certain gap configurations. Normally, only sphere gaps are used for voltage measurements. In certain cases uniform field gaps and rod gaps are also used, but their accuracy is less. The spark gap breakdown, especially the sphere gap breakdown, is independent of the voltage waveform and hence is highly suitable for all types of waveforms from d.c. to impulse voltages of short rise times (rise time $\geq 0.5\,\mu$ s). As such, sphere gaps can be used for radio frequency a.c. voltage peak measurements also (up to 1 MHz).

Sphere Gap Measurements

Sphere gaps can be arranged either (*i*) vertically with lower sphere grounded, or (*ii*) horizontally with both spheres connected to the source voltage or one sphere grounded. In horizontal configurations, it is generally arranged such that both spheres are symmetrically at high voltage above the ground. The two spheres used are identical in size and shape. The schematic arrangement is shown in Figs. 7.18a and 7.18b. The voltage to be measured is applied between the two spheres and the distance

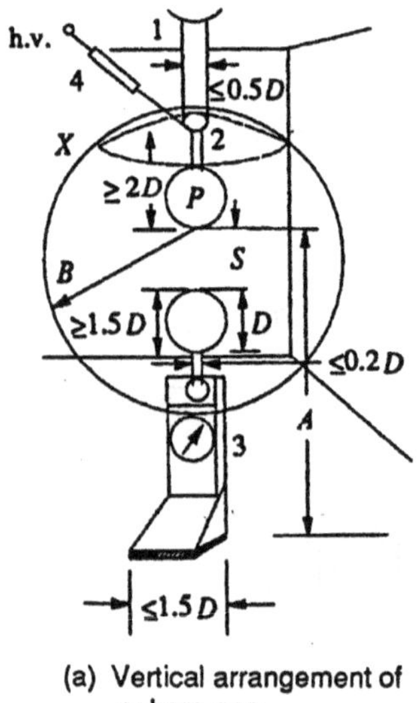

1 — Insulator support
2 — Sphere shank
3 — Operating gear and motor for changing gap distance
4 — H.V. connection
P — Sparking point
D — Diameter of the sphere
S — Spacing
A — Height of P above earth
B — Radius of the clearance from external structures
X — High voltage lead should not pass through this plane within a distance B from P

(a) Vertical arrangement of sphere gap

Fig. 7.18a Sphere gap for voltage measurement

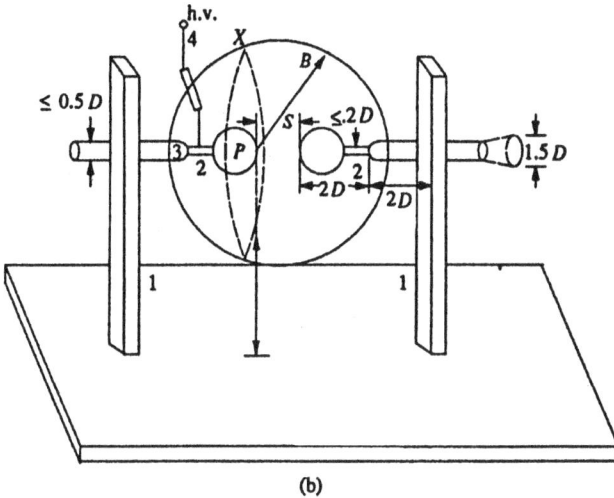

Fig. 7.18b Horizontal arrangement of sphere gap
(Legend as in Fig. 7.18a)

or spacing S between them gives a measure of the sparkover voltage. A series resistance is usually connected between the source and the sphere gap to (*i*) limit the breakdown current, and (*ii*) to suppress unwanted oscillations in the source voltage when breakdown occurs (in case of impulse voltages). The value of the series resistance may vary from 100 to 1000 kilo ohms for a.c. or d.c. voltages and not more than 500 Ω in the case of impulse voltages.

In the case of a.c. peak value and d.c. voltage measurements, the applied voltage is uniformly increased until sparkover occurs in the gap. Generally, a mean of about five breakdown values is taken when they agree to within ±3%.

In the case of impulse voltages, to obtain 50% flashover voltage, two voltage limits, differing by not more than 2% are set such that on application of lower limit value either 2 or 4 flashovers take place and on application of upper limit value 8 or 6 flashovers take place respectively. The mean of these two limits is taken as 50% flashover voltage. In any case, a preliminary sparkover voltage measurement is to be made before actual measurements are made.

The flashover voltage for various gap distances and standard diameters of the spheres used are given in Tables 7.3 and 7.4 respectively. The values of sparkover voltages are specified in BS : 358, IEC Publication 52 of 1960 and IS : 1876 of 1962. The clearances necessary are shown in Figs. 7.18a and 7.18b for measurements to be within ±3%. The values of A and B indicated in the above figures are given in Table 7.5.

Table 7.3 Peak value of sparkover voltage in kV for a.c., d.c. voltages of either polarity, and for full negative standard impulse voltages (one sphere earthed) (a) and positive polarity impulse voltages and impulse voltages with long tails (b) at temperature: 25°C and pressure: 760 torr

Gap spacing (cm)	Sphere diameter (cm)															
	5		10		15		25		50		100		150		200	
	A	B	A	B	A	B	A	B	A	B	A	B	A	B	A	B
0.5	17.4	17.4	16.9	16.8	16.9	16.9										
1.0	32.0	32.0	31.7	31.7	31.4	31.4	31.2	31.4								
1.5	44.7	45.5	44.7	45.1	44.7	45.1	44.7	44.7								
2.0	57.5	58.0	58.0	58.0	58.0	58.0	58.0	58.0								
2.5			71.5	71.5	71.5	71.5	71.5	71.5	71.5	71.5						
3.0			85.0	85.0	85.0	85.0	85.0	85.0	85.0	85.0						
3.5			95.5	96.0	97.0	97.0	97.0	97.0	97.0	97.0						
4.0			106.0	108.0	108.0	110.0	110.0	110.0	110.0	110.0						
5.0			(123.0)	(127.0)	127.0	132.0	135.0	136.0	136.0	136.0						
7.5					(181.0)	(187.0)	195.0	196.0	199.0	199.0						
10.0							257	268	259	259	262	262	262	262	262	262
12.5							277	294	315	317						
15.0							(309)	(331)	367	374	383	384	384	384	384	384
17.5							(336)	(362)	413	425						
20.0									452	472	500	500	500	500	500	500
25.0									520	545	605	610				
30.0									(575)	(610)	700	715	730	735	735	740
35.0									(725)	(755)	785	800				
40.0											862	885	940	950	960	965
45.0											925	965				
50.0											1000	1020	1110	1130	1160	1170
75.0											(1210)	(1260)	1420	1460	1510	1590
100.0															1870	1900

Table 7.4 Sphere gap sparkover voltages in kV (peak) in air for a.c., d.c., and impulse voltage of either polarity for symmetrical sphere gaps at temperature: 20°c and pressure: 760 torr

Gap spacing (cm)	Sphere diameter (cm)								Remarks
	5	10	15	25	50	100	150	200	
0.5	17.5	16.9	16.5						For spacings less than 0.5 D, the accuracy is ± 3% and for spacings ≥ 0.5 D, the accuracy is ± 5%.
1.0	32.2	31.6	31.3	31.0					
1.5	46.1	45.8	45.5	45.0					
2.0	58.3	59.3	59.2	59.0					
2.5	69.4	72.4	72.9	73.0					
3.0	(79.3)	84.9	85.8	86.0					
4.0		107.0	111.0	113.0	112.0				
5.0		128.0	134.0	138.0	138.0	137.0	137.0	137.0	
8.0		(177)	194.0	207.0	214.0	266.0	267.0	267.0	
10.0				248.0	263.0				
12.0				286.0	309.0				
14.0				320.0	353.0				
16.0				352.0	394.0				
18.0					452.0				
20.0					495.0	504.0	511.0	511.0	
25.0					558.0	613.0	628.0	632.0	
30.0						744.0	741.0	746.0	
35.0						812.0	848.0	860.0	
40.0						902.0	950.0	972.0	
50.0						1070.0	1140.0	1180.0	
60.0						(1210)	1320.0	1380.0	
70.0							1490.0	1560.0	
80.0							(1640)	1730.0	
90.0								1900.0	
100.0								2050.0	

Sphere Gap Construction and Assembly

Sphere gaps are made with two metal spheres of identical diameters D with their shanks, operating gear, and insulator supports (Fig. 7.18a or b). Spheres are generally made of copper, brass, or aluminium; the latter is used due to low cost. The standard diameters for the spheres are 2, 5, 6.25, 10, 12.5, 15, 25, 50, 75, 100, 150, and 200 cm. The spacing is so designed and chosen such that flashover occurs near the sparking point P. The spheres are carefully designed and fabricated so that their surfaces are smooth and the curvature is uniform. The radius of curvature measured with a spherometer at various points over an area enclosed by a circle of 0.3 D around the sparking point should not differ by more than ±2% of the nominal value. The surface of the sphere should be free from dust, grease, or any other coating. The surface should be maintained clean but need not be polished. If excessive pitting occurs due to repeated sparkovers, they should be smoothened. The dimensions of the shanks used, the grading ring used (if necessary) with spheres, the ground clearances, etc. should follow the values indicated in Figs. 7.18a and 7.18b and Table 7.5. The high voltage conductor should be arranged such that it does not affect the field configuration. Series resistance connected should be outside the shanks at a distance 2D away from the high voltage sphere or the sparking point P.

Irradiation of sphere gap is needed when measurements of voltages less than 50 kV are made with sphere gaps of 10 cm diameter or less. The irradiation may be obtained from a quartz tube mercury vapour lamp of 40 W rating. The lamp should be at a distance B or more as indicated in Table 7.5.

Table 7.5 Clearances for Sphere Gaps

D (cm)	Value of A		Value of B (min)
	Max	Min	
up to 6.25	7 D	9 D	14S
10 to 15	6 D	8 D	12S
25	5 D	7 D	10S
50	4 D	6 D	8S
100	3.5 D	5 D	7S
150	3 D	4 D	6S
200	3 D	4 D	6S

A and B are clearances as shown in Figs. 7.18a and 7.18b.
D = diameter of the sphere; S = spacing of the gap; and $S/D \leq 0.5$.

Factors Influencing the Sparkover Voltage of Sphere Gaps

Various factors that affect the sparkover voltage of a sphere gap are:
  (i) nearby earthed objects,
  (ii) atmospheric conditions and humidity,
  (iii) irradiation, and
  (iv) polarity and rise time of voltage waveforms.

Detailed investigations of the above factors have been made and analysed by Craggs and Meek[1], Kuffel and Abdullah[2], Kuffel[15], Davis and Boulder[16], and several other investigators. Only a few important factors are presented here.

(l) Effect of nearby earthed objects

The effect of nearby earthed objects was investigated by Kuffel[14] by enclosing the earthed sphere inside an earthed cylinder. It was observed that the sparkover voltage is reduced. The reduction was observed to be

$$\Delta V = m \log (B/D) + C \tag{7.21}$$

where,
$$\Delta V = \text{percentage reduction,}$$
$$B = \text{diameter of earthed enclosing cylinder,}$$
$$D = \text{diameter of the spheres,}$$
$$S = \text{spacing, and } m \text{ and } C \text{ are constants.}$$

The reduction was less than 2% for $S/D \leq 0.5$ and $B/D \geq 0.8$. Even for $S/D \approx 1.0$ and $B/D \geq 1.0$ the reduction was only 3%. Hence, if the specifications regarding the clearances are closely observed the error is within the tolerances and accuracy specified. The variation of breakdown voltage with A/D ratio is given in Figs. 7.19a and b for a 50 cm sphere gap. The reduction in voltage is within the accuracy limits, if S/D is kept less than 0.6. A in the above ratio A/D is the distance from sparking point to horizontal ground plane (also shown in Fig. 7.19).

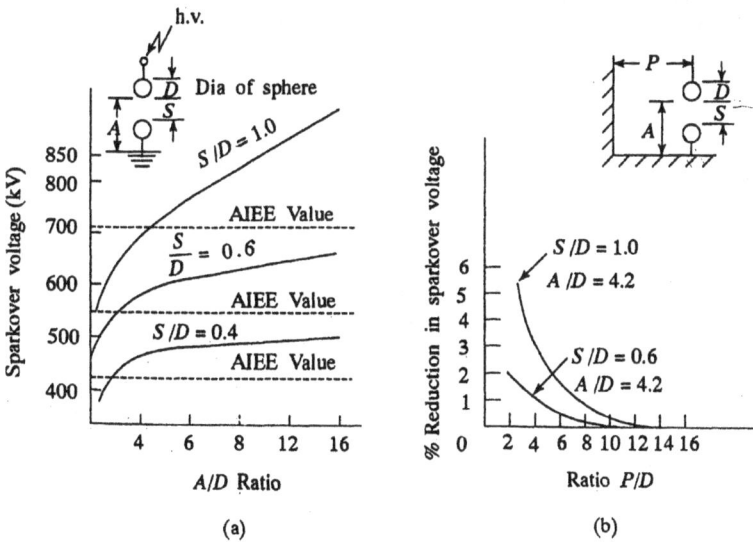

Fig. 7.19   Influence of ground planes on sparkover voltage

(ii) Effect of atmospheric conditions

The sparkover voltage of a spark gap depends on the air density which varies with the changes in both temperature and pressure. If the sparkover voltage is V under test conditions of temperature T and pressure p torr and if the sparkover voltage is V_0 under standard conditions of temperature $T = 20°C$ and pressure $p = 760$ torr, then

$$V = kV_0$$

where k is a function of the air density factor d, given by

$$d = \frac{p}{760}\left(\frac{293}{273+T}\right)$$ (7.22)

The relationship between d and k is given in Table 7.6.

Table 7.6 Relation between Correction Factor k and Air Density Factor d

d	0.70	0.75	0.80	0.85	0.90	0.95	1.0	1.05	1.10	1.15
k	0.72	0.77	0.82	0.86	0.91	0.95	1.0	1.05	1.09	1.12

The sparkover voltage increases with humidity. The increase is about 2 to 3% over normal humidity range of 8 g/m^3 to 15 g/m^3. The influence of humidity on sparkover voltage of a 25 cm sphere gap for 1 cm spacing is presented in Fig. 7.20. It can be seen that the increase in sparkover voltage is less than 3% and the variation between a.c. and d.c. breakdown voltages is negligible (< 0.5%). Hence, it may be concluded that (*i*) the humidity effect increases with the size of spheres and is maximum for uniform field gaps, and (*ii*) the sparkover voltage increases with the partial pressure of water vapour in air, and for a given humidity condition, the change in sparkover voltage

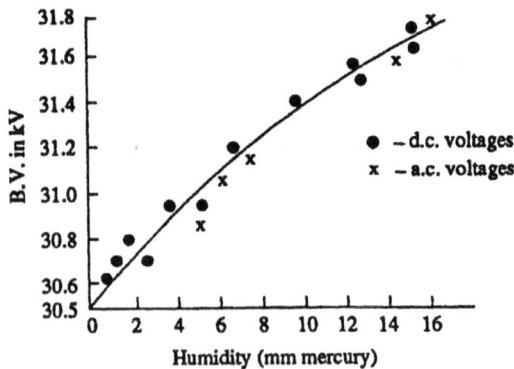

Fig. 7.20   Influence of humidity on d.c. and a.c. breakdown voltages (25 cm dia sphere gap, 1 cm spacing)

increases with the gap length. As the change in sparkover voltage with humidity is within 3%, no correction is normally given for humidity.

(iii) Effect of Irradiation

Illumination of sphere gaps with ultra-violet or x-rays aids easy ionization in gaps. The effect of irradiation is pronounced for small gap spacings. A reduction of about 20% in sparkover voltage was observed for spacings of 0.1 D to 0.3 D for a 1.3 cm sphere gap with d.c. voltages. The reduction in sparkover voltage is less than 5% for gap spacings more than 1 cm, and for gap spacings of 2 cm or more it is about 1.5%. Hence, irradiation is necessary for smaller sphere gaps of gap spacing less than 1 cm for obtaining consistent values.

(iv) Effect of polarity and waveform

It has been observed that the sparkover voltages for positive and negative polarity impulses are different. Experimental investigation showed that for sphere gaps of 6.25 to 25 cm diameter, the difference between positive and negative d.c. voltages is not more than 1%. For smaller sphere gaps (2 cm diameter and less) the difference was about 8% between negative and positive impulses of 1/50 µ s waveform. Similarly, the wave front and wave tail durations also influence the breakdown voltage. For wave fronts of less than 0.5 µ s and wave tails less than 5 µ s the breakdown voltages are not consistent and hence the use of sphere gap is not recommended for voltage measurement in such cases.

Uniform Field Electrode Gaps

Sphere gaps, although widely used for voltage measurements, have only limited range with uniform electric field. Hence, it is not possible to ensure that the sparking always takes place along the uniform field region. Rogowski (see Craggs and Meek[1]) presented a design for uniform field electrodes for sparkover voltages up to 600 kV. The sparkover voltage in a uniform field gap is given by

$$V = AS + B\sqrt{S}$$

where A and B are constants, S is the gap spacing in cm, and V is the sparkover voltage.

Typical uniform field electrodes are shown in Fig. 7.21. The constants A and B were found to be 24.4 and 7.50 respectively at a temperature $T = 25°C$ and pressure = 760 torr. Since the sparking potential is a function of air density, the sparkover voltage for any given air density factor d (see Eq. 7.22) is modified as

$$V = 24.4 \, dS + 7.50 \, \sqrt{dS} \qquad (7.23)$$

Bruce (see Craggs et al.[1] and Kuffel et al.[2]) made uniform field electrodes with a sine curve in the end region. According to Bruce, the electrodes with diameters of 4.5, 9.0, and 15.0 in. can be used for maximum voltages of 140, 280, and 420 kV respectively. For the Bruce profile, the constants A and B are respectively 24.22 and 6.08. Later, it was found that with humidity the sparkover voltage increases, and the relationship for sparkover voltage was modified as

$$V = 6.66 \, \sqrt{dS} + [24.55 + 0.41(0.1e - 1.0)]dS \qquad (7.24)$$

where, $V =$ sparkover voltage, kV_{peak} (in $kV_{d.c.}$).

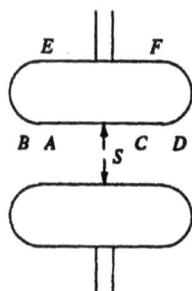

AC, EF — Flat portion (≥ S)

Curvature A to B and C to D ≥ 108

Curvature B to E and D to F
continuously increasing

AB — Flat portion
BC — Sine curve
CD — arc of a circle with centre at O

$XY - OC \sin\left(\dfrac{\pi}{2} \cdot \dfrac{BX}{BO}\right)$

(a) Electrodes for 300 kV (rms)
    spark gap

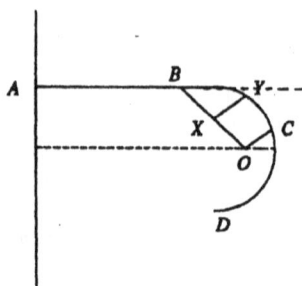

(b) Bruce profile (half contour)

Fig. 7.21   Uniform field electrode spark gap

S = spacing between the electrodes, cm,

d = air density factor, and

e = vapour pressure of water in air (mm Hg).

The constants A and B differ for a.c., d.c., and impulse voltages. A comparison between the sparkover voltages (in air at a temperature of 20°C and a pressure of 760 torr) of a uniform field electrode gap and a sphere gap is given in Table 7.7. From this table it may be concluded that within the specified limitations and error limits, there is no significant difference among the sparkover voltages of sphere gaps and uniform field gaps.

Table 7.7   Sparkover Voltages of Uniform Field Gaps and Sphere Gaps at $t = 20°C$ and $p = 760$ torr

Gap spacing (cm)	Sparkover voltage with uniform field electrodes as measured by			Sphere gap sparkover voltage (kV)
	Ritz (kV)	Bruce (kV)	Schumann (kV)	
0.1	4.54		4.50	4.6
0.2	7.90	7.56	8.00	8.0
0.5	17.00	16.41	17.40	17.0
1.0	31.35	30.30	31.70	31.0
2.0	58.70	57.04	59.60	58.0
4.0	112.00	109.00	114.00	112.0

6.0	163.80	160.20	166.20	164.0
8.0	215.00	211.00	216.80	215.0
10.0	265.00	261.1	266.00	265.0
12.0	315.00	311.6		312.0

The sparkover voltage of uniform field electrode gaps can also be found from calculations. However, no such calculation is available for sphere gaps. In spite of the superior performance and accuracy, the uniform field spark gap is not usually used for measurement purposes, as very accurate finish of the electrode surfaces and careful alignment are difficult to obtain in practice.

Rod Gaps

A rod gap is also sometimes used for approximate measurement of peak values of power frequency voltages and impulse voltages. IEEE recognise that this method gives an accuracy within ±8%. The rods will be either square edged or circular in cross-section. The length of the rods may be 15 to 75 cm and the spacing varies from 2 to 200 cm. The sparkover voltage, as in other gaps, is affected by humidity and air density. The power frequency breakdown voltage for 1.27 cm square rods in air at 25°C and at a pressure of 760 torr with the vapour pressure of water of 15.5 torr is given in Table 7.8. The humidity correction is given in Table 7.9. The air density correction factor can be taken from Table 7.6.

Table 7.8 Sparkover Voltage for Rod Gaps

Gap spacing (cm)	Sparkover voltage (kV)	Gap spacing (cm)	Sparkover voltage (kV)
2	26	30	172
4	47	40	225
6	62	50	278
8	72	60	332
10	81	70	382
15	102	80	435
20	124	90	488
25	147	100	537

The rods are 1.27 cm square edged at $t = 27°C$, $p = 760$ torr, and vapour pressure of water = 15.5 torr.

Table 7.9 Humidity Correction for Rod Gap Sparkover Voltages

Vapour pressure of water (torr)	2.54	5	10	15	20	25	30
Correction factor %	− 16.5	− 13.1	− 6.5	− 0.5	4.4	7.9	10.1

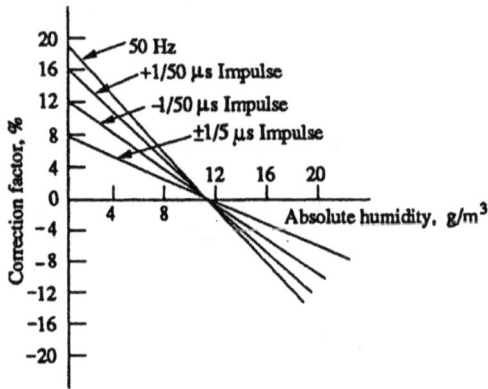

Fig. 7.22 Correction factor for rod gaps

In case of impulse voltage measurements, the IEC and IEEE recommend horizontal mounting of rod gaps on insulators at a height of 1.5 to 2.0 times the gap spacing above the ground. One of the rods is usually earthed. For 50% flashover voltages, the procedure followed is the same as that for sphere gaps. Corrections for humidity for $1/50~\mu$ s impulse and $1/50~\mu$ s impulse waves of either polarity are given in Fig. 7.22. The sparkover voltages for impulse waves are given in Table 7.10.

Table 7.10 Sparkover Voltages of Rod Gaps for Impulse Voltages at Temperature = 20°c, Pressure = 760 torr and Humidity = 11 g/cm²

Gap length (cm)	$1/5~\mu$ s wave (kV)		$1/50~\mu$ s wave (kV)	
	ositive	Negative	Positive	Negative
5	60	66	56	61
10	101	111	90	97
20	179	208	160	178
30	256	301	226	262
40	348	392	279	339
50	431	475	334	407
60	513	557	397	470
80	657	701	511	585
100	820	855	629	703

7.2.7 Potential Dividers for Impulse Voltage Measurements

Potential or voltage dividers for high voltage impulse measurements, high frequency a.c. measurements, or for fast rising transient voltage measurements are usually either resistive or capacitive or mixed element type. The low voltage arm of the divider is usually connected to a fast recording oscillograph or a peak reading instrument through a delay cable. A schematic diagram of a potential divider with its terminating equipment is given in Fig. 7.23. Z_1 is usually a resistor or a series of resistors in case of a resistance potential divider, or a single or a number of capacitors in case of a capacitance divider. It can also be a combination of both resistors and capacitors. Z_2 will be a resistor or a capacitor or an R-C impedance depending upon the type of the divider. Each element in the divider, in case of high voltage dividers, has a self-resistance or capacitance. In addition, the resistive elements have residual inductances, a terminal stray capacitance to ground, and terminal to terminal capacitances.

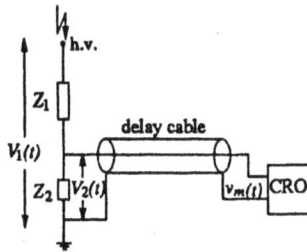

Fig. 7.23 Schematic diagram of a potential divider with a delay cable and oscilloscope

The lumped-circuit equivalent of a resistive element is already shown in Fig. 7.7, and the equivalent circuit of the divider with inductance neglected is of the form shown in Fig. 7.8a. A capacitance potential divider also has the same equivalent circuit as in Fig. 7.7a, where C_s will be the capacitance of each elemental capacitor, C_g will be the terminal capacitance to ground, and R will be the equivalent leakage resistance and resistance due to dielectric loss in the element. When a step or fast rising voltage is applied at the high voltage terminal, the voltage developed across the element Z_2 will not have the true waveform as that of the applied voltage. The cable can also introduce distortion in the waveshape. The following elements mainly constitute the different errors in the measurement:

(*i*)   residual inductance in the elements;
(*ii*) stray capacitance occurring
    (*a*) between the elements,
    (*b*) from sections and terminals of the elements to ground, and
    (*c*) from the high voltage lead to the elements or sections;
(*iii*) the impedance errors due to

(a)   connecting leads between the divider and the test objects, and

(b)   ground return leads and extraneous current in ground leads; and

(iv) parasitic oscillations due to lead and cable inductances and capacitance of high voltage terminal to ground.

The effect to residual and lead inductances becomes pronounced when fast rising impulses of less than one microsecond are to be measured. The residual inductances damp and slow down the fast rising pulses. Secondly, the layout of the test objects, the impulse generator, and the ground leads also require special attention to minimize recording errors. These are discussed in Sec. 7.4.

Resistance Potential Divider for Very Low Impulse Voltages and Fast Rising Pulses

A simple resistance potential divider consists of two resistances R_1 and R_2 in series ($R_1 \gg R_2$) (see Fig. 7.24). The attenuation factor of the divider or the voltage ratio is given by

$$a = \frac{V_1(t)}{V_2(t)} = 1 + \frac{R_1}{R_2}. \qquad (7.25)$$

The divider element R_2, in practice, is connected through the coaxial cable to the oscilloscope. The cable will generally have a surge impedance Z_0 and this will come in parallel with the oscilloscope input impedance (R_m, C_m). R_m will generally be greater than one megaohm and C_m may be 10 to 50 picofarads. For high frequency and impulse voltages (since they also contain high frequency fundamental and harmonics), the ratio in the frequency domain will be given by

$$a = \frac{V_1}{V_2} = 1 + \frac{R_1}{(R_2/1 + j\omega R_2 C_m)} \qquad (7.26)$$

Hence, the ratio is a function of the frequency. To avoid the frequency dependance of the voltage ratio a, the divider is compensated by adding an additional capacitance C_1 across R_1. The value of C_1, to make the divider independent of the frequency, may be obtained from the relation,

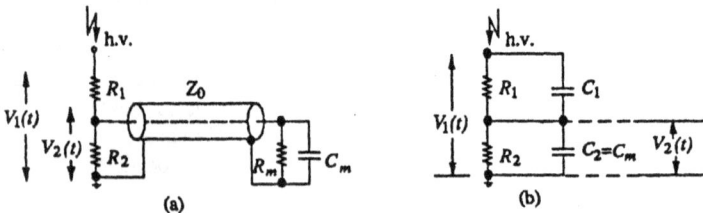

Resistance potential divider
with surge cable and
oscilloscope terminations

Comepensated resistance
potential divider

Fig. 7.24 a & b   Resistance potential dividers

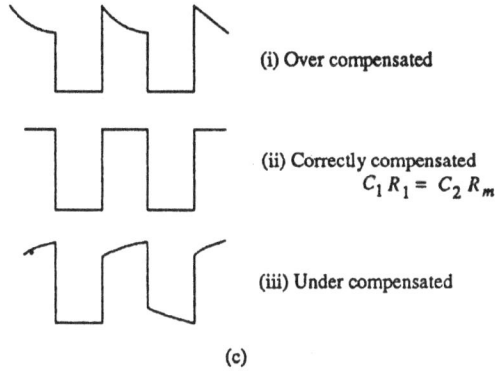

(i) Over compensated

(ii) Correctly compensated
$$C_1 R_1 = C_2 R_m$$

(iii) Under compensated

(c)

Fig. 7.24c Output of compensated resistance voltage divider for different degrees of compensation

$$\frac{R_1}{R_2} = \frac{C_m}{C_1} \qquad (7.27)$$

or, $\qquad\qquad R_1 C_1 = R_2 C_m$

meaning that the time constant of both the arms should be the same. This compensation is used for the construction of high voltage dividers and probes used with oscilloscopes. Usually, probes are made with adjustable values of C_m so that the value of C_m can include any stray capacitance including that of a cable, etc. A typical high voltage probe with a four nanosecond rise time rated for 40 kV (peak) has an input impedance of 100 MΩ in parallel with 2.7 pF. The output waveforms of a compensated divider are shown in Fig. 7.24c with over and under compensation for a square wave input. In Fig. 7.24 c(i) is shown the waveform of an R-C divider when C_1 is too large or overcompensated, while in Fig. 7.24 c(iii) is shown the waveform when C_1 is small or under compensated. For the exponential slope or for the rising portion of the wave, the time constant $\tau = [R_1 R_2/(R_1 + R_2)](C_1 + C_m)$. This will be too large when the value of C_1 is greater than that required for correct compensation, i.e. $R_1 C_1 = R_2 C_m$ and hence an overshoot with an exponential decay occurs as shown in Fig. 7.24 c(i). For under compensation, the charging time is too high and as such an exponential rise occurs as shown in Fig. 7.24 c(iii). The schematic circuit of a compensated oscilloscope probe is shown in Fig. 7.25.

Potential Dividers Used for High Voltage Impulse Measurements

In a resistance potential divider, R_1 and R_2 are considered as resistors of small dimensions in the previous section. For voltages above 100 kV, R_1 is no longer small in dimension and is usually made of a number of sections. Hence the divider is no longer a small resistor of lumped parameters, but has to be considered as an equivalent distributed network with its terminal to ground capacitances and inter-sectional series capacitances as shown in Fig. 7.26. The total series resistance R_1 is made of n resistors of value R'_1 and $R = nR'_1$. C_g is the terminal to ground capacitance of each of the

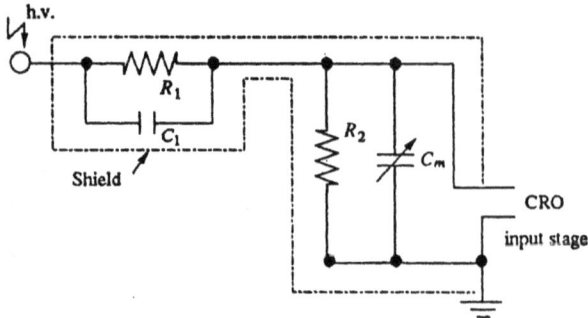

Fig. 7.25 Schematic circuit arrangement of a cathode ray
oscilloscope voltage divider probe

resistor elements R'_1, and C_s is the capacitance between the terminals of each section. The inductance of each element (L'_1) is not shown in the figure as it is usually small compared to the other elements (i.e. R'_1, C_s and C_g). This type of divider produces a non-linear voltage distribution along its length and also acts like an R-C filter for applied voltages. The output of such a divider for various values of C_g/C_s ratio is shown in Fig. 7.27 for a step input. By arranging guard rings at various elemental points, the equivalent circuit can be modified as shown in Fig. 7.28, where C_h

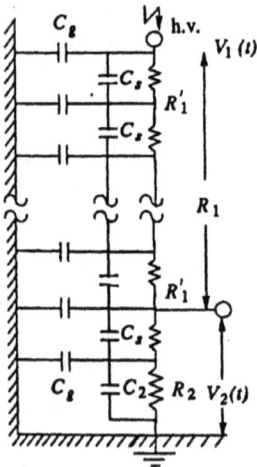

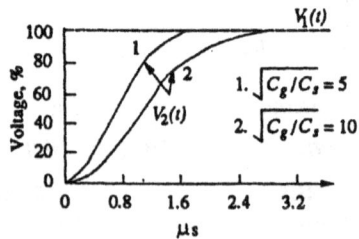

Fig. 7.26 Resistance potential
divider with inter-sectional
and ground capacitances

Fig. 7.27 Output of the divider shown in
Fig. 7.26 for a step input

represents the stray capacitance introduced between the high voltage lead and the guard elements. This reduces the distortion introduced by the original divider.

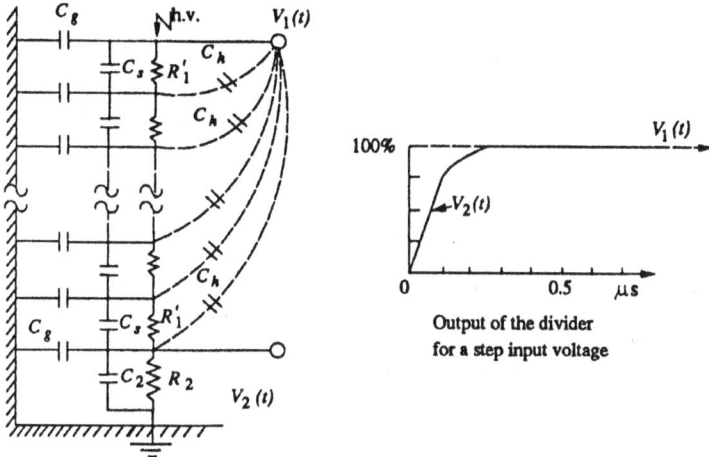

Fig. 7.28 Equivalent circuit of a resistance potential divider with shield and guard rings

Capacitance Voltage Dividers

Capacitance voltage dividers are ideal for measurement of fast rising voltages and pulses. The capacitance ratio is independent of the frequency, if their leakage resistance is high enough to be neglected. But usually the dividers are connected to the source voltage through long leads which introduce lead inductances and residual resistances. Also, the capacitance used for very high voltage work is not small in dimension and hence cannot be considered as a lumped element. Therefore, the output of the divider for high frequencies and impulses is distorted as in the case of resistance dividers.

Pure Capacitance Dividers

A pure capacitance divider for high voltage measurements and its electrical equivalent network without stray elements is shown in Fig. 7.29. The ratio of the divider

$$a = \frac{V_1(t)}{V_2(t)} = 1 + \frac{C_2}{C_1} \tag{7.28}$$

Capacitance C_1 is formed between the h.v. terminal of the source (impulse generator) and that of the test object or any other point of measurement. The CRO is located within the shielded screen surrounding capacitance C_2. C_2 includes the capacitance used, the lead capacitance, input capacitance of the CRO, and other

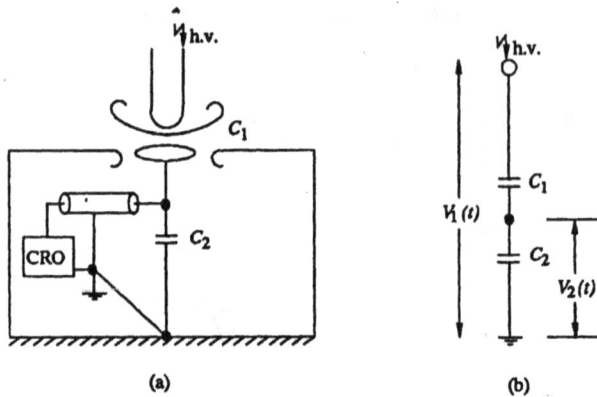

Fig. 7.29 Capacitance voltage divider for very high voltages and its electrical equivalent circuit

ground capacitances. The advantage of this connection is that the loading on the source is negligible; but a small disturbance in the location of C_2 or h.v. electrode or the presence of any stray object nearby changes the capacitance C_1, and hence the divider ratio is affected.

In many cases a standard air or compressed gas capacitor is used which has coaxial cylindrical construction. Accurate ratios that could be calculated up to 1000 : 1 have been achieved for a maximum impulse voltage of 350 kV, and the upper frequency limit is about 10 MHz. For smaller or moderate high voltages (up to 100 kV) capacitance dividers are built with an upper frequency limit of 200 MHz.

Another type of design frequently used is to make C_1 to consist of a number of capacitors C'_1 in series for the given voltage V_1. In such cases the equivalent circuit is similar to that of a string insulator unit used in transmission lines (Fig. 7.30). The voltage distribution along the capacitor chain is non-linear and hence causes distribution of the output wave. But the ratio error is constant and is independent on frequency as compared to resistance dividers. A simplified equivalent circuit is shown in Fig. 7.30 b, which can be used if $C_1 \ll C_2$ and $C_g \ll C_1$. The voltage ratio is

$$a = \frac{V_1(t)}{V_2(t)} \approx \left[1 + \frac{C_2}{C_1}\right]\left[1 + \frac{C_g}{6C_1}\right] \tag{7.29}$$

This ratio is constant and gives an error of less than 5% when $C_1 = 3C_g$. This equivalent circuit is quite satisfactory up to 1 MHz.

Field Controlled Voltage Dividers

The electrostatic or capacitive field distribution of a shield or guard ring placed over a resistive divider to enforce a uniform field in the neighbourhood and along the divider may be adopted for high voltage measurements. The schematic diagram is shown in Fig. 7.31 and its equivalent circuit is same as that given in Fig. 7.28. The shield is of the form of a cone. R_1 is a non-linear resistance in the sense the resistance per unit length is not the same but is variable. The main advantage is that the

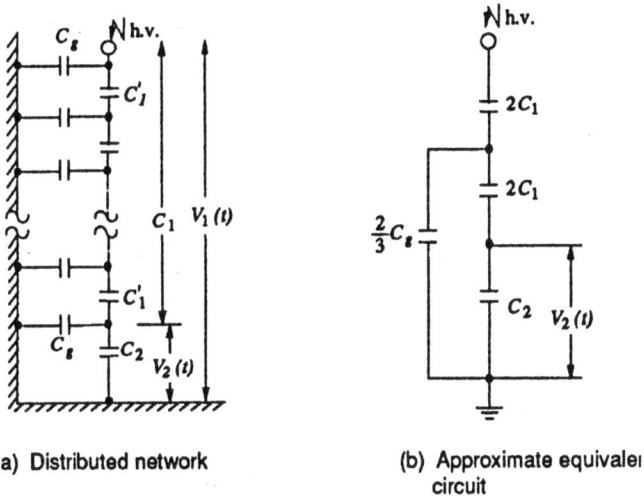

(a) Distributed network

(b) Approximate equivalent circuit

Fig. 7.30   Capacitance voltage divider with distributed network and its equivalent circuit

capacitance per unit length is small and hence loading effect is reduced. Sometimes the parallel resistance R_2 together with the lead inductance and shunt capacitances cause oscillations as shown in Fig. 7.32a. The oscillations can be reduced by adding a damping resistance R_d as shown in Fig. 7.31. Such dividers are constructed for very

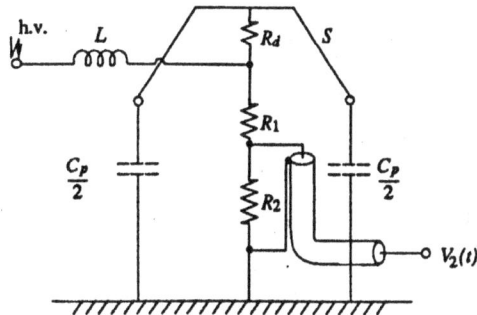

R_d — Damping resistor

L — Lead inductance

C_P — Capacitance of the shield to ground

S — Shield

Fig. 7.31   Field controlled resistance divider with a damping resistor

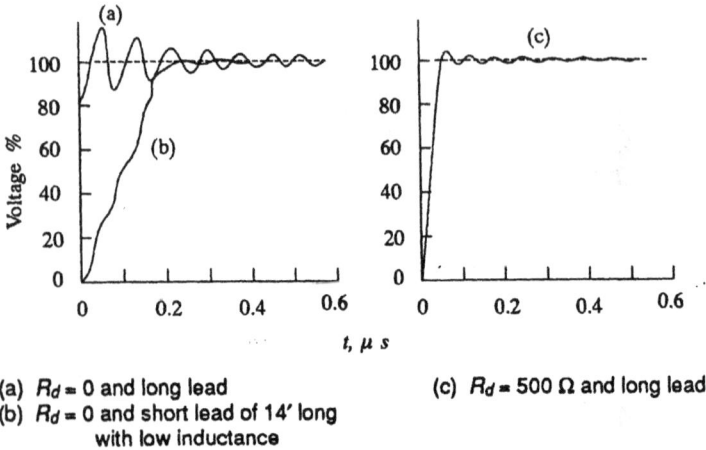

(a) $R_d = 0$ and long lead
(b) $R_d = 0$ and short lead of 14' long
with low inductance

(c) $R_d = 500 \Omega$ and long lead

Fig. 7.32 Step response of field controlled voltage divider of Fig. 7.31

high voltages (up to 2 MV) with response times less than 30 ns. The resistance column, R_1 is made of woven resistance of 20 kilo ohms. The step response of such a divider is shown in Fig. 7.32, with and without a damping resistor. With a proper damping resistor (R_4) the response time is much less and the overshoot is reduced.

Mixed R-C Potential Dividers

Mixed potential dividers use $R\text{-}C$ elements in series or in parallel. One method is to connect capacitance in parallel with each R'_1 element. This is successfully employed

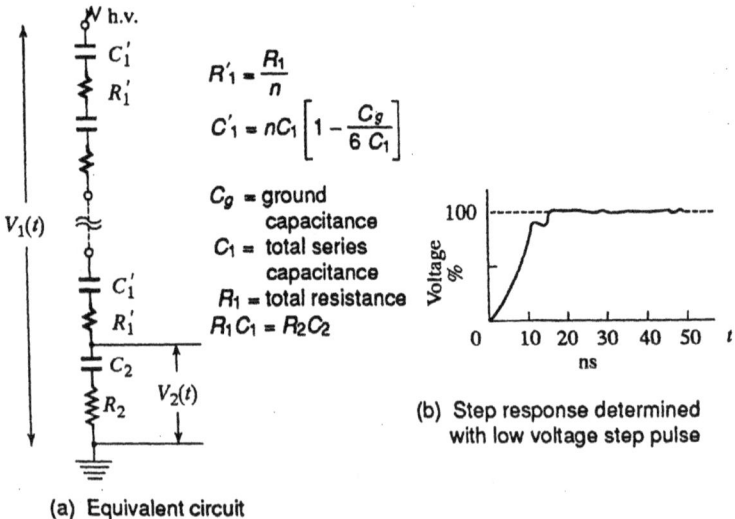

$$R'_1 = \frac{R_1}{n}$$

$$C'_1 = nC_1 \left[1 - \frac{C_g}{6\,C_1} \right]$$

C_g = ground capacitance
C_1 = total series capacitance
R_1 = total resistance
$R_1 C_1 = R_2 C_2$

(a) Equivalent circuit

(b) Step response determined with low voltage step pulse

Fig. 7.33 Equivalent circuit of a series R-C voltage divider and its step response

for voltage dividers of rating 2 MV and above. A better construction is to make an R-C series element connection. The equivalent circuit of such a construction is shown in Fig. 7.33. Such dividers are made for 5 MV with response times less than 30 n s. The low voltage arm R_2 is given "L peaking" by connecting a variable inductance L in series with R_2. The step response of the divider and the schematic connection of low voltage arm are shown in Fig. 7.34. However, for a correctly designed voltage divider L peaking will not be necessary.

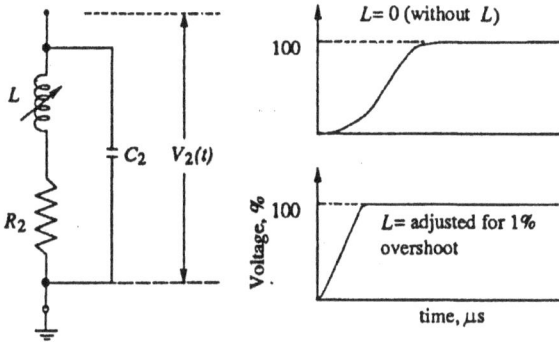

Fig. 7.34 L peaking in low voltage arm and step response of the divider with L peaking

R-C Potential dividers for 2 MV rating and above

Voltage dividers used for measuring more than one million volts attenuate the measuring signal to a value in the range 100 V to few hundreds of volts. The criteria required to assess the dividers are: (i) the shape of the voltage in the test arrangement should be transferred without any distortion to the L.V. side, (ii) simple determination of transfer behaviour should be ensured, and (iii) they should be suitable for multipurpose use, i.e. for use with a.c. power frequency voltages, switching impulse voltages as well as with lightning impulse voltages. This condition necessitates that the dividers should have broad bandwidths. The above requirements are generally met by (a) optimally damped R-C dividers, or (b) under damped or low damped R-C dividers. The high voltage arm of such dividers consists of series R-C units while the secondary arm is usually an R-C series or parallel circuit. In case of optimally damped dividers, $R_1 = 4\sqrt{L_1/C_g}$, where L_1 is the inductance of the high voltage lead and the H.V. portion of the divider, and C_g is the equivalent capacitance to ground. Usually this resistance will be 400 to 1000 ohms. On the other hand, for low or underdamped dividers, R_1 will be equal to 0.25 to 1.5 times $\sqrt{L/C_1}$ where L is the inductance for the complete measuring loop and C_1 is the capacitance of the H.V. part of the divider. In this case, the normal value of R_1 lies between 50 and 300 ohms. The step response of the two types of dividers mentioned above is shown in Fig. 7.35. In actual practice, because of the large time constant $(R_d + R_1)C_1$, the optimal damped divider affects the voltage shape at the test object. Standard lightning impulses sometimes cannot be generated to the correct standard specifications. As such, R-C potential dividers are

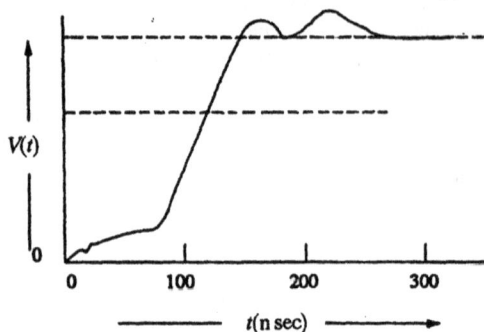

(i) Optimally damped

Response time : 50 n sec
Front time   : 50 n sec
Overshoot   : ≈ 3%
Parameters   : R_1 = 1000 Ω
                       C_1 = 360 pF

Damping
  resistance   : 500 Ω
$(R_d + R_1) C_1 = R_2 C_2$

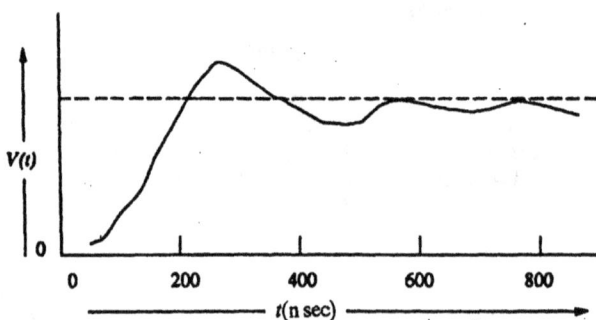

(ii) Underdamped

Response time : 4 n sec
Front time   : 110 n sec
Overshoot   : ≈ 30%
Parameters   : R_1 = 256 Ω
                       C_1 = 400 pF

Damping
  resistance : 0 Ω
$R_1 C_1 = R_2 C_2$

Fig. 7.35   Step response of a 4 MV R-C divider

$t\,(\mu sec)$

| 0 | 0.5 | 1.0 | 1.5 | 2.0 | 2.5 |

(A)

Amplitude 0

(B)

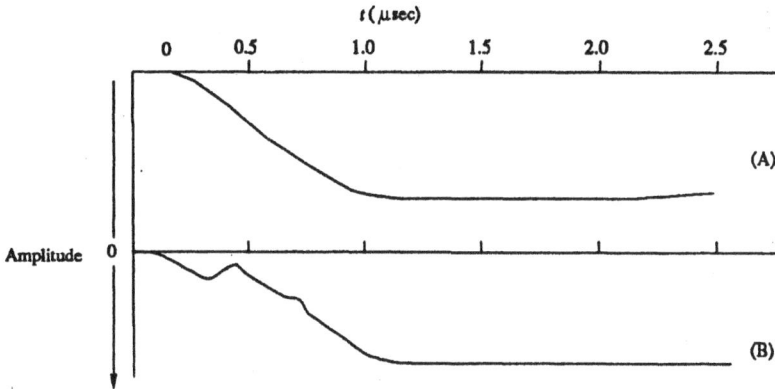

Fig. 7.36 Record of the front portion of a lightning impulse wave with under-damped (curve A) and optimally damped (curve B) dividers for a negative polarity wave when both dividers are connected in parallel

not suitable for measurements with test objects of very low capacitance. The low or underdamped R-C divider acts as a load capacitance and a voltage divider, and is suitable for applications over a broad bandwidth, i.e. a.c., switching impulses, lightning impulses, chopped waves etc. Underdamped R-C dividers are also suitable for measurement of steep fronted impulse waves. A typical record of lightning impulse wave (1.2/50 μ s wave) obtained using both the above types of dividers is shown in Fig. 7.36. It may be noted that even though the step response is poor in the case of underdamped dividers, they can be used to measure the standard impulse wave to a better accuracy.

Different Connections Employed with Potential Dividers

Different arrangements and connections of voltage or potential dividers with a cathode ray oscilloscope are shown in Figs. 7.37 and 7.38.

A simple arrangement of a resistance divider is shown in Fig. 7.37 a. The possible errors are (i) $R_2 \neq Z_0$ (surge impedance of the cable), (ii) capacitance of the cable and CRO shunting R_2 and hence introducing distortion, (iii) attenuation or voltage drop in surge cable Z_0, and (iv) ground capacitance effect. These errors are already discussed in Sec. 7.2.7. To avoid reflections at the junction of the cable and R_2, R_2 is varied and adjusted to give the best possible step response. When a unit function voltage is applied to the circuit shown in Fig. 7.37 b, the effect of the cable is to take a fraction of the voltage $[C_1/(C_1 + C_2)]$ into it and cause reflections at the input end. In the beginning the cable acts like a resistance of value $= Z_0$ the surge impedance, but later behaves like a capacitor of value equal to the total capacitance of the cable. This behaviour introduces distortion and is compensated by using a split capacitor connection as shown in Fig. 7.37 c with $(C_1 + C_2) = (C_3 + C_k)$ [C_k = capacitance of the cable]. On the other hand if $C_k/(C_1 + C_2 + C_k) = 0.1$, the error will be less than 1.5%.

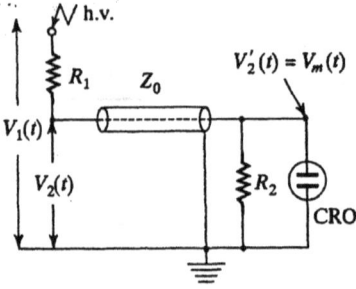

(a) Resistance potential
divider with surge cable
and CRO

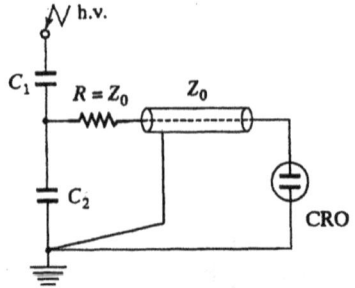

(b) Capacitance divider
with surge cable
and CRO

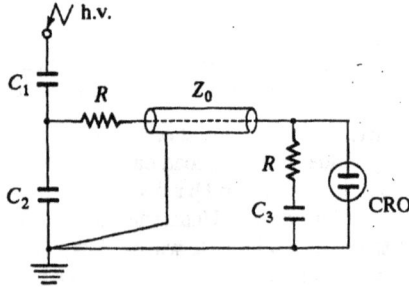

(c) Split capacitor arrangement $R = Z_0$

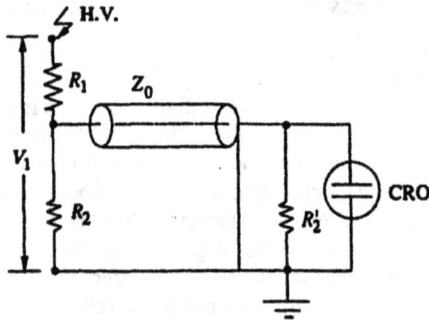

(d) Resistance potential divider with surge cable and CRO.
Voltage ratio, $V_1/V_2 = 1 + (R_1/R_2) + (R_1/R_2')$ where R_2'/Z_0

Fig. 7.37 Potential divider arrangements

The arrangements for mixed potential dividers are shown in Fig. 7.38. The arrangement shown in Fig. 7.38a is modified and improved in the arrangement of Fig. 7.38b. With

$$R_1 C_1 = \frac{C_2 Z_0 (C_1 + C_2 + C_k)}{(C_1 + C_2)} \qquad (7.30)$$

$$Z_0 = R_3 + \left(\frac{R_1 R_2}{R_1 + R_2} \right), \text{ and } R_1 C_1 = R_2 C_3 \qquad (7.31)$$

the response is greatly improved. The arrangement shown in Fig. 7.38c is simple and gives the desired impedance matching.

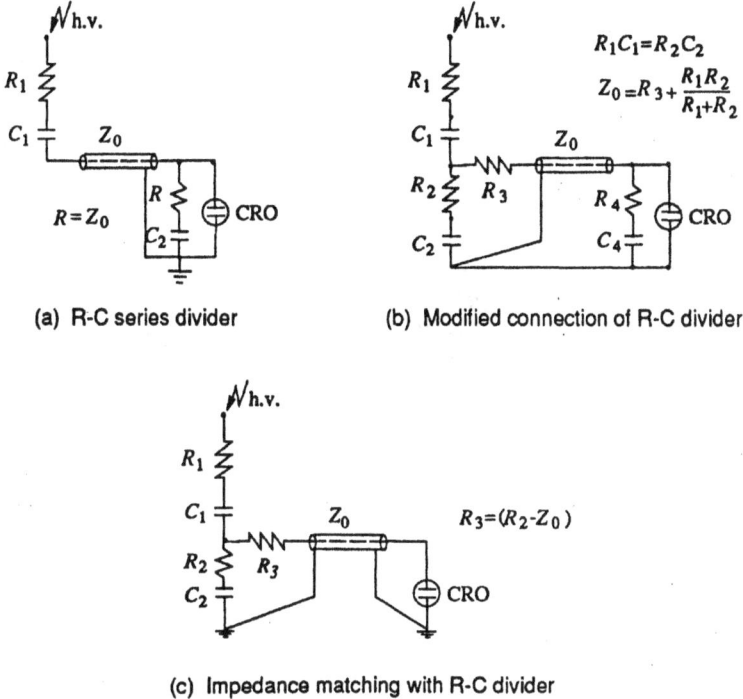

(a) R-C series divider   (b) Modified connection of R-C divider

(c) Impedance matching with R-C divider

Fig. 7.38   Mixed potential divider arrangements

Low voltage arms of the measuring system connected to voltage dividers

The mode of connection and the layout arrangement of the secondary arm of the divider is very critical for the distortionless measurement of fast transients. The L.V. arm of the divider itself introduces large distortions if not properly connected. Different corrections employed for connecting the L.V. arm to the measuring instrument via the signal cables are shown in Figs. 7.37 and 7.38. The signal cable Z_0 may be assumed to be loss-free so that the surge impedance, $Z_0 = \sqrt{L/C}$ is independent of the frequency and the travel time for the signal, $T_0 = \sqrt{LC}$ (refer to Chapter 8 for details). In the case of resistance dividers, the cable matching is achieved by having a pure resistance, $R_2 = Z_0$ at the end of the cable. The surge cable Z_0 and the resistance R_2 form an integral part of the cable system. Typically, Z_0 has values of 50 or 75 ohms. In actual practice, signal cables do have losses due to skin effect at high frequencies and hence Z_0 becomes a complex quantity. Thus, the matching of R_2 with Z_0 should be done at high frequencies or with a step input as indicated earlier. In the case of long cables, the cable resistance including that of the shield wire should be taken as a part of the matching resistance. The divider ratio in the case of the connection shown in Fig. 7.37 is

$$a = V_1/V_2 = 1 + R_1/R_2 + R_1/R'_2 \text{ and}$$

$$R'_2 = Z_0 \tag{7.32}$$

For the capacitance dividers, the signal cable cannot be completely matched. A low ohmic resistance connected in parallel with C_2 would load the L.V arm and hence, the output gets decreased. Connection of a resistance $R = Z_0$ at the input end (see figures 7.37 and 7.38) will make the voltage across the CRO the same as that across C_2. The transient voltage ratio, at $t = 0$ is given as

$$a = V_1/V_2 = 1 + C_2/C_1 \text{ and}$$

$$\text{effective} = 1 + (C_2 + C_k)/C_1 \text{ for } t \gg 2T_0 \tag{7.33}$$

Where C_k is the cable capacitance.

Thus, an initial overshoot of $\Delta V = C_k/(C_1 + C_2)$ will appear. This will be either small or negligible for medium and low cable lengths, and for high values of capacitance C_2. This error can be avoided and the response improved in the case of R-C dividers by using the arrangements shown in Fig. 7.38.

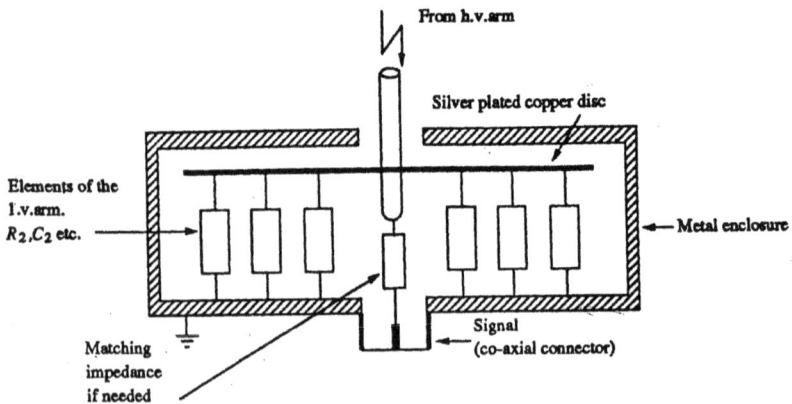

Fig. 7.39   L.V. arm layout for voltage dividers

Usually, the L.V. arms are made co-axial and are enclosed in metal boxes that are solidly grounded. The series resistors used in R-C divider forms an integral part of the divider's L.V. arm. Further, all the L.V. arm capacitors and inductors should have a very low inductance. A typical L.V. arm arrangement is shown in Fig. 7.39.

7.2.8 Peak Reading Voltmeters for Impulse Voltages

Sometimes it is enough if the peak value of an impulse voltage wave is measured; its waveshape might already be known or fixed by the source itself. This is highly useful in routine impulse testing work. The methods are similar to those employed for a.c. voltage crest value measurements. The instrument is normally connected to the low voltage arms of the potential dividers described in Sec. 7.2.7. The basic circuit along with its equivalent circuit and the response characteristic is shown in Fig. 7.40. The circuit consists of only valve rectifiers.

Diode D conducts for positive voltages only. For negative pulses, the diode has to be connected in reverse. When a voltage impulse $v(t)$ appears across the low voltage arm of the potential divider, the capacitor C_m is charged to the peak value of the pulse. When the amplitude of the signal starts decreasing the diode becomes reverse biased and prevents the discharging of the capacitor C_m. The voltage developed across C_m is measured by a high impedance voltmeter (an electrostatic voltmeter or an electrometer). As the diode D has finite forward resistance, the voltage to which C_m is charged will be less than the actual peak of the signal, and is modified by the R-C network of the diode resistance and the measuring capacitance C_m. The error is shown in Fig. 7.40c. The error can be estimated if the waveform is known. The actual forward

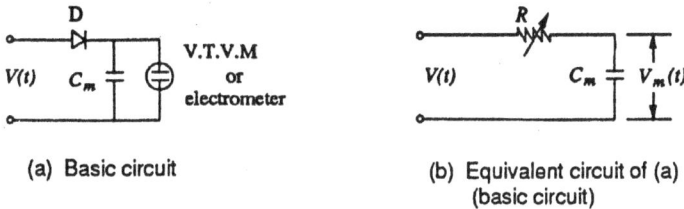

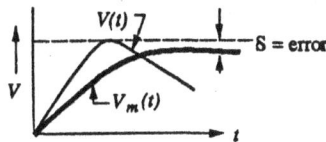

(a) Basic circuit

(b) Equivalent circuit of (a)
     (basic circuit)

(c) Waveforms of input and meter voltages
     showing error δ (diode reverse resistance
     is assumed to be infinite)

Fig. 7.40 A peak reading voltmeter and its equivalent circuit (R-C approximation)

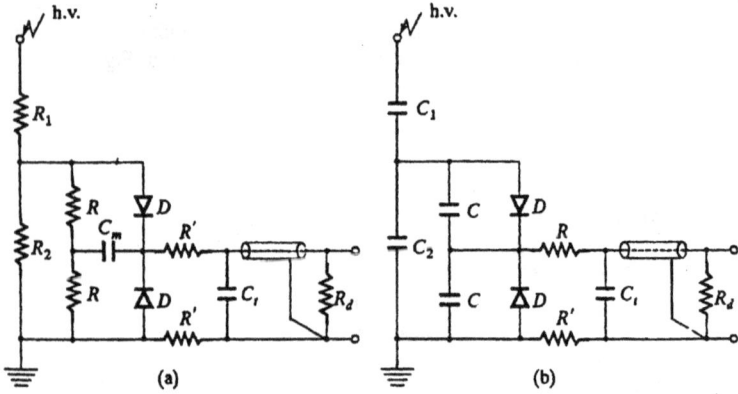

Fig. 7.41 Peak reading voltmeter for either polarity with
(a) resistance divider, and (b) capacitance divider

resistance of the diode D (dynamic value) is difficult to estimate, and hence the meter is calibrated using an oscilloscope. Peak voltmeters for either polarity employing resistance dividers and capacitance dividers are shown in Fig. 7.41. In this arrangement, the voltage of either polarity is transferred into a proportional positive measuring signal by a resistive or capacitive voltage divider and a diode circuit. An active network with feedback circuit is employed in commercial instruments, so that the fast rising pulses can also be measured. Instruments employing capacitor dividers require discharge resistance across the low voltage arm to prevent the build-up of d.c. charge.

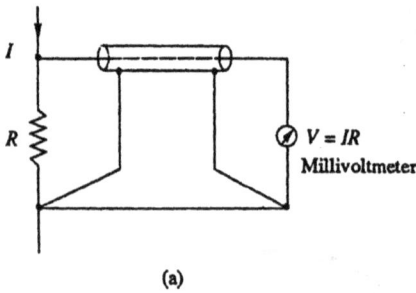

Low ohmic shunt with a
millivoltmeter for current
measurement

CC — Current terminals
PP — Potential terminals
R — Ohmic element

Four terminal shunt

(In 3 terminal construction the
bottom C and P terminals are
made common)

Fig. 7.42 Calibrated ohmic shunt for d.c. current measurements

7.3 MEASUREMENT OF HIGH d.c., a.c. AND IMPULSE CURRENTS

In power systems, it is often necessary to measure high currents, arising due to short circuits. For conducting temperature rise and heat run tests on power equipments like conductors, cables, circuit breakers, etc., measurement of high currents is required. During lightning discharges and switching transients also, large magnitudes of impulse and switching surge currents occur, which require special measuring techniques at high potential levels.

7.3.1 Measurement of High Direct Currents

High magnitude direct currents are measured using a resistive shunt of low ohmic value. The voltage drop across the resistance is measured with a millivoltmeter. The value of the resistance varies usually between 10 $\mu\Omega$ and 13 mΩ. This depends on the heating effect and the loading permitted in the circuit. High current resistors are usually oil immersed and are made as three or four terminal resistances (see Fig. 7.42). The voltage drop across the shunt is limited to a few millivolts (< 1 Volt) in power circuits.

Hall Generators for d.c. Current Measurements

The principle of the "Hall effect" is made use of in measuring very high direct currents. If an electric current flows through a metal plate located in a magnetic field perpendicular to it, Lorenz forces will deflect the electrons in the metal structure in a direction normal to the direction of both the current and the magnetic field. The charge displacement generates an emf in the normal direction, called the "Hall voltage". The Hall voltage is proportional to the current i, the magnetic flux density B, and the reciprocal of the plate thickness d; the proportionality constant R is called the "Hall coefficient".

$$V_H = R \frac{B_i}{d} \tag{7.34}$$

For metals the Hall coefficient is very small, and hence semi-conductor materials are used for which the Hall coefficient is high.

In large current measurements, the current carrying conductor is surrounded by an iron cored magnetic circuit, so that the magnetic field intensity $H = (I/\delta)$ is produced in a small air gap in the core. The Hall element is placed in the air gap (of thickness δ), and a small constant d.c. current is passed through the element. The schematic arrangement is shown in Fig. 7.43. The voltage developed across the Hall element in the normal direction is proportional to the d.c. current I. It may be noted that the Hall coefficient R depends on the temperature and the high magnetic field strengths, and suitable compensation has to be provided when used for measurement of very high currents.

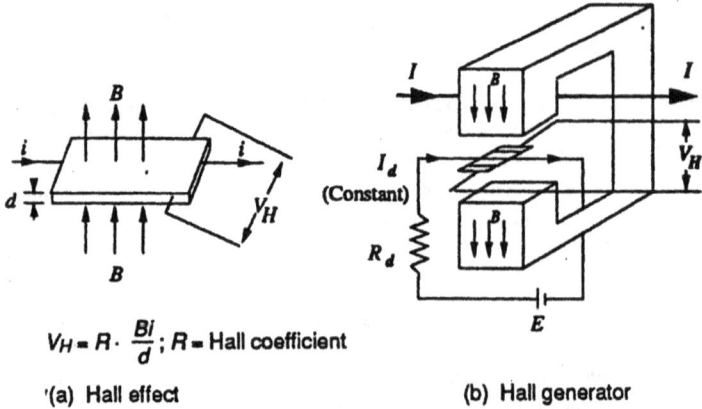

$$V_H = R \cdot \frac{Bi}{d} \; ; R = \text{Hall coefficient}$$

(a) Hall effect   (b) Hall generator

Fig. 7.43   Hall generator for measuring high d.c. currents

7.3.2 Measurement of High Power Frequency Alternating Currents

Measurement of power frequency currents are normally done using current transformers only, as use of current shunts involves unnecessary power loss. Also the
current transformers provide electrical isolation from high voltage circuits in power
systems. Current transformers used for extra high voltage (EHV) systems are quite
different from the conventional designs as they have to be kept at very high voltages
from the ground. A new scheme of current transformer measurements introducing

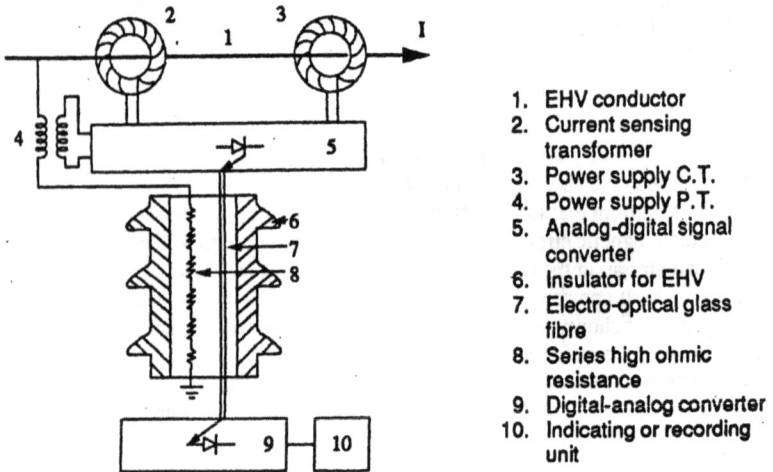

1. EHV conductor
2. Current sensing
   transformer
3. Power supply C.T.
4. Power supply P.T.
5. Analog-digital signal
   converter
6. Insulator for EHV
7. Electro-optical glass
   fibre
8. Series high ohmic
   resistance
9. Digital-analog converter
10. Indicating or recording
    unit

Fig. 7.44   Current transformer with electro-optical signal converter for EHV systems

electro-optical technique is described in Fig. 7.44. A voltage signal proportional to the measuring current is generated and is transmitted to the ground side through an electro-optical device. Light pulses proportional to the voltage signal are transmitted by a glass-optical fibre bundle to a photodetector and converted back into an analog voltage signal. Accuracies better than ±0.5% have been obtained at rated current as well as for high short circuit currents. The required power for the signal converter and optical device are obtained from suitable current and voltage transformers as shown in the Fig. 7.44.

7.3.3 Measurement of High Frequency and Impulse Currents

In power system applications as well as in other scientific and technical fields, it is often necessary to determine the amplitude and waveforms of rapidly varying high currents. High impulse currents occur in lightning discharges, electrical arcs and post arc phenomenon studies with circuit breakers, and with electric discharge studies in plasma physics. The current amplitudes may range from a few amperes to few hundred kiloamperes. The rate of rise for such currents can be as high as 10^6 to 10^{12} A/s, and rise times can vary from few microseconds to few nano seconds. In all such cases the sensing device should be capable of measuring the signal over a wide frequency band. The methods that are frequently employed are (i) resistive shunts, (ii) magnetic potentiometers or probes, and (iii) the Faraday and Hall effect devices.

The accuracy of measurement varies from 1 to 10%. In applications where only peak value measurement is required, peak reading voltmeters described in Sec. 7.2.8 may be employed with a suitable shunt.

Resistive shunts

The most common method employed for high impulse current measurements is a low ohmic pure resistive shunt shown in Fig. 7.45. The equivalent circuit is shown in Fig. 7.45b. The current through the resistive element R produces a voltage drop $v(t) = i(t)R$. The voltage signal generated is transmitted to a CRO through a coaxial cable of surge impedance Z_0. The cable at the oscilloscope end is terminated by a resistance $R_i = Z_0$

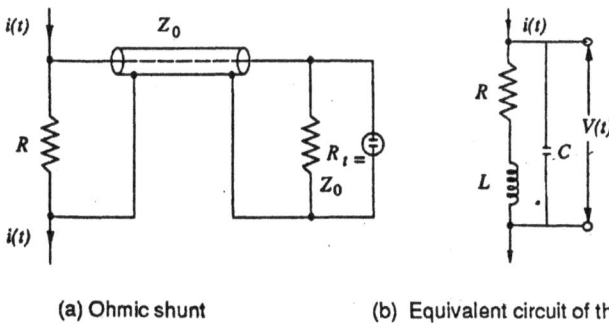

(a) Ohmic shunt   (b) Equivalent circuit of the shunt

Fig. 7.45   Calibrated low ohmic shunt and its equivalent circuit for impulse current measurements

to avoid reflections. The resistance element, because of its large dimensions will have a residual inductance L and a terminal capacitance C. The inductance L may be neglected at low frequencies (ω), but becomes appreciable at higher frequencies (ω) when ωL is of the order of R. Similarly, the value of C has to be considered when the reactance $1/\omega C$ is of comparable value. Normally L and C become significant above a frequency of 1 MHz. The resistance value usually ranges from 10 $\mu\Omega$ to few milliohms, and the voltage drop is usually about a few volts. The value of the resistance is determined by the thermal capacity and heat dissipation of the shunt.

The voltage drop across the shunt in the complex frequency domain may be written as:

$$V(s) = \frac{(R + Ls)}{(1 + RCs + LCs^2)} I(s) \qquad (7.35)$$

where s is the complex frequency or Laplace transform operator and $V(s)$ and $I(s)$ are the transformed quantities of the signals $v(t)$ and $i(t)$. With the value of C neglected it may be approximated as:

$$V(s) = (R + Ls)I(s) \qquad (7.36)$$

It may be noted here that the stray inductance and capacitance should be made as small as possible for better frequency response of the shunt. The resistance shunt is usually designed in the following manner to reduce the stray effects.

(a) Bifilar flat strip design,

(b) coaxial tube or Park's shunt design, and

(c) coaxial squirrel cage design.

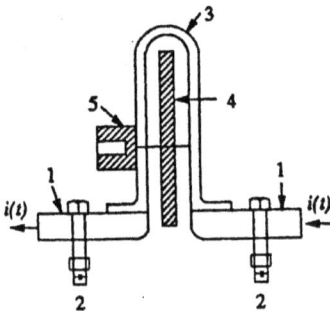

(a) Schematic arrangement

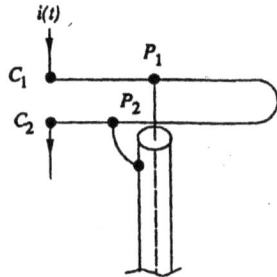

(b) Connection for potential and current terminals

1. Metal base
2. Current terminals (C_1 and C_2)
3. Bifilar resistance strip
4. Insulating spacer (teflon or bakelite)
5. Coaxial UHF connector

P_1, P_2 — Potential terminals

Fig. 7.46 Bifilar flat strip resistive shunt

(a) Bifilar Strip Shunt

The bifilar design (Fig. 7.46) consists of resistor elements wound in opposite directions and folded back, with both ends insulated by a teflon or other high quality insulation. The voltage signal is picked up through a ultra high frequency (UHF) coaxial connector. The shunt suffers from stray inductance associated with the resistance element, and its potential leads are linked to a small part of the magnetic flux generated by the current that is measured. To overcome these problems, coaxial shunts are chosen.

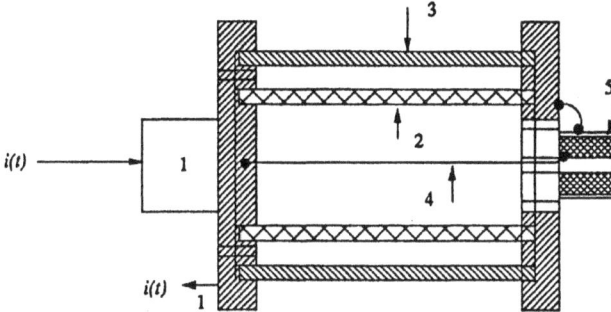

1. Current terminals
2. Coaxial cylindrical resistive element
3. Coaxial cylindrical return conductor (copper or brass tube)
4. Potential pick up lead
5. UHF coaxial connector

Fig. 7.47 Schematic arrangement of a coaxial ohmic shunt

(b) Coaxial Tubular or Park's Shunt

In the coaxial design (Fig. 7.47) the current is made to enter through an inner cylinder or resistive element and is made to return through an outer conducting cylinder of copper or brass. The voltage drop across the resistive element is measured between the potential pick-up point and the outer case. The space between the inner and the outer cylinder is air and hence acts like a pure insulator. With this construction, the maximum frequency limit is about 1000 MHz and the response time is a few nanoseconds. The upper frequency limit is governed by the skin effect in the resistive element. The equivalent circuit of the shunt is given in Fig. 7.48. The step response and the frequency response are shown in Fig. 7.49. The inductance L_0 shown in Fig. 7.48 may be written as:

$$L_0 = \frac{\mu dl}{2\pi r} \qquad (7.37)$$

where,
$\mu = \mu_0 \mu_r$; the magnetic permeability, $\mu_0 = 4\pi \times 10^{-9}$
Vs/A cm is the magnetic field constant of vacuum

$d = $ thickness of the cylindrical tube,

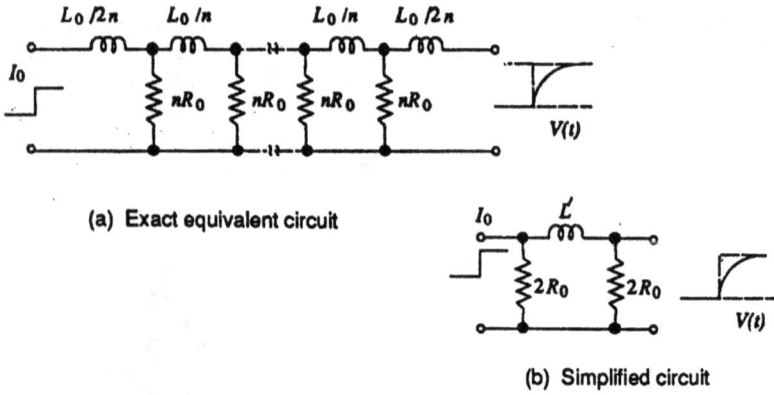

(a) Exact equivalent circuit

(b) Simplified circuit

L_0 — Inductance
R_0 — d.c. resistance
n — Number of sections per unit length

$L' - 0.43 L_0$

Fig. 7.48　Simplified and exact equivalent circuits of a coaxial tubular shunt

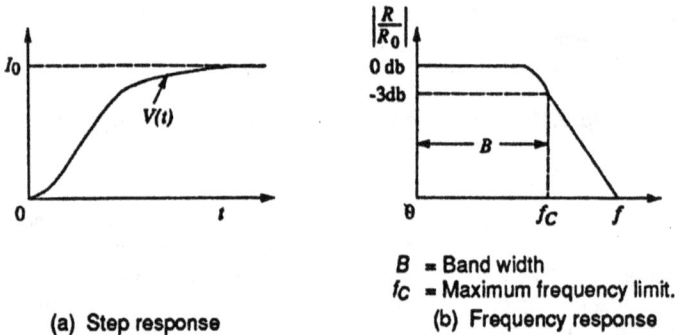

B = Band width
f_C = Maximum frequency limit.

(a) Step response

(b) Frequency response

Fig. 7.49　Step and frequency responses of a coaxial tubular shunt

$l =$　length of the cylindrical tube, and
$r =$　radius of the cylindrical tube

The effective resistance is given by

$$R = \frac{V(t)}{I_0} = R_0 \theta(\omega t) \qquad (7.38)$$

where, $R_0 =$ the d.c. resistance; $L_0 =$ inductance for d.c. currents and $\theta(\omega t)$ is the theta function of type 3 and is equal to

$$[1 + 2 \sum_{n=1}^{\infty} (-1)^n \exp(-n^2 \omega t)]$$

in which $\omega = \dfrac{(\pi^2 R_0)}{L_0}$

$V(t)$ = signal developed; and I_0 is the step current.

The effective impedance of the shunt for any frequency f according to Silsbee is given by:

$$Z = \frac{R_0(1 + j)\delta}{\sinh\,[(1 + j)\delta]} \qquad (7.39)$$

where, $R_0 =$ d.c. resistance Ω,

$\delta = 2\pi\,d\sqrt{(f\mu)/\rho}$,

$\rho =$ resistivity of the material, Ω-cm,

$d =$ thickness of the tube, cm,

$f =$ frequency, Hz, and

$\mu =$ permeability as defined earlier.

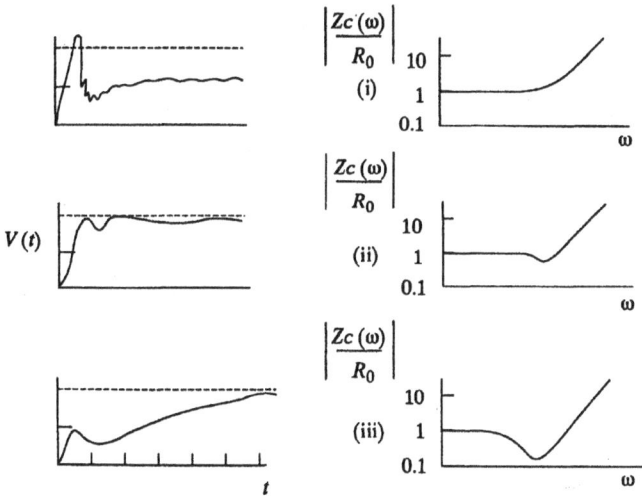

(a) Step response   (b) Frequency response

(i)   number of rods too small
(ii) ideal number of rods
(iii) number of rods too high

Fig. 7.50   Response of squirrel cage shunt for different number of rods

The simplified equivalent circuit shown in Fig. 7.48 is convenient to calculate the rise time of the shunt. The rise time accordingly is given by,

$$T = 0.237 \frac{\mu d^2}{\rho}$$

and the bandwidth is given by

$$B = \frac{1.46\,R}{L_0} = \frac{1.46\,\rho}{\mu d^2} \qquad (7.40)$$

The coaxial tubular shunts were constructed for current peaks up to 500 kA; shunts constructed for current peaks as high as 200 kA with di/dt of about 5×10^{10} A/s have induced voltages less than 50 V and the voltage drop across the shunt was about 100 V.

(c) Squirrel Cage Shunts

In certain applications, such as post arc current measurements, high ohmic value shunts which can dissipate larger energy are required. In such cases tubular shunts are not suitable due to their limitations of heat dissipation, larger wall thickness, and the skin effect. To overcome these problems, the resistive cylinder is replaced by thick rods or strips, and the structure resembles the rotor construction of double squirrel cage induction motor. The equivalent circuit for squirrel cage construction is different, and complex. The shunts show peaky response for step input, and a compensating network has to be designed to get optimum response. In Fig. 7.50, the step response (Fig. 7.50a) and frequency response (Fig. 7.50b) characteristics are given. Rise times of better than 8 n s with bandwidth more than 400 MHz were obtained for this type of shunts. A typical R-C compensating network used for these shunts is shown in Fig. 7.51.

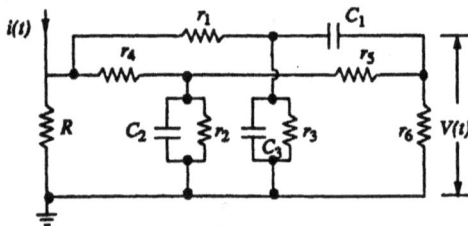

R — Shunt resistance
$r_1 - r_6$ — Resistors and capacitors in compensating double T network and $C_1 - C_6$

Fig. 7.51 Compensating network for squirrel cage shunts

(d) Materials and Technical Data for the Current Shunts

The important factor for the materials of the shunts is the variation of the resistivity of the material with temperature. In Table 7.11 physical properties of some materials with low temperature coefficient, which can be used for shunt construction are given.

Table 7.11 Properties of Resistive Materials

Property	Material				
	Constantan	Manganin	Nichrome	German silver	Ferro-alloy
Resistivity ρ at 20°C (Ω-m)	0.49×10^{-6}	0.43×10^{-6}	1.33×10^{-6}	0.23×10^{-6}	0.49×10^{-6}
Temperature coefficient per °C(10^{-6})	30	20	20	≈ 50	40
Density at 20°C kg/litre	8.9	8.4	8.1	≈ 7.5	8.8
Specific heat kilo calories/ kg °C	0.098	0.097	0.11	≈ 0.1	≈ 0.1

The importance of the skin effect has been pointed out in the coaxial shunt design. The skin depth d for a material of conductivity σ at any frequency f is given by

$$d = \frac{1}{\sqrt{\pi f \mu \sigma}} \qquad (7.41)$$

Skin depth, d, is defined as the distance or depth from the surface at which the magnetic field intensity is reduced to '$1/e$' ($e = 2.718 \ldots$) of the surface value for a given frequency f. Materials of low conductivity σ (high resistivity materials) have large skin depth and hence exhibit less skin effect.

It may be stated that low ohmic shunts of coaxial type or squirrel cage type construction permit measurements of high currents with response times less than 10 n s.

Measurement of High Impulse Currents Using Magnetic Potentiometers (Rogowski Coils) and Magnetic Links

If a coil is placed surrounding a current carrying conductor, the voltage signal induced in the coil is $v_i(t) = M dI(t)/dt$ where M is the mutual inductance between the conductor and the coil, and $I(t)$ is the current flowing in the conductor. Usually, the coil is wound on a nonmagnetic former of toroidal shape and is coaxially placed surrounding the current carrying conductor. The number of turns on the coil is chosen to be large, to get enough signal induced. The coil is wound cross-wise to reduce the leakage inductance. Usually an integrating circuit (see Fig. 7.52) is employed to get the output signal voltage proportional to the current to be measured. The output voltage is given by

$$V_m(t) = \frac{1}{CR} \int_0^t v_i(t) = \frac{M}{CR} I(t) \qquad (7.42)$$

Rogowski coils with electronic or active integrator circuits have large bandwidths (about 100 MHz). At frequencies greater than 100 MHz the response is affected by

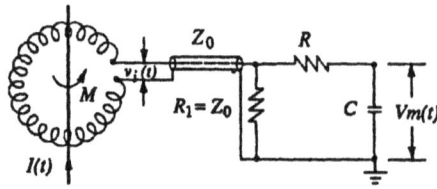

$V_i(t)$ — Induced voltage in the coil = $M \dfrac{d[I(t)]}{dt}$

Z_0 — Coaxial cable of surge impedance Z_0

R-C — Integrating network

Fig. 7.52 Rogowski coil for high impulse current measurements

the skin effect, the capacitance distributed per unit length along the coil, and due to the electromagnetic interferences. However, miniature probes having nanosecond response time are made using very few turns of copper strips for UHF measurements.

Magnetic Links

Magnetic links are short high retentivity steel strips arranged on a circular wheel or drum. These strips have the property that the remanent magnetism for a current pulse of 0.5/5 μ s is same as that caused by a d.c. current of the same value. Hence, these can be used for measurement of peak value of impulse currents. The strips will be kept at a known distance from the current carrying conductor and parallel to it. The remanent magnetism is then measured in the laboratory from which the peak value of the current can be estimated. These are useful for field measurements, mainly for estimating the lightning currents on the transmission lines and towers. By using a number of links, accurate measurement of the peak value, polarity, and the percentage oscillations in lightning currents can be made.

Other Techniques for Impulse Current Measurements

(a) Hall Generators
Hall generators described earlier can be used for a.c. and impulse current measurements also. The bandwidth of such devices was found to be about 50 MHz with suitable compensating devices and feedback. The saturation effect in magnetic core can be minimized, and these devices are successfully used for post arc and plasma current measurements.

(b) Faraday Generator or Ammeter
When a linearly polarized light beam passes through a transparent crystal in the presence of a magnetic field, the plane of polarization of the light beam undergoes rotation.

The angle of rotation α is given by:

$$\alpha = VBl \tag{7.43}$$

where, $V =$ a constant of the crystal which depends on the wavelength of the light,

$B =$ magnetic flux density, and

$l =$ length of the crystal.

To measure the waveform of a large current in a EHV system an arrangement shown in Fig. 7.53 may be employed. A beam of light from a stabilized light source is passed through a polarizer P_1 to fall on a crystal F placed parallel to the magnetic field produced by the current I. The light beam undergoes rotation of its plane of polarization. After passing through the analyser, the beam is focused on a photo-multiplier, the output of which is fed to a CRO. The output beam is filtered through a filter M, which allows only the monochromatic light. The relation between the oscillograph display and the current to be measured are complex but can be determined. The advantages of this method are that (i) there is no electric connection between the source and the device, (ii) no thermal problems even for large currents of several kiloamperes, and (iii) as the signal transmission is through an optical system, no insulation problems or difficulties arise for EHV systems. However, this device does not operate for d.c. currents.

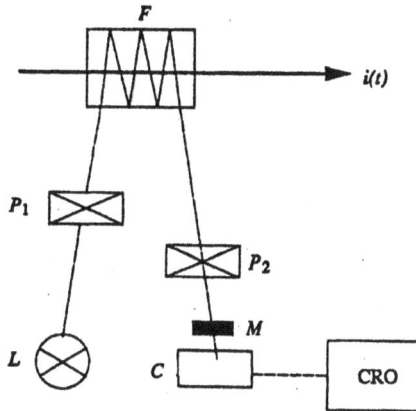

L — Light source	F — Crystal	C — Photo-multiplier
P_1 — Polarizer	CRO — Recording oscillograph	
P_2 — Analyser	M — Filter	

Fig. 7.53 Magneto-optical method of measuring impulse currents

7.4 CATHODE RAY OSCILLOGRAPHS FOR IMPULSE VOLTAGE AND CURRENT MEASUREMENTS

When waveforms of rapidly varying signals (voltages or currents) have to be measured or recorded, certain difficulties arise. The peak values of the signals in high voltage measurements are too large, may be several kilovolts or kiloamperes. There-fore, direct measurement is not possible. The magnitudes of these signals are scaled down by voltage dividers or shunts to smaller voltage signals. The reduced signal $V_m(t)$ is normally proportional to the measured quantity. The procedure of transmit-ting the signal and displaying or recording it is very important. The associated electromagnetic fields with rapidly changing signals induce disturbing voltages, which have, to be avoided. The problems associated in the above procedure are discussed in this section.

7.4.1 Cathode Ray Oscillographs for Impulse Measurements

Modern oscillographs are sealed tube hot cathode oscilloscopes with photographic arrangement for recording the waveforms. The cathode ray oscilloscope for impulse work normally has input voltage range from 5 mV/cm to about 20 V/cm. In addition, there are probes and attenuators to handle signals up to 600 V (peak to peak). The bandwidth and rise time of the oscilloscope should be adequate. Rise times of 5 n s and bandwidth as high as 500 MHz may be necessary.

Sometimes high voltage surge test oscilloscopes do not have vertical amplifier and directly require an input voltage of 10 V. They can take a maximum signal of about 100 V (peak to peak) but require suitable attenuators for large signals.

Oscilloscopes are fitted with good cameras for recording purposes. Tektronix model 7094 is fitted with a lens of 1 : 1.2 polaroid camera which uses 10,000 ASA film which possesses a writing speed of 9 cm/n s.

With rapidly changing signals, it is necessary to initiate or start the oscilloscope time base before the signal reaches the oscilloscope deflecting plates, otherwise a portion of the signal may be missed. Such measurements require an accurate initia-tion of the horizontal time base and is known as triggering. Oscilloscopes are normally provided with both internal and external triggering facility. When external triggering is used, as with recording of impulses, the signal is directly fed to actuate the time

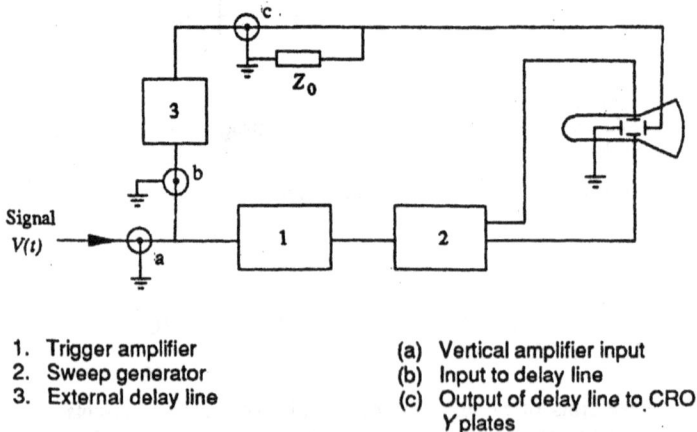

1. Trigger amplifier	(a) Vertical amplifier input
2. Sweep generator	(b) Input to delay line
3. External delay line	(c) Output of delay line to CRO Y plates

Fig. 7.54a Block diagram of a surge test oscilloscope (older arrangement)

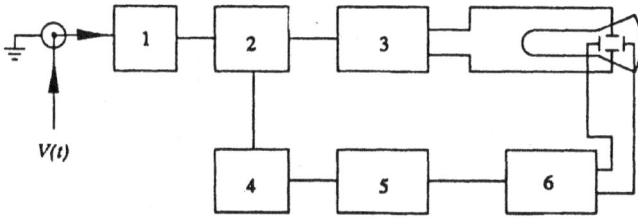

1. Plug-in amplifier
2. Y amplifier
3. Internal delay line
4. Trigger amplifier
5. Sweep generator
6. X amplifier

Fig. 7.54b Simplified block diagram of surge test oscilloscopes
(recent schemes)

base and then applied to the vertical or Y deflecting plates through a delay line. The delay is usually 0.1 to 0.5 μ s. The delay is obtained by:

(1) A long interconnecting coaxial cable 20 to 50 m long. The required triggering is obtained from an antenna whose induced voltage is applied to the external trigger terminal.

(2) The measuring signal is transmitted to the CRO by a normal coaxial cable. The delay is obtained by an externally connected coaxial long cable to give the necessary delay. This arrangement is shown in Fig. 7.54.

(3) The impulse generator and the time base of the CRO are triggered from an electronic tripping device. A first pulse from the device starts the CRO time base and after a predetermined time a second pulse triggers the impulse generator.

7.4.2 Instrument Leads and Arrangement of Test Circuits

It is essential that leads, layout, and connections from the signal sources to the CRO are to be arranged such that the induced voltages and stray pick-ups due to electromagnetic interference are avoided. For slowly varying signals, the connecting cables behave as either capacitive or inductive depending on the load at the end of the cable. For fast rising signals, however, the cables have to be accounted as transmission lines with distributed parameters. A travelling wave or signal entering such a cable encounters the surge impedance of the cable. To avoid unnecessary reflections at the cable ends, it has to be terminated properly by connecting a resistance equal to the surge impedance of the cable. In cables, the signal travels with a velocity less than that of light which is given by:

$$v = \frac{C}{\sqrt{\varepsilon_r \mu_r}}$$

where $C = 3 \times 10^8$ m/s and ε_r and μ_r are the relative permittivity and relative permeability respectively of the cable materials. Therefore the cable introduces a finite propagation time

$$t = \frac{1}{v} \times \text{ length of the cable}$$

Measuring devices such as oscilloscopes have finite input impedance, usually about 1 to 10 MΩ resistance in parallel with a 10 to 50 pF capacitance. This impedance at high frequencies ($f \approx 100$ MHz) is about 80Ω and thus acts as a load at the end of a surge cable. This load attenuates the signal at the CRO end.

Cables at high frequencies are not lossless transmission lines. Because of the ohmic resistance loss in the conductor and the dielectric loss in the cable material, they introduce attenuation and distortion to the signal. Cable distortion has to be eliminated as far as possible. Cable shields also generate noise, voltages due to ground loop currents and due to the electromagnetic coupling from other conductors. In Fig. 7.55, the ground loop currents and their path are indicated. To eliminate these noise voltages multiple shielding arrangement as shown in Fig. 7.56 may have to be used.

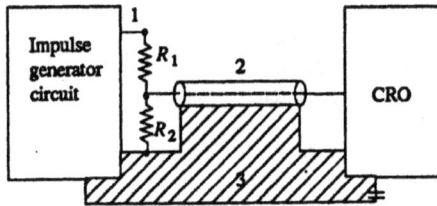

1. Potential divider
2. Coaxial signal cable
3. Ground loop

Fig. 7.55 Ground loops in impulse measuring systems

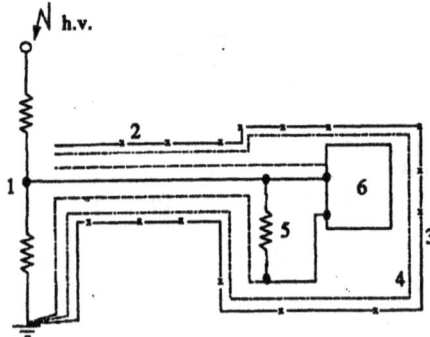

1. Potential divider
2. Triple shielded cable
3. Outer shield enclosure
4. Inner shielded enclosure
5. Terminating impedance
6. CRO

Fig. 7.56 Impulse measurements using multiple shield enclosures and signal cable

Another important factor is the layout of power and signal cables in the impulse testing laboratories. Power and interconnecting cables should not be laid in a zig-zag manner and should not be cross connected. All power cables and control cables have to be arranged through earthed and shielded conduits. A typical schematic layout is shown in Fig. 7.57. The arrangement should provide for branched wiring from the cable tree and should not form loops. Where environmental conditions are so severe that true signal cannot be obtained with all countermeasures, electro-optical techniques for transmitting signal pulses may have to be used.

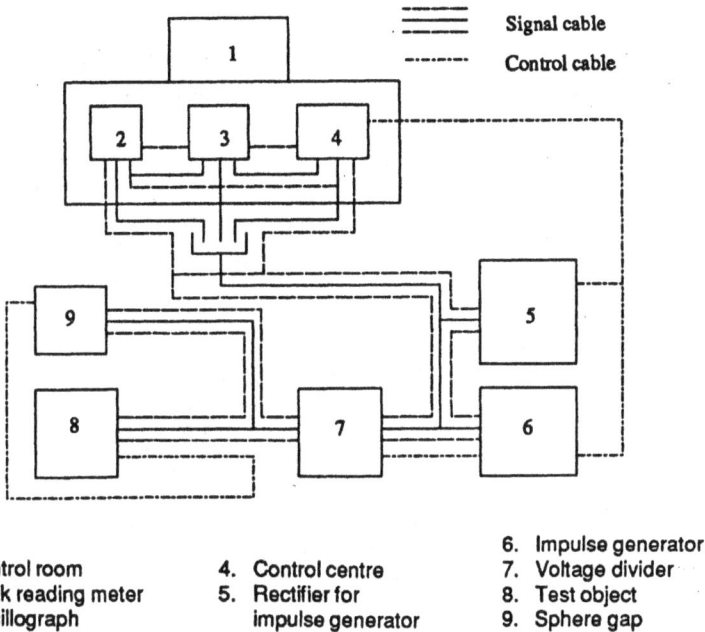

Fig. 7.57   Layout of an impulse testing laboratory with control and signal cables

1. Control room
2. Peak reading meter
3. Oscillograph
4. Control centre
5. Rectifier for impulse generator
6. Impulse generator
7. Voltage divider
8. Test object
9. Sphere gap

QUESTIONS

Q.7.1 Discuss the different methods of measuring high d.c. voltages. What are the limitations in each method ?

Q.7.2 Describe the generating voltmeter used for measuring high d.c. voltages. How does it compare with a potential divider for measuring high d.c. voltages ?

Q.7.3 Compare the relative advantages and disadvantages of using a series resistance microammeter and a potential divider with an electrostatic voltmeter for measuring high d.c. voltages ?

Q.7.4 Why are capacitance voltage dividers preferred for high a.c. voltage measurements ?

Q.7.5 What is capacitance voltage transformer ? Explain with phasor diagram how a tuned capacitance voltage transformer can be used for voltage measurements in power systems.

Q.7.6 Explain the principle and construction of an electrostatic voltmeter for very high voltages. What are its merits and demerits for high voltage a.c. measurements ?

Q.7.7 Give the basic circuit for measuring the peak voltage of (a) a.c. voltage, and (b) impulse voltage. What is the difference in measurement technique in the above two cases ?

Q.7.8 Explain how a sphere gap can be used to measure the peak value of voltages. What are the parameters and factors that influence such voltage measurement ?

Q.7.9 Compare the use of uniform field electrode spark gap and sphere gap for measuring peak values of voltages.

Q.7.10 What are the conditions to be satisfied by a potential divider to be used for impulse work ?

Q.7.11 Give the schematic arrangement of an impulse potential divider with an oscilloscope connected for measuring impulse voltages. Explain the arrangement used to minimise errors.

Q.7.12 What is a mixed potential divider ? How is it used for impulse-voltage measurements ?

Q.7.13 Explain the different methods of high current measurements with their relative merits and demerits.

Q.7.14 What are the different types of resistive shunts used for impulse current measurements ? Discuss their characteristics and limitations.

Q.7.15 What are the requirements of an oscillograph for impulse and high frequency measurements in high voltage test circuits ?

Q.7.16 Explain the necessity of earthing and shielding arrangements in impulse measurements and in high voltage laboratories. Give a sketch of the multiple shielding arrangements used for impulse voltage and current measurements.

Q.7.17 A generating voltmeter is to read 250 kV with an indicating meter having a range of $(0 - 20)$ μA calibrated accordingly. Calculate the capacitance of the generating voltmeter when the driving motor rotates at a constant speed of 1500 r.p.m.

Q.7.18 The effective diameter of the moving disc of an electrostatic voltmeter is 15 cm with an electrode separation of 1.5 cm. Find the weight in gms that is necessary to be added to balance the moving plate when measuring a voltage of 50 kV d.c. Derive the formula used. What is the force of attraction between the two plates when they are balanced ?

Q.7.19 A compensated resistance divider has its high voltage arm consisting of a series of resistance whose total value is 25 kilo-ohms shunted by a capacitance of 400 pF. The L.V. arm has a resistance of 75 ohms. Calculate the capacitance needed for the compensation of this divider.

Q.7.20 What are the usual sources of errors in measuring high impulse voltages by resistance potential dividers? How are they eliminated? An impulse resistance divider has a high voltage arm with a 5000 ohm resistance and the L.V. arm with a 5 ohm resistance. If the oscilloscope is connected to the secondary arm through a cable of surge impedance 75 ohms, determine, (i) the terminating resistance, and (ii) the effective voltage ratio.

Q.7.21 A mixed R-C divider has its h.v. arm consisting of a capacitance of 400 pF in series with a resistance of 100 ohms. The L.V. arm has a resistance of 0.175 ohm in series with a capacitance C_2. What should be the L.V. arm capacitance for correct compensation? The divider is connected to a CRO through a measuring cable of 75 ohms surge impedance. What should be the values of R_4 and C_4 (see Fig. 7.38(b)) in the matching impedance? Determine the voltage ratio of the divider.

WORKED EXAMPLES

Example 7.1: A generating voltmeter has to be designed so that it can have a range from 20 to 200 kV d.c. If the indicating meter reads a minimum current of 2 μA and maximum current of 25 μA, what should the capacitance of the generating voltmeter be ?

Solution: Assume that the driving motor has a synchronous speed of 1500 rpm.

$$I_{rms} = \frac{VC_m}{\sqrt{2}} \omega$$

where,

$V =$ applied voltage,

$C_m =$ capacitance of the meter, and

$\omega =$ angular speed of the drive

Substituting,

$$2 \times 10^{-6} = \frac{20 \times 10^3 \times C_m}{\sqrt{2}} \times \frac{1500}{60} \times 2 \pi$$

∴ $$C_m = 0.9 \text{ p.F}$$

At $$200 \text{ kV}, I_{rms} = \frac{200 \times 10^3 \times 0.9 \times 10^{-12} \times 1500}{\sqrt{2} \times 60} 2\pi$$

$$= 20.0 \text{ μA}$$

The capacitance of the meter should be 0.9 pF. The meter will indicate 20 kV at a current 2 μA and 200 kV at a current of 20 μA.

Example 7.2: Design a peak reading voltmeter along with a suitable micro-ammeter such that it will be able to read voltages, up to 100 kV (peak). The capacitance potential divider available is of the ratio 1000 : 1.

Solution: Let the peak reading voltmeter be of the Haefely type shown in Fig. 7.17a.

Let the micro-ammeter have the range $0 - 10$ μA.

The voltage available at the C_2 arm $= 100 \times 10^3 \times \frac{1}{1000}$

$$= 100 \text{ V (peak)}$$

The series resistance R in series with the micro ammeter

$$= \frac{100}{10 \times 10^{-6}}$$

$$= 10^7 \Omega$$

$C_S R = 1 \text{ to } 10 \text{ s}$

Taking the higher value of 10 s, $C_S = \frac{10}{10^7}$

$$= 1 \text{ μF}$$

The values of C_S and R are 1 μF and $10^7 \Omega$.

Example 7.3: Calculate the correction factors for atmospheric conditions, if the laboratory temperature is 37°C, the atmospheric pressure is 750 mm Hg, and the wet bulb temperature is 27°C.

Solution: Air density factor, $d = \dfrac{p}{760} \dfrac{293}{(273 + t)}$

At $t = 37°C$

$$d = \frac{750}{760} \frac{293}{310}$$

$$= 0.9327$$

From Table 7.6 air density correction factor $K = 0.9362$. From Fig. 10.1, the absolute humidity (by extrapolation) corresponding to the given temperature is 18 g/m^3. From Fig. 10.2, the humidity correction factor for 50 Hz (curve a) is 0.925.

(Note: No humidity correction is needed for sphere gaps.)

Example 7.4: A resistance divider of 1400 kV (impulse) has a high voltage arm of 16 kilo-ohms and a low voltage arm consisting 16 members of 250 ohms, 2 watt resistors in parallel. The divider is connected to a CRO through a cable of surge impedance 75 ohms and is terminated at the other end through a 75 ohm resistor. Calculate the exact divider ratio.

Solution: h.v. arm resistance, $R_1 = 16{,}000$ ohms

l.v. arm resistance, $R_2 = \dfrac{250}{16}$ ohms

Terminating resistance, $R_2' = 75$ ohms

hence, the divider ratio, $a = 1 + R_1/R_2 + R_1/R_2'$

$$= 1 + 16{,}000 \times 16/250$$

$$= 1 + 16{,}000/75$$

$$= 1 + 1024 + 213.3 = 1238.3$$

Example 7.5: The H. V. arm of an R-C, divider has 15 numbers of 120 ohm resistors with a 20 pF capacitor to ground from each of the junction points. The L.V. arm resistance is 5 ohms. Determine the capacitance needed in the L.V. arm for correct compensation.

Solution: Ground capacitance per unit = $C_g' = 20$ pF

Effective ground capacitance = $C_e = (2/3)\, C_g$

$$= 2/3\, (15 \times 20)$$

(Refer Fig. 7.34)

$$= 200\ \text{pF}$$

This capacitance is assumed to be between the center tap of the H.V. arm and the ground as shown in Fig. 7.28.

Here, $R_1/2 = 15 \times 120/2 = 900$ ohms

$R_2 = 5$ ohms.

Then, the effective time constant of the divider,

$$= (R_1/2 \cdot (2/3)\, C_g) = R_1 C_e/2$$

$$= ((900 \times 200 \times 10^{-12})/2s = 90 \text{ n s}$$

Making the L.V. arm time constant to be the same as that of the H.V arm; the capacitance required for compensation is calculated as:

$$R_2 C_2 = 90 \text{ n sec.}$$

$$C_2 = 90/5 \text{ nF} = 18 \times 10^{-9} \text{F}$$

Example 7.6: A coaxial shunt is to be designed to measure an impulse current of 50 kA. If the bandwidth of the shunt is to be at least 10 MHz and if the voltage drop across the shunt should not exceed 50 V, find the ohmic value of the shunt and its dimensions.

Solution: Resistance of the shunt (max) $R = \dfrac{50}{50 \times 10^2}$

$$= 1 \text{ m}\Omega$$

Taking the simplified equivalent circuit of the shunt as given in Fig. 7.48(b)

Bandwidth $\qquad\qquad B = \dfrac{1.46R}{L_0} = 10 \text{ MHz}$

or, $\qquad\qquad L_0 = \dfrac{1.46R}{B} = \dfrac{1.46 \times 10^{-2}}{10 \times 10^6}$

$$= 1.46 \times 10^{-10} \text{ H}$$

$$\text{or } 0.146 \text{ n H}$$

d, the thickness of the cylindrical resistive tube is taken from the consideration of the bandwidth as

$$B = \dfrac{1.46\rho}{\mu d^2}$$

where,

$\qquad\qquad\qquad \rho =$ resistivity of the material,

$\qquad\qquad \mu = \mu_0 = 4\pi \times 10^{-7}$ H/m, and

$\qquad\qquad\qquad d =$ thickness of the tube in metres

Let $\qquad\qquad\qquad r =$ radius of the resistive tube,

$\qquad\qquad\qquad l =$ length of the resistive tube,

$\qquad\qquad\qquad d =$ thickness of the resistive tube, and

$\qquad\qquad\qquad \rho =$ resistivity of the tube material.

Then the bandwidth $\qquad B = \dfrac{1.46\rho}{\mu d^2}$

where, $\qquad\qquad \mu = \mu_0 \mu_r \approx \mu_0$

Substituting $\qquad\qquad B = 10^7$ Hz

$$\rho = 30 \times 10^{-8} \ \Omega m$$

$$\mu_0 = 4\mu \times 10^{-7}$$

$$d = \sqrt{\frac{1.46\rho}{\mu B}}$$

$$= \sqrt{\frac{1.46 \times 30 \times 10^{-8}}{4\pi \times 10^{-7} \times 10^7}}$$

$$= 0.187 \times 10^{-8} \text{ m}$$

$$= 0.187 \text{ mm}$$

Let the length l be taken as 10 cm or 10^{-1} m;

then,

$$R = \frac{\rho l}{A} = \frac{\rho l}{(2\pi r)d} = 1 \text{ m}\Omega$$

or,

$$r = \frac{\rho l}{2\pi R d}$$

$$= \frac{30 \times 10^{-8} \times 10^{-1}}{2\pi \times 10^{-3} \times 0.187 \times 10^{-3}}$$

$$= 25.5 \times 10^{-3} \text{ m}$$

$$\text{or } 25.5 \text{ mm.}$$

For the return conductor the outer tube can be taken to have a length = 10 cm, radius = 30 mm, and thickness = 1 mm, and it can be made from copper or brass.

Example 7.7: A Rogowski coil is to be designed to measure impulse currents of 10 kA having a rate of change of current of 10^{11} A/s. The current is read by a VTVM as a potential drop across the integrating circuit connected to the secondary. Estimate the values of mutual inductance, resistance, and capacitance to be connected, if the meter reading is to be 10 V for full-scale deflection.

Solution: $V_m(t) = \dfrac{M}{CR} I(t)$ for $\dfrac{1}{CR} << \omega$ (Eq. 7.42),

taking the peak values

$$\frac{M}{CR} = \frac{V_m(t)}{I(t)} = \frac{10}{10^4} = 10^{-3}$$

The time interval of the change of current assuming sinusoidal variation is

$$\frac{10^4}{10^{11}} = 10^{-7} \text{ s} = \frac{1}{4} \text{ of a cycle}$$

∴ frequency

$$= \frac{10^7}{4} \text{ Hz}$$

and,

$$\omega = 2\pi f = \frac{\pi}{2} \times 10^7$$

Taking

$$\frac{1}{CR} = \frac{\omega}{10\pi} = \frac{10^6}{2}$$

$$CR = \frac{2}{10^6}$$

$$M = 10^{-3}CR = 10^{-3}\frac{2}{10^6}$$

$$= 2 \times 10^{-9} \text{ H or 2 n H}.$$

Taking $R = 2 \times 10^3 \, \Omega$,

$$C = \frac{CR}{R} = 2 \times 10^{-6}/20 \times 10^2$$

$$= 10^{-9} \text{ F or 1000 pF}$$

(It should be noted that for a given frequency, $X_c \ll R$; otherwise the low frequency response will be poor. Here X_c at $f = 10^7/4$ is 60Ω only.)

Example 7.8: If the coil in Example 7.7 is to be used for measuring impulse current of 8/20 μ s wave and of the same peak current, what should be the R-C integrating circuit.

Solution: In this case, the lowest frequency to be read should be at least $\frac{1}{3}$ to $\frac{1}{5}$ of the lowest frequency component present in the waveform.

The frequency corresponding to the tail time is

$$\frac{1}{20 \times 10^{-6}} = 50 \text{ kHz}$$

∴ lowest frequency to be read is

$$50 \times \frac{1}{5} = 10 \text{ kHz}$$

$$\therefore \omega = 2\pi \times 10^4 \text{ radians}$$

Taking

$$\frac{1}{CR} = \frac{\omega}{10\pi}$$

$$= \frac{2\pi \times 10^4}{10\pi} = 2 \times 10^3$$

$$M = 10^{-3} \times \frac{1}{2 \times 10^3} = 0.5 \times 10^{-6} \text{ H, or 0.5 } \mu\text{H}$$

Taking

$$R = 2 \times 10^3 \Omega \text{ as before,}$$

$$C = \frac{0.5 \times 10^{-3}}{2 \times 10^3}$$

$$= 0.25 \, \mu\text{F}$$

(X_c at a frequency of 10 kHz is about 60Ω which is very much less than R.)

REFERENCES

1. Craggs, J.D. and Meek, J.M., *High Voltage Laboratory Techniques*, Butterworths Scientific Publications, London (1964).
2. Kuffel, E., and Abdullah, M., *High Voltage Engineering*, Pergamon Press, Oxford (1970).
3. Schwab, A.J., *High Voltage Measurement Technique*, M.I.T. Press, Cambridge, Massachusetts (1972).
4. Bowlder, G.W., *Measurement in High Voltage Test Circuits*, Pergamon Press, Oxford (1975).
5. Begamudre, R.D., "*E.H.V. A.C. Transmission Engineering*, Wiley Eastern, New Delhi (1986).
6. Hylten Cavallius, N., *High Voltage Laboratory Planning*, Haefely and Co., Basel, Switzerland (1988).
7. "Method of voltage measurement by means of sphere gaps (one sphere earthed)" *IS* : 1876-1961.
8. "Methods of impulse voltage testing", *IS* : *2070-1962*.
9. "Methods of high voltage testing", *IS* : *2071-1962*.
10. "On sphere gap measurements", *BS* : *358-1960*.
11. "Recommendations for voltage measurements by sphere gaps", *IEC Publication No. 52-1960*.
12. Trump, G.J. and Van De Graaff *et al.*, "Generator voltmeter of high voltage sources", *Rev. Sci. Instr.*, **11**, 54 (1940).
13. Hamwell and Van Voorhis, "An electrostatic generating voltmeter", *Rev. Sci. Instr.*, **4**, 540 (1933).
14. Kuffel, E., "The influence of nearby earth objects and polarity of voltage on d.c. breakdown of sphere gaps", *Proc. IEE, Part A*, **108**, 302 (1961).
15. Kuffel, E., "The effect of irradiation on breakdown voltage of sphere gaps in air under d.c. and a.c. voltages", *Proc. IEE, Part C*, **106**, 133 (1956).
16. Davis *et al.*, "Measurement of high voltages with special reference to peak voltages", *Proc. IEE*, **68**, 1222 (1930).
17. Spharpe *et al.*, "Crest voltmeters", *Tr. AIEE*, **35**, 99 (1916).
18. Rabus, W., "Peak voltmeters", *Z. Elektrot*, **3**, 7 (1950).
19. Haefely and Co., "*H.V. meter for peak and rms value measurements*", Druckschirf, BD., 6589 Basel (1967).
20. Rabus, W., "Measurements of surges by V.T.V.M. and electrostatic voltmeters", *ETZ(A)*, **75**, 6761 (1953).
21. Blalock *et al.*, "A capacitive voltage divider for UHV outdoor testing", *Tr. IEEE PAS*, PAS-89, 1404 (1970).
22. Ziegler, "Highly stable 150 kV voltage divider", *Tr. IEEEIM*, **IM-19**, 395 (1970).
23. High voltage testing techniques, *IEC Publication No. 60-1962*.
24. Koshrt, F., "Hall generators for high d.c. current measurements", *ETZ (A)*, **77**, 487 (1956).
25. Thomas, R.F., "Fast light pulse measuring schemes", *Tr. IEEEIM*, **IM-17**, 12 (1968).
26. Mckibbin, F., "Use of delta modulation in pulse transmission system", *EE Trans.*, **6**, 55 (1970).
28. Witt, R., "Response of low resistance shunts for impulse currents", *Eleteckric*, **47**, 54 (1960).
29. Schwab, A., "Low ohmic resistors for impulse currents", *Transaction 12, of H.V. Laboratory*, University of Karlsruhe, 1972.

30. Schwab, A., *ETZ(A)*, **87**, 181 (1966).
31. Heumann, K., "Magnetic potentiometer of high precision", *Tr. IEEEIM*, **IM-15**, 242 (1966).
32. Ficchi, R.C., *Electrical interference*, McGraw-Hill, New York (1963).
33. Thomas, R.T. *et al.*, "High impulse current and voltage measurements", *Tr. IEEE IM*, **IM-19**, 102 (1970).
34. Cassidy, E.C. *et al.*, "Electro-optical H.V. pulse measurement techniques", *Tr. IEEE IM*, **IM-19**, 395 (1970).
35. Zaengel, W., "Impulse voltage dividers and leads", *Bull, SEV*, **61**, 1003 (1970).
36. Faser, K., "Transient behaviour of damped capacitive voltage dividers of million-volts", *Tr.IEEE PAS*, **93**.116 (1974).
37. Hylten Cavallius, N., et al., "A new approach to minimise response errors in the measurement of high voltages", *Tr. IEEE PAS, Vol. PAS-102, 2077* (1983).
38. Hylten Cavallius, N., et al., "Response errors of shunts", *Int. Symp. on High Voltage Engineering, Paper No. 61.05*, Athens, Greece (1983).

8

Overvoltage Phenomenon and Insulation Coordination in Electric Power Systems

It is essential for electrical power engineers to reduce the number of outages and preserve the continuity of service and electric supply. Therefore, it is necessary to direct special attention towards the protection of transmission lines and power apparatus from the chief causes of overvoltages in electric systems, namely lightning overvoltages and switching overvoltages. Lightning overvoltage is a natural phenomenon, while switching overvoltages originate in the system itself by the connection and disconnection of circuit breaker contacts or due to initiation or interruption of faults. Switching overvoltages are highly damped short duration overvoltages. They are "temporary overvoltages" of power frequency or its harmonic frequency either sustained or weakly damped and originate in switching and fault clearing processes in power systems. Although both switching and power frequency overvoltages have no common origin, they may occur together, and their combined effect is important in insulation design. Probability of lightning and switching overvoltages coinciding together is very small and hence can be neglected. The magnitude of lightning voltages appearing on transmission lines does not depend on line design and hence lightning performance tends to improve with increasing insulation level, that is with system voltage. On the other hand, switching overvoltages are proportional to operating voltage. Hence, there is a system operating voltage at which the emphasis changes from lightning to switching surge design, this being important above 500 kV. In the range of 300 kV to 765 kV, both switching overvoltages and lightning overvoltages have to be considered, while for ultra high voltages (> 700 kV), perhaps switching surges may be the chief condition for design considerations.

For the study of overvoltages a basic knowledge of the origin of overvoltages, surge phenomenon, and its propagation is desirable. The present chapter is therefore devoted to a summary of the above topics.

8.1 NATURAL CAUSES FOR OVERVOLTAGES — LIGHTNING PHENOMENON

Lightning phenomenon is a peak discharge in which charge accumulated in the clouds discharges into a neighbouring cloud or to the ground. The electrode separation, i.e. cloud-to-cloud or cloud-to-ground is very large, perhaps 10 km or more. The mechanism of charge formation in the clouds and their discharges is quite a complicated and uncertain process. Nevertheless, a lot of information has been collected since the last fifty years and several theories have been put forth for explaining the phenomenon. A summary of the various processes and theories is presented in this section.

8.1.1 Charge Formation in the Clouds

The factors that contribute to the formation or accumulation of charge in the clouds are too many and uncertain. But during thunderstorms, positive and negative charges become separated by the heavy air currents with ice crystals in the upper part and rain in the lower parts of the cloud. This charge separation depends on the height of the clouds, which range from 200 to 10,000 m, with their charge centres probably at a distance of about 300 to 2000 m. The volume of the clouds that participate in lightning flashover are uncertain, but the charge inside the cloud may be as high as 1 to 100 C. Clouds may have a potential as high as 10^7 to 10^8 V with field gradients ranging from 100 V/cm within the cloud to as high as 10 kV/cm at the initial discharge point. The energies associated with the cloud discharges can be as high as 250 kWh. It is believed that the upper regions of the cloud are usually positively charged, whereas the lower region and the base are predominantly negative except the local region, near the base and the head, which is positive. The maximum gradient reached at the ground level due to a charged cloud may be as high as 300 V/cm, while the fair weather gradients are about 1 V/cm. A probable charge distribution model is given in Fig. 8.1 with the corresponding field gradients near the ground.

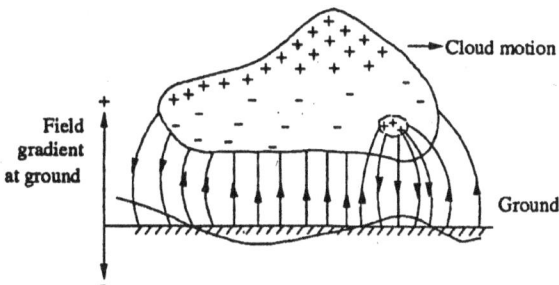

Fig. 8.1 Probable field gradient near the ground corresponding to the probable charge distribution in a cloud

According to the Simpson's theory (Fig. 8.2) there are three essential regions in the cloud to be considered for charge formation. Below region A, air currents travel above 800 cm/s, and no raindrops fall through. In region A, air velocity is high enough to break the falling raindrops causing a positive charge spray in the cloud and negative charge in the air. The spray is blown upwards, but as the velocity of air decreases, the positively charged water drops recombine with the larger drops and fall again. Thus region A, eventually becomes predominantly positively charged, while region B above it, becomes negatively charged by air currents. In the upper regions in the cloud, the temperature is low (below freezing point) and only ice crystals exist. The impact of air on these crystals makes them negatively charged, thus the distribution of the charge within the cloud becomes as shown in Fig. 8.2.

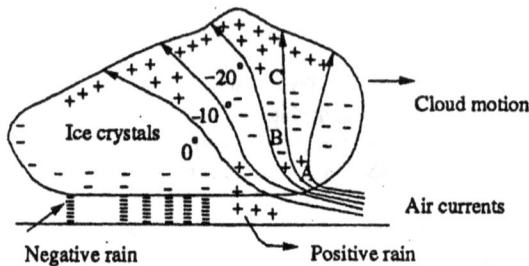

Fig. 8.2 Cloud model according to Simpson's theory

However, the above theory is obsolete and the explanation presented is not satisfactory. Recently, Reynolds and Mason proposed modification, according to which the thunder clouds are developed at heights 1 to 2 km above the ground level and may extend up to 12 to 14 km above the ground. For thunder clouds and charge formation air currents, moisture and specific temperature range are required.

The air currents controlled by the temperature gradient move upwards carrying moisture and water droplets. The temperature is 0°C at about 4 km from the ground and may reach – 50°C at about 12 km height. But water droplets do not freeze as soon as the temperature is 0°C. They freeze below – 40°C only as solid particles on which crystalline ice patterns develop and grow. The larger the number of solid sites or nuclei present, the higher is the temperature (> – 40°C) at which the ice crystals grow. Thus in clouds, the effective freezing temperature range is around – 33°C to – 40°C. The water droplets in the thunder cloud are blown up by air currents and get super cooled over a range of heights and temperatures. When such freezing occurs, the crystals grow into large masses and due to their weight and gravitational force start moving downwards. Thus, a thunder cloud consists of supercooled water droplets moving upwards and large hail stones moving downwards.

When the upward moving supercooled water droplets act on cooler hail stone, it freezes partially, i.e. the outer layer of the water droplets freezes forming a shell with water inside. When the process of cooling extends to inside warmer water in the core, it expands, thereby splintering and spraying the frozen ice shell. The splinters being fine in size are moved up by the air currents and carry a net positive charge to the upper region of the cloud. The hail stones that travel downwards carry an equivalent negative charge to the lower regions of the cloud and thus negative charge builds up in the bottom side of the cloud.

According to Mason, the ice splinters should carry only positive charge upwards. Water being ionic in nature has concentration of H^+ and OH^- ions. The ion density depends on the temperature. Thus, in an ice slab with upper and lower surfaces at temperatures T_1 and T_2, $(T_1 < T_2)$, there will be a higher concentration of ions in the lower region. However, since H^+ ions are much lighter, they diffuse much faster all over the volume. Therefore, the lower portion which is warmer will have a net negative charge density, and hence the upper portion, i.e. cooler region will have a net positive charge density. Hence, it must be appreciated, that the outer shells of the freezed water droplets coming into contact with hail stones will be relatively cooler (than their inner core—warmer water) and therefore acquire a net positive charge. When the shell splinters, the charge carried by them in the upward direction is positive.

According to the Reynold's theory, which is based on experimental results, the hail packets get negatively charged when impinged upon by warmer ice crystals. When the temperature conditions are reversed, the charging polarity reverses. However, the extent of the charging and consequently the rate of charge generation was found to disagree with the practical observations relating to thunder clouds. This type of phenomenon also occurs in thunder clouds.

Rate of Charging of Thunder Clouds

Mason considered thunder clouds to consist of a uniform mixture of positive and negative charges. Due to hail stones and air currents the charges separate vertically. If λ is a factor which depends on the conductivity of the medium, there will be a resistive leakage of charge from the electric field built up, and this should be taken into account for cloud charging.

Let E be the electric field intensity, v be the velocity of separation of charges, and ρ the charge density in the cloud. Then, the electric field intensity E is given by

$$\frac{dE}{dt} + \lambda E = \rho v \tag{8.1}$$

Hence
$$E = \frac{\rho v}{\lambda} [1 - \exp(-\lambda t)] \tag{8.2}$$

This equation assumes initially $E = 0$ at $t = 0$, the start of charge separation, i.e. there is no separation initially.

Let Q, be the separated charge and Q_g be the generated charge, then

$$\rho = \frac{Q_g}{Ah} \tag{8.3a}$$

and
$$E = \frac{Q_s}{A\varepsilon_0} \tag{8.3b}$$

where ε_0 is the permittivity of the medium, A is the cloud area and h is the height of the charged region.

From Eq. (8.2), on substitution

$$Q_g = \frac{Q_s h}{v[1 - \exp(-\lambda t)]} = \frac{M}{v[1 - \exp(-\lambda t)]} \qquad (8.4)$$

where $M = Q_s \cdot h =$ the electric moment of the thunder-storm.
The average values observed for thunder-clouds are:

time constant $= \dfrac{1}{\lambda} = 20$ s

electric moment $M = 110$ C-km and

time for first lightning flash to appear, $t = 20$ s

The velocity of separation of charges, $v = 10$ to 20 m/s.

Substituting these values, we get

$$Q_g = \frac{20,000}{v} \text{ C}$$

$$= \frac{20,000}{20} \text{ C} = 1000 \text{ C for } v = 20 \text{ m/s}$$

Calculations using Mason's theory show that a maximum charge transfer of $3 \times 10^{-3} T$ esu/cm^2 of contact surface for a contact period of 0.01 s, where T is the temperature difference.

The theory and observations of Reynolds *et al.*, gave values of 5×10^{-9} esu per crystal impact for a temperature difference of 5°C. Mason's theory seems to give much higher values, yet it explains the phenomenon satisfactorily.

8.1.2 Mechanism of Lightning Strokes

When the electric field intensity at some point in the charge concentrated cloud exceeds the breakdown value of the moist ionized air ($\approx$ 10 kV/cm), an electric streamer with plasma starts towards the ground with a velocity of about 1/10 times that of the light, but may progress only about 50 m or so before it comes to a halt emitting a bright flash of light. The halt may be due to insufficient build-up of electric charge at its head and not sufficient to maintain the necessary field gradient for further progress of the streamer. But after a short interval of about 100 μs, the streamer again starts out repeating its performance. The total time required for such a stepped leader to reach the ground may be 20 ms. The path may be quite lustruous, depending on the local conditions in air as well as the electric field gradients. Branches from the initial leader may also be formed. Since the progress of this leader stroke is by a series of jumps, it is referred as stepped leader. The picture of a typical leader stroke taken with a Boy's camera is shown in Fig. 8.3.

The lightning stroke and the electrical discharges due to lightning are explained based on the "streamer" or "kanel" theory for spark discharges in long gaps with non-uniform electric fields. The lightning consists of few separate discharges starting from a leader discharge and culminates in return strokes or main discharges. The velocity of the leader stroke of the first discharge may be 1.5×10^7 cm/s, of the succeeding leader strokes about 10^8 cm/s, and of the return strokes may be 1.5×10^9 to 1.5×10^{10} cm/s (about 0.05 to 0.5 times the velocity of light).

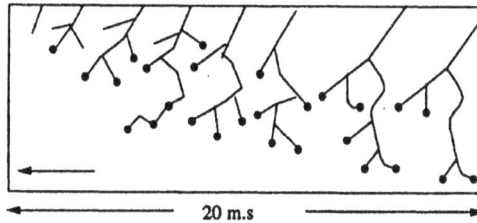

Fig. 8.3 Propagation of a stepped leader stroke from a cloud
(• Bright tips recorded)

After the leader touches the ground, the return stroke follows. As the leader moves towards the ground, positive charge is directly accumulated under the head of the stroke or canal. By the time the stroke reaches the ground or comes sufficiently near the ground, the electrical field intensity on the ground side is sufficiently large to build up the path. Hence, the positive charge returns to the cloud neutralizing the negative charge, and hence a heavy current flows through the path. The velocity of the return or main stroke ranges from 0.05 to 0.5 times the velocity of light, and currents will be of the order of 1000 to 250,000 A. The return strokes vanish before they reached the cloud, suggesting that the charge involved is that conferred to the stroke itself. The duration of the main or return stroke is about 100 μ s or more. The diameters of the return strokes were estimated to be about 1 to 2 cm but the corona envelop may be approximately 50 cm. The return strokes also may develop branches but the charges in the branches are neutralized in succession so that their further progress is arrested. A Boy's camera picture of return stroke is shown in Fig. 8.4.

Fig. 8.4 Development of the main or return stroke

After the completion of the return stroke, a much smaller current of 100 to 1000 A may continue to flow which persists approximately 20 ms. Due to these currents the initial breakdown points in the cloud are considerably reduced and discharges concentrate towards this point. Therefore, additional reservoirs of charge become available due to penetration of a cloud mass known as preferred paths and lead to repeated strokes. The leader strokes of the repeated strokes progress with much less velocity ($\approx 1\%$ of that of light) and do not branch. This stroke is called continuous leader, and return stroke for this leader follows with much less current. The interval between the repeated strokes may be from 0.6 ms to 500 ms with an average of 30 ms. Multiple strokes may last for 1 s. The total duration of the lightning may be more than 1 s. The current from the ground by the main return stroke may have a peak value of 250,000 A,

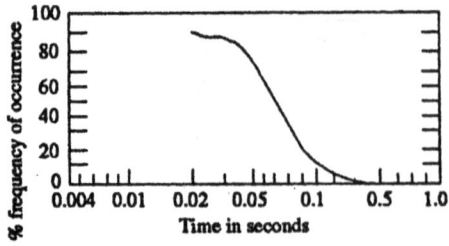

Fig. 8.5 Time interval between successive strokes (ERA average curve)

Fig. 8.6 Number of strokes in lightning discharges (ERA average curve)

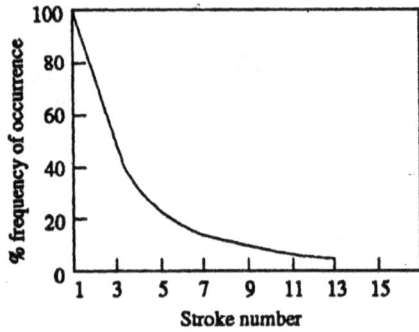

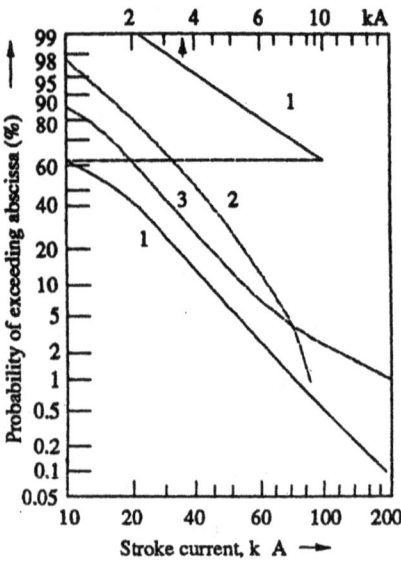

1. AIEE committee (1950)
2. Anderson (1968)
3. CIGRE (1972)

Fig. 8.7 Cumulative distributions of lightning stroke currents (peak values)

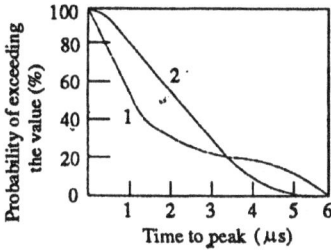

Fig. 8.8 Time to peak of lightning stroke currents

1. McEachron (1941)
2. Anderson (1968)

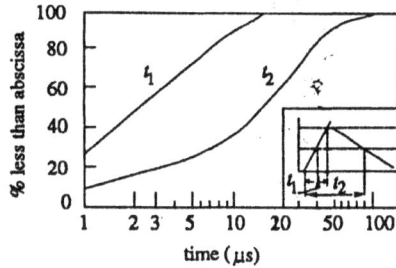

Fig. 8.9 Wavefront and wave tail times of lightning strokes (ref: Muller Hiller Brand, 1965)

and rates of rise may be as high as 100 kA/μ s or 10^{11} A/s. The time intervals between successive strokes, the number of successive strokes, the duration of lightning discharges the discharge current, the rate of rise of current, and wavefront and wave tail times and their probability distribution are given in Figs. 8.5 to 8.10.

8.1.3 Parameters and Characteristics of the Lightning Strokes

The parameters and characteristics of lightning include the amplitude of the currents, the rate of rise, the probability distribution of the above, and the waveshapes of the lightning voltages and currents.

Typical oscillograms of the lightning current and voltage waveshapes on a transmission line are shown in Figs. 8.11 and 8.12. The lightning current oscillograms indicate an initial high current portion which is characterized by short front times up to 10 μ s. The high current peak may last for some tens of microseconds followed by a long duration low current portion lasting

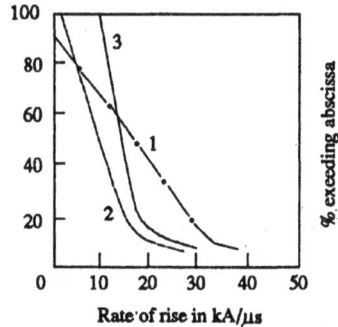

Fig. 8.10 Rate of rise of current of lightning strokes

(ref: Westinghouse T and D reference book)
1. Bergen—43 records on transmission tower
2. Norinder—magnetic field measurements
3. McEachron—strokes on Empire State Building by CRO measurements

for several milliseconds. This last portion is normally responsible for damages (thermal damage). Lightning currents are usually measured either directly from high towers or buildings or from the transmission tower legs. The former gives high values and does not represent typical currents that occur on electrical transmission lines, and the latter gives inaccurate values due to non-uniform division of current in legs and

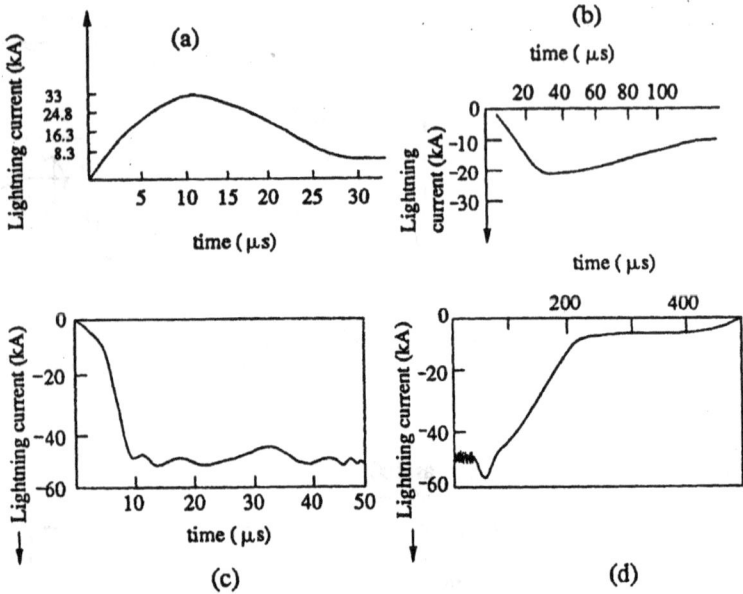

Fig. 8.11 Typical lightning current oscillograms

  (a) to a capacitive balloon (CIGRE)
  (b) on Empire State Building (McEachron)
  (c) and (d) on transmission line tower (Berger)
  ref: Westinghouse T and D reference book

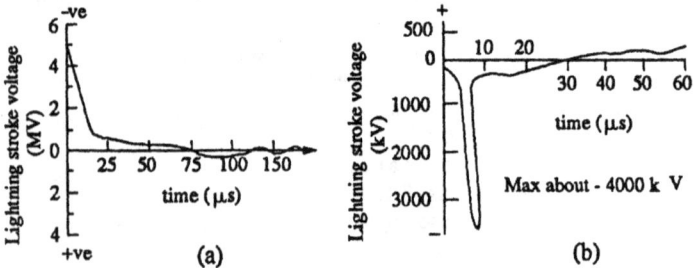

Fig. 8.12 Typical lightning stroke voltage on a transmission line without ground wire

  ref: Bell et al.,'Transactions AIEE, Vol. 50, 1931

the presence of ground wires and adjacent towers. Measurements made by several investigators and committees indicated the large strokes of currents (> 100 kA) are possible (Fig. 8.7). It was shown earlier that tall objects attract a large portion of high current strokes, and this would explain the shift of the frequency distribution curves towards higher currents.

Other important characteristics are time to peak value and its rate of rise. From the field data, it was indicated that 50% of lightning stroke currents have a rate of rise greater than 7.5 kA/µ s, and for 10% strokes it exceeded 25 kA/µ s. The duration of the stroke currents above half the value is more than 30 µ s.

Measurements of surge voltages indicated that a maximum voltage, as high as 5,000 kV, is possible on transmission lines, but on the average, most of the lightning strokes give rise to voltage surges less than 1000 kV on lines. The time to front of these waves varies from 2 to 10 µ s and tail times usually vary from 20 to 100 µ s. The rate of rise of voltage, during rising of the wave may be typically about 1 MV/µ s.

Lightning strokes on transmission lines are classified into two groups—the direct strokes and the induced strokes. When a thunder cloud directly discharges on to a transmission line tower or line wires it is called a direct stroke. This is the most severe form of the stroke. However, for bulk of the transmission systems the direct strokes are rare and only the induced strokes occur.

When the thunderstorm generates negative charge at its ground end, the earth objects develop induced positive charges. The earth objects of interest to electrical engineers are transmission lines and towers. Normally, it is expected that the lines are unaffected because they are insulated by string insulators. However, because of high field gradients involved, the positive charges leak from the tower along the insulator surfaces to the line conductors. This process may take quite a long time, of the order of some hundreds of seconds. When the cloud discharges to some earthed object other than the line, the transmission line is left with a huge concentration of charge (positive) which cannot leak suddenly. The transmission line and the ground will act as a huge capacitor charged with a positive charge and hence overvoltages occur due to these induced charges. This would result in a stroke and hence the name "induced lightning stroke".

Sometimes, when a direct lightning stroke occurs on a tower, the tower has to carry huge impulse currents. If the tower footing resistance is considerable, the potential of the tower rises to a large value, steeply with respect to the line and consequently a flashover may take place along the insulator strings. This is known as back flashover.

8.1.4 Mathematical Model for Lightning

During the charge formation process, the cloud may be considered to be a non-conductor. Hence, various potentials may be assumed at different parts of the cloud. If the charging process is continued, it is probable that the gradient at certain parts of the charged region exceeds the breakdown strength of the air or moist air in the cloud. Hence, local breakdown takes place within the cloud. This local discharge may finally lead to a situation wherein a large reservoir of charges involving a considerable mass of cloud hangs over the ground, with the air between the cloud and the ground as a dielectric. When a streamer discharge occurs to ground by first a leader stroke, followed by main strokes with considerable currents flowing, the lightning stroke may be thought to be a current source of value I_0 with a source impedance Z_0 discharging to earth. If the stroke strikes an object of impedance Z, the voltage built across it may be taken as

$$V = IZ$$

$$= I_0 \frac{ZZ_0}{Z + Z_0} \tag{8.5}$$

$$= I_0 \frac{Z}{1 + \dfrac{Z}{Z_0}} \tag{8.6}$$

The source impedance of the lightning channels are not known exactly, but it is estimated to be about 1000 to 3000 Ω. The objects of interest to electrical engineers, namely, transmission line, etc. have surge impedances less than 500 Ω (overhead lines 300 to 500 Ω, ground wires 100 to 150 Ω, towers 10 to 50 Ω, etc.). Therefore, the value Z/Z_0 will usually be less than 0.1 and hence can be neglected. Hence, the voltage rise of lines, etc. may be taken to be approximately $V = I_0 Z$, where I_0 is the lightning stroke current and Z the line surge impedance.

If a lightning stroke current as low as 10,000 A strikes a line of 400 Ω surge impedance, it may cause an overvoltage of 4000 kV. This is a heavy overvoltage and causes immediate flashover of the line conductor through its insulator strings.

In case a direct stroke occurs over the top of an unshielded transmission line, the current wave tries to divide into two branches and travel on either side of the line. Hence, the effective surge impedance of the line as seen by the wave is $Z_0/2$ and taking the above example, the overvoltage caused may be only $10,000 \times (400/2) = 2000$ kV. If this line were to be a 132 kV line with an eleven 10 inch disc insulator string, the flashover of the insulator string will take place, as the impulse flashover voltage of the string is about 950 kV for a 2 μ s front impulse wave.

The incidence of lightning strikes on transmission lines and sub-stations is related to the degree of thunderstorm activity. It is based on the level of "Thunderstorm days" (TD) known as "Isokeraunic Level" defined as the number of days in a year when thunder is heard or recorded in a particular location. But this indication does not often distinguish between the ground strikes and the cloud-to-cloud strikes. If a measure of ground flashover density (N_g) is obtained, then the number of ground flashovers can be computed from the TD level. From the past records and the past experience, it is found that

$$N_g = (0.1 \text{ to } 0.2) \text{ TD/strokes/km}^2\text{-year.}$$

It is reported that TD is between 5 and 15 in Britain, Europe and Pacific west of North America, and is in the range of 30 to 50 in Central and Eastern states of U.S.A. A much higher level is reported from South Africa and South America. No literature is available for the different regions in India, but a value of 30 to 50 may be taken for the coastal areas and for the central parts of India.

8.1.5 Travelling Waves on Transmission Lines

Any disturbance on a transmission line or system such as sudden opening or closing of a line, a short circuit or a fault results in the development of over voltages or over currents at that point. This disturbance propagates as a travelling wave to the ends of the line or to a termination, such as, a sub-station. Usually these travelling waves are high frequency disturbances and travel as waves. They may be reflected, transmitted, attenuated or distorted during propagation until the energy is absorbed. Long transmission lines are to be considered as electrical networks with distributed electrical elements. In Fig. 8.13, a typical two-wire transmission line is shown along with the distributed electrical elements R, L, C and G.

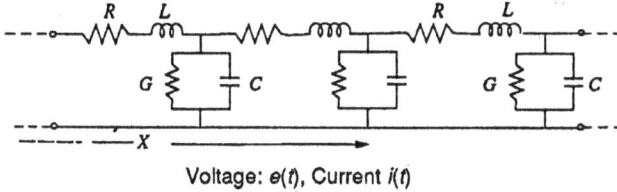

Voltage: $e(t)$, Current $i(t)$

R — Resistance per unit length C — Capacitance per unit length
L — Inductance per unit length G — Leakage conductance per unit length

Fig. 8.13   Distributed characteristic of a long transmission line

The propagation of any travelling wave, say a voltage wave can be analysed by considering an elemental length of the line dx. The voltage drop in the positive x-direction in the elemental length dx due to the inductance and resistance is

$$dV = \frac{\delta V}{\delta x} \cdot dx = iRdx + \frac{\delta \psi}{\delta t} \tag{8.7}$$

Here, $\delta \psi$ is the change of flux linkages and is equal to $i.L.dx$, where i the current through the line.

$$\therefore \qquad dV = iR \cdot dx + L \frac{\delta}{\delta t} (i \cdot dx)$$

$$= \left(R + L \frac{\delta}{\delta t} \right) i \cdot dx \tag{8.8}$$

The shunt current through the leakage conductance (G) and capacitance (C) is

$$di = \frac{\delta i}{\delta x} \cdot dx = VG \cdot dx + \frac{\delta}{\delta t} (\Delta \varphi) \tag{8.9}$$

Here, $\Delta \varphi$ is the change in electrostatic field flux and is equal to $VC \cdot dx$, where V is the potential at the point x.

$$di = VG \cdot dx + \frac{\delta}{\delta t} (VC \cdot dx)$$

$$= \left(G + C \frac{\delta}{\delta t} \right) V \cdot dx$$

Hence, the above equations can be written as

$$\frac{dV}{dX} = R + L \left(\frac{di}{dt} \right) \text{ and } \frac{di}{dx} = G + C \left(\frac{dV}{dt} \right)$$

Taking Laplace transform with respect to the time variable t, the equations can be put in the operation form as

$$\frac{\delta V}{\delta x} = (R + Ls) \cdot i = Z \cdot i,$$

and

$$\frac{\delta i}{\delta x} = (G + Cs) V = YV$$

where,

$$Z = (R + Ls) \text{ and } Y = (G + Cs)$$

Eliminating i and V from above equations and differentiating w.r.t. x, we get

$$\frac{\delta^2 V}{\delta x^2} = Z \frac{\delta i}{\delta x}$$

$$= YZV = \gamma^2 V \qquad (8.10)$$

$$\frac{\delta^2 i}{\delta x^2} = Y \frac{\delta V}{\delta x}$$

$$= YZi = \gamma^2 i \qquad (8.10a)$$

where the product $YZ = \gamma^2 = RG + (RC + LG)s + LCs^2$

The above two equations are called wave equations or telegraphic equations. The solutions for the above equations can be written in the form :

$$V = e^{\gamma x} f_1(t) + e^{-\gamma x} f_2(t) \qquad (8.11)$$

and

$$i = -\sqrt{Y/Z} \, [e^{\gamma x} f_1(t) - e^{-\gamma x} f_2(t)] \qquad (8.11a)$$

where $f_1(t)$ and $f_2(t)$ are any arbitrary functions that satisfy the boundary conditions. The operator γ is simplified as

$$\gamma = [(R + Ls)(G + Cs)]^{1/2} = \frac{1}{\sqrt{LC}} (s + R/L)^{1/2} (s + G/C)^{1/2}$$

$$= 1/v = [(s + \alpha)^2 - \beta^2]^{1/2} \qquad (8.12)$$

where,

$$v = 1/\sqrt{LC} = \text{propagation velocity} \qquad (8.12a)$$

$$\alpha = 1/2 \, (R/L + G/C) = \text{attenuator constant} \qquad (8.12b)$$

and

$$\beta = 1/2 \, (R/L - G/C) = \text{wavelength constant} \qquad (8.12c)$$

Also

$$\sqrt{Y/Z} = [G + Cs/R + Ls]^{1/2}$$

$$= \sqrt{C/L} \, [(s + \alpha - \beta)/(s + \alpha + \beta)]^{1/2} = Y(s)$$

$Y(s)$ is called the surge admittance, the reciprocal of which

$$Z(s) = \sqrt{L/C} \, [(s + \alpha + \beta)/(s + \alpha - \beta)]^{1/2} \qquad (8.13)$$

$Z(s)$ is called the surge impedance of the transmission line.

8.1.5.1 Classification of Transmission Lines

Transmission lines are usually classified as
- (a) lines with no loss or ideal lines,
- (b) lines without distortion or distortionless lines,
- (c) lines with small losses, and
- (d) lines with infinite and finite length defined by all the four parameters.

(a) Ideal lines : A line is said to be an ideal line if $R = 0$ and $G = 0$. In this case, the surge impedance of the line is $Z = L/C$ and the surge admittance of the line is $Y = C/L$. The solution for the voltage and current waves for this type of line is obtained as

$$V = f_1(t + x/v) + f_2(t - x/v) \tag{8.14}$$

$$i = Y f_2(t - x/v) - f_2(t + x/v) \tag{8.14a}$$

The function $f_2(t - x/v)$ represents the forward travelling wave and the function $f_1(t + x/v)$ represents the backward travelling wave. The propagation velocity for either wave is given as

$$v = 1/\sqrt{L/C}$$

•(b) Distortionless lines : If, for any line $R/L = G/C = \alpha$ then, $\gamma = \sqrt{ZY} = LC(Ps + \alpha)$ and the surge impedance is

$$Z(s) = \sqrt{L/C} \tag{8.15}$$

For these conditions, the solution for the wave equations (8.10) and (8.10a) will be modified as

$$V = \exp(ax/v) f_1(t + x/v) + \exp(-ax/v) f_2(t - x/v) \tag{8.16}$$

and
$$i = (1/Z) [\exp(-ax/v) f_2(t - x/v) - \exp(ax/v)$$
$$f_1(t - x/v) \tag{8.17}$$

The voltage and current waves for an ideal line represented by equations (8.14) and (8.14a) will be of the same shape. However, for a distortionless line, their magnitudes will decrease by the factor $\exp(\pm ax/v)$, which is the reduction with respect to the distance x. These solutions can be rearranged by putting

$$\lambda_1 = t + x/v \text{ and } \lambda_2 = t - x/v$$

Under these conditions,

$$
\begin{aligned}
V &= \exp[\alpha(\lambda_1 - t)] \cdot f_1(\lambda_1) + \exp[\alpha(\lambda_2 - t)] \cdot f_2(\lambda_2) \\
&= \exp(-\alpha t) [\exp(\lambda_1) f_1(\lambda_1) + \exp(\lambda_2) f_2(\lambda_2)] \\
&= \exp(-\alpha t) [f_3(\lambda_1) + f_4(\lambda_2)] \\
&= \exp(-\alpha t) [f_3(t + x/v) + f_4(t - x/v)] \tag{8.16a}
\end{aligned}
$$

similarly,
$$i = \frac{\exp(-\alpha t)}{Z} [f_4(t - x/v) - f_3(t + x/v)] \tag{8.17a}$$

Thus the attenuation can be either with respect to the distance, x or the time, t. Equations (8.16) and (8.17) are more useful if the voltage distribution at $t = 0$ (initial condition) is specified.

(c) Line with small losses : In this type of lines, the time constants of the line are large i.e. R/L and G/C are small. Then γ can be approximated to be equal to $(s + \alpha/v)$, and $Y(s)$ to be equal to $\sqrt{C/L} (1 - \beta/s)$. Under these conditions, the solutions for the voltage and the current waves become

$$V = \exp(\alpha x/v) \cdot f_1(t + x/v) + \exp(-\alpha x/v) \cdot f_2(t + x/v)$$

and
$$i = Y(s) \cdot V \tag{8.18}$$

$$= Y(s) \, [\exp(-\alpha x/v) \cdot f_2'(t - x/v) + \exp(\alpha x/v)$$

$$f_1(t + x/v)] + V\beta \, [\exp(\alpha x/v) \int_{-x/v}^{t} f_1 \; (t + x/v) \, dt -$$

$$\exp(-\alpha x/v) \cdot \int_{x/v}^{t} f_2(t - x/v) \cdot dt \qquad (8.19)$$

The voltage equation (8.18) is similar to equation (8.14) and the current equation (8.19) is similar to equation (8.14a), along with the other expression containing the time integral of the functions f_1 and f_2.

In a line with small losses, voltage solution shows that the voltage wave is the same as that in the case of distortionless line. However, the solution for the current wave differs by an amount equal to the energy loss in the resistance of the line. Thus, this solution valid only for small intervals of time, i.e. for small values of $(t + x/v)$.

(d)   Exact solution for lines of finite or infinite length defined by all the four parameters : The exact solution of the wave equation of this type of lines is quite complex and is normally of little practical importance. However, some of the inferences that can be drawn are:

   (i)   the current and the voltage wave are dissimilar
   (ii)   the attenuation and distortion due to normal line resistance and leakage conductance are of little consequence
   (iii)   the surge impedance $Z(s) = e(s)/i(s)$ is a complex function and is not uniquely defined

8.1.5.2 Attenuation and Distortion of Travelling Waves

As a travelling wave moves along a line, it suffers both attenuation and distortion. The decrease in the magnitude of the wave as it propagates along the line is called attenuation. The elongation or change of wave shape that occurs is called distortion. Sometimes, the steepness of the wave is reduced by distortion. Also, the current and voltage wave shapes become dissimilar even though they may be the same initially. Attenuation is caused due to the energy loss in the line and distortion is caused due to the inductance and capacitance of the line. The energy loss may be in the conductor resistance as modified by the skin effect, changes in ground resistance, leakage resistance and non-uniform ground resistances etc. The changes in the inductance are due to the skin effect, the proximity effect and the non-uniform distribution effect of currents, and the nearness to steel structures such as transmission towers. The variation is capacitance is due to capacitance change in the insulation nearest to the ground structures etc. If the wave shapes remain approximately the same, then the surge impedance can be taken to be constant, in which case the attenuation can be estimated. The other factor that contributes for the attenuation and distortion is the corona on the lines. For distortionless lines, the attenuation is approximated as a loss function $\varphi(V)$, considering that the attenuation is due to the energy lost per unit length of the line in the resistance as the wave travels. It can be shown that

$$\varphi(V) = -C \frac{dV^2}{dt} \qquad (8.20)$$

For different line conditions, $\varphi(V)$ and attenuation are as follows:

(i) For lines having all the parameters R, L, G and C

$$\varphi(V) = [(RC + LG)/L]V^2 \tag{8.21}$$

and $\qquad dV/dt = -\alpha V$, where $\alpha = (1/2)\,[(R/L) + (G/C)]$ $\qquad$ (8.22)

From the equation (8.20), α is called the attenuation factor.

$$\text{Hence } \quad V = V_0 \exp(-\alpha t) \tag{8.23}$$

where the initial voltage at $t = 0$ is taken as V_0.

(ii) *The Skilling formula:* If φV is assumed to be equal to $\beta(V - V_C)$, where V_C is critical corona voltage; then

$$dV/dt = -\beta/2c\,((V - V_c/V))$$

and if the initial voltage at $t = 0$ is taken as V_0, then

$$(V_0 - V) + V_c \ln\,[(V_0 - V_c)/(V - V_c)] = (\beta/2c)t \tag{8.24}$$

(iii) *The quadratic formula :* If φV is assumed to vary as

$$(V - V_c)^2, \text{ then, } dV/dt = (-\gamma/2c)\,[(V_0 - V_c)/(V - V_c)]^2$$

Integrating the above equation, we get

$$[(V_0 - V) \cdot V_c/(V_0 - V_c)\,(V - V_c)] + \ln\,[(V_0 - V_c)/(V - V_c)] = \gamma t/2c \tag{8.25}$$

(iv) *The Foust and Manger formula :* Here, φV is assumed to be equal to λV^3, so that

$$\frac{dV}{dt} = (-\lambda/2c)V^2$$

It follows from the above equation that

$$V = V_0/(1 + KV_0 t) \tag{8.26}$$

where $\qquad K = \lambda/2c.$

The Foust and Manger formula is simpler for the purpose of calculation and the exceptional attenuation obtained in equation (8.21) can be used easily for all mathematical operations.

(a) *Attenuation due to corona :* The effect of corona is to reduce the crest of the voltage wave under propagation, limiting the peak value to the critical corona voltage. Hence, the excess voltage above the critical voltage will cause power loss by ionising the surrounding air. This mechanism is explained as follows: the travelling wave is divided into a number of sections corresponding to different voltage levels, each voltage level corresponding to a different velocity of propagation since each lamination ionises a different diameter of the air layer surrounding the conductor and hence have different capacitances. Hence, a distortion is caused in the wave shape. This explanation ignores the power loss due to corona. The mechanism of corona power loss as explained by Skilling is as follows :

The charges liberated by the ionisation of air surrounding the conductor takes such positions on the conductor so as to make the critical field intensity (gradient) for air to reach values that cannot be exceeded. The supply of space charge to the above regions continue as long as the voltage is increasing and the energy is supplied. After the crest of wave is reached and the wave is trailing, the space charge remains constant in magnitude. The only energy loss caused during this period is due to the diffusion of ions, a process which is very slow. Based on this explanation Skilling gives the formula for corona power loss as $P = K(V - V_c)^2$, where V_c is the critical corona voltage and K is a constant.

8.1.5.3 Reflection and Transmission of Waves at Transition Points

Whenever there is an abrupt change in the parameters of a transmission line, such as an open circuit or a termination, the travelling wave undergoes a transition, part of the wave is reflected or sent back and only a portion is transmitted forward. At the transition point, the voltage or current wave may attain a value which can vary from zero to two times its initial value. The incoming wave is called the incident wave and the other waves are called the reflected and transmitted waves at the transition point. Such waves are formed according to the Kirchhoff's laws and they satisfy the line differential equations. In Fig. 8.14, is shown a typical general transition point.

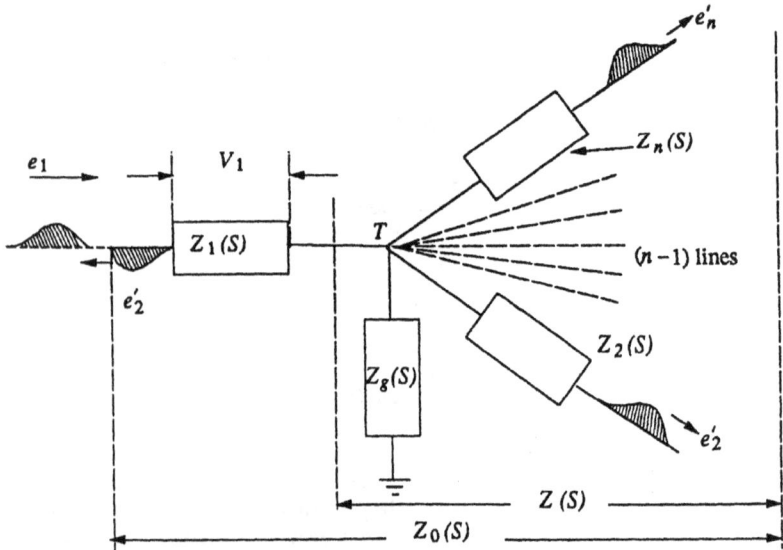

Fig. 8.14 Transition point *(T)* and the propagation of the wave at the transition point

Let the transformed equations for line impedances be $Z_1(s)$, $Z_2(s)$ ... $Z_n(s)$, and the impedance to the ground $Z_g(s)$ be the total impedance as seen from the transition point $Z(s)$ and looking beyond the transition point, $Z_0(s)$. Taking the transition point T as reference, the distance along the line away from the transition point or the line is taken as negative, so that the incoming wave towards the transition point is counted as travelling in the positive direction. Taking the lines as lossless lines, the relations can be written as follows :

$$
\left.
\begin{array}{l}
e_1/i_1 \;= Z_1 \text{ (incident wave)}, \\
e_1'/i_1' = - Z_1 \text{ (reflected wave)} \\
e''_k/i''_k = Z_k \text{ (transmitted wave)}
\end{array}
\right\}
\qquad (8.27)
$$

and

In the above equation, e, i etc. are the unprimed quantities for the incident waves, single primed quantities e', i' etc. are for the reflected waves and the double primed quantities e'', i'' etc. are for the transmitted waves.

At the junction point T, $i_0 = i_1 + i'_1$

and

$$
e_0 = \; e_1 + e'_1 = Z_0(s) \cdot i_0 \qquad (8.28)
$$

From equations (8.27) and (8.28), the reflected voltage wave is

$$
e'_1 = \; [\{Z_0(s) - Z_1\}/\{Z_0(s) + Z_1\}] \cdot e_1 \qquad (8.29)
$$

and the reflected current wave is

$$
i'_1 = \; [\{Z_0(s) - Z_1\}/\{Z_0(s) + Z_1\}] \cdot i_1 \qquad (8.30)
$$

The junction voltage e_0 and the total current i_0 are given by

$$
e_0 = \; [2Z_0(s)/(Z_0(s) + Z_1)] \cdot e_1 \text{ and } i_0 = [2e_1/(Z_0(s) + Z_1)]
$$
$$
(8.31)
$$

The coefficients multiplying the incident quantities are called the reflection coefficients.

The impedance of any line is given by $Z_k(s) + Z_k$.

Hence the total impedance $Z_0(s) = Z_1(s) + Z(s)$

i.e.

$$
Z_0(s) = \; Z_1(s) + 1/\left[(1/Z_g(s)) + \left(\sum_{k=2}^{n} (1/Z_k(s) + Z_k)\right)\right] \quad (8.32)
$$

Solving the above equations, the junction potential e_0 can be written as

$$
e_0 = Z(s) \cdot i_0 = [2Z(s)/Z_0(s) + Z_1)] \cdot e_1
$$

and the ground current, $i_g = e_1/Z_g(s)$

$$
= \; 2[\{Z(s)/(Z_g(s)) \cdot \{e_1/(Z_0(s) + Z_1)\}] \qquad (8.33)
$$

The transmitted voltage and current waves through any line K $(2 < K < n)$ can be expressed as

$$
e''_k = \; [\{2Z(s)/(Z_0(s) + Z_1)\} \cdot \{Z_K/(Z_K(s) + Z_K)\}] \cdot e_1
$$
$$
(8.34)
$$

and

$$
i_k = \; [\{2Z(s)/(Z_0(s) + Z_1)\} \cdot \{e_1/(Z_K(s) + Z_K)\}]
$$

The voltage drop V_K across any lumped $Z_K(s)$ is given by

$$
V_K = \; [\{2Z(s)/(Z_0(s) + Z_1)\} \cdot \{Z_K(s)'(Z_K(s) + Z_K)\}] \cdot e_1
$$
$$
(8.35)
$$

All the above functions are in the transformed form as functions of 's' (the Laplace transform operator), and the inverse transform gives the desired time functions.

In simpler cases where the junction consists of two impedances only, the reflection and transmission coefficients become simpler and are given by reflection coefficient

$$\gamma = (Z_2 - Z_1)/(Z_2 + Z_1)$$

and, the transmission coefficient is $(1 + \gamma)$ for voltage waves. The reflected and transmitted waves are then given by

$$e' = \gamma e, i' = -\gamma i, \ e'' = (1 + \gamma) \cdot e \text{ and } i'' = (1 - \gamma) \cdot i \qquad (8.36)$$

It may be easily verified that

$$e/i = Z_1, e'/i' = -Z_1, \text{ and } e''/i'' = Z_2 \qquad (8.37)$$

Solutions for lines terminated with lumped impedances can be easily worked out and the solutions for few cases are given in section 8.1.6. The above analysis of the travelling waves indicate how the waves are modified at transition points. There can be doubling effect at the junction point which contributes to the over voltages in power system networks. In many overvoltage calculations, travelling wave analysis will be useful for overvoltage and over current calculations.

(a) Successive reflections and lattice diagrams: In many problems involving short cable lengths, or lines tapped at intervals, the travelling waves encounter successive reflections at the transition point. It is exceedingly difficult to calculate the multiplicity of these reflections and in his book, Bewley has given the lattice or time-space diagrams from which the motion of reflected and transmitted waves and their positions at every instant can be obtained. The principles observed in the lattice diagrams are as follows:

(i)   all waves travel downhill, i.e. into the positive time

(ii)   the position of the wave at any instant is given by means of the time scale at the left of the lattice diagram

(iii)   the total potential at any instant of time is the superposition of all the waves which arrive at that point until that instant of time, displaced in position from each other by time intervals equal to the time differences of their arrival

(iv)   attenuation is included so that the amount by which a wave is reduced is taken care of and

(v)   the previous history of the wave, if desired can be easily traced. If the computation is to be carried out at a point where the operations cannot be directly placed on the lattice diagram, the arms can be numbered and the quantity can be tabulated and computed.

The above comprehensive description can be understood by considering the example shown in Fig. 8.15.

In the arrangement shown in the figure, there are two junctions 1 and 2. The travel times for the waves are different through Z_1, Z_2, and Z_3. The lines with surge impedances Z_1, Z_2, and Z_3 are connected on either side of the junctions. Let α and β be the attenuation coefficients for the two sections Z_2 and Z_3. Let a_1 and a'_1 be the reflection coefficients for the waves approaching from the left and the right at junction 1, and a_2 and a'_2 be the corresponding reflection coefficients at junction 2. Similarly, let b_1 and b'_1 be the transmission coefficients for the waves that approach from the left and the right at junction 1, and the corresponding coefficients be b_2 and b'_2 at junction 2. To construct the lattice diagram, the position 0 is taken when the wave coming from Z_1 reaches junction 1. Junction 2 is taken to scale at the time interval

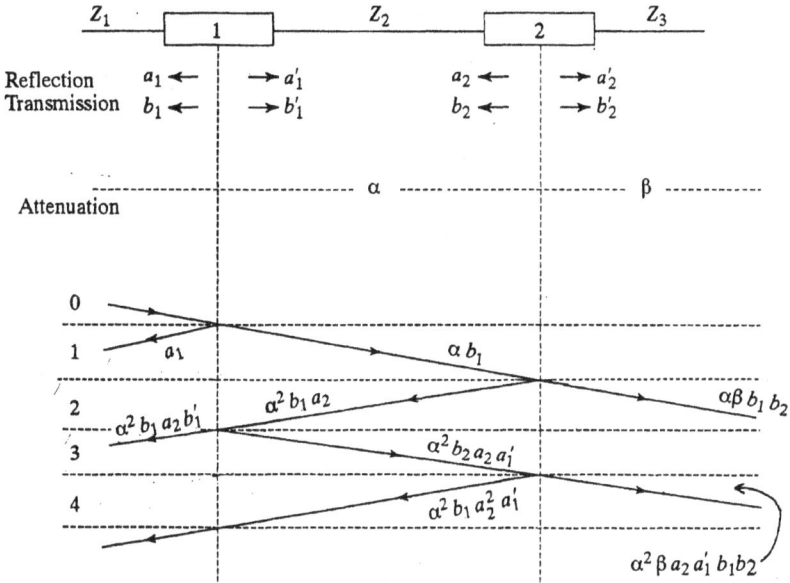

Fig. 8.15   Reflection lattice of a travelling wave

equal to the travel time through the line Z_2 between the junctions 1 and 2. The diagram is drawn by choosing a suitable time scale. The reflection and the transmission factors are marked as shown in the figure. The process of calculation is indicated on the slope of the lines in the diagram. The process can be continued for up to the required time interval. A numerical example illustrating the use of the above technique for the computation is given under worked examples (Example 8.6).

8.1.6   Behaviour of Rectangular Travelling Wave [Unit Step Function $U(t)$] at Transition Points—Typical Cases

Reflection and transmission of a travelling wave at junction points of unequal impedances in a transmission line are of great importance in transmission systems. Depending on the type of impedance at transition points, the travelling wave is modified, and sometimes a voltage rise or build-up of voltage can occur. The following cases are of practical importance and as such are discussed here. The solution is obtained using the Laplace Transforms rather than using operational calculus, as many of the readers may not be familiar with the Heaviside Operational Calculus.

Case (i): Open ended transmission line of surge impedance Z:
Let the voltage of the travelling wave incident on the line be

$$e = E U(t)$$

then, $\qquad Z_1 = Z \text{ and } Z_2 = \infty$

$\therefore$ coefficient of reflection $\quad \Gamma = \dfrac{(Z_2 - Z_1)}{(Z_2 + Z_1)}$

$$= \dfrac{(1 - Z_1/Z_2)}{(1 + Z_1/Z_2)}$$

Substituting $\qquad \Gamma = \dfrac{(1 - Z/\infty)}{(1 + Z/\infty)}$

$$= 1$$

$\therefore$ voltage of the reflected wave, $e' = \Gamma e = e = E\,U(t)$ and the voltage of the transmitted wave,

$$e'' = (1 + \Gamma)e = 2e = 2\,E\,U(t)$$

Hence the voltage at the open end rises to double its value.

Case (ii): Short circuited line:

Voltage of the wave, $e = E\,U(t)$

   Surge impedances $Z_1 = Z$ and $Z_2 = 0$

$\therefore$ coefficient of reflection, $\Gamma = \dfrac{0 - Z}{0 + Z}$

$$= -1$$

$\therefore$ voltage of the reflected wave, $\quad e' = \Gamma \cdot e$

$$= -E\,U(t)$$

The voltage of the transmitted wave, $e'' = (1 + \Gamma)e = 0$

Further i', the magnitude of the reflected current wave

$$= \left| -\dfrac{e'}{Z} \right|$$

$$= \dfrac{E\,U(t)}{Z} \text{ (the magnitude of the incident}$$

$$\text{current wave)}$$

The total current at the junction point

$$i_0 = (i + i') = 2i$$

Thus, the current at the junction point rises to double the value of the incident current wave.

Case (iii): Line terminated with a resistance equal to the
*   surge impedance of the line:*

In this case, $Z_1 = Z$ and $Z_2 = R = Z$

   $\therefore$ coefficient of reflection,

$$\Gamma = \dfrac{(R - Z)}{(R + Z)} = 0$$

   $\therefore$ voltage of the reflected wave, e' is $\Gamma e = 0$

The voltage of the transmitted wave is $(1 + \Gamma)e = e$

Thus, there is no reflected wave. There is no discontinuity of the line, and the travelling wave proceeds without reflection and disappears. It is very important to note that there will be no reflections at the junction, if a transmission line or cable is terminated with a resistance equal to the surge impedance of the line or cable.

Case (iv): Line terminated with a capacitor:

In this case, $e = E\,U(t)$, $Z_1 = Z$, and $Z_2 = \dfrac{1}{Cs}$

where s is the Laplace transform operator.

The coefficient of reflection $\Gamma = \dfrac{(Z_2 - Z_1)}{(Z_2 + Z_1)}$

$$= \frac{\left(\dfrac{1}{Cs} - Z\right)}{\left(\dfrac{1}{Cs} + Z\right)}$$

$$= \frac{(1 - CZs)}{(1 + CZs)}$$

The voltage of the reflected wave, $e' = \Gamma e$

Taking the Laplace transform, $e'(s) = \dfrac{(1 - CZs)}{(1 + CZs)}\dfrac{E}{s}$

$$= \left[1 - \frac{2CZs}{1 + CZs}\right]\frac{E}{s}$$

$$= \left[\frac{1}{s} - \frac{2CZs}{1 + CZs}\right]E$$

Taking the inverse transform

$$e' = [1 - 2\exp(-2\exp(-t/CZ)]\,E\,U(t)$$

The voltage of the Laplace transformed transmitted wave,

$$e''(s) = (1 + \Gamma)\,e(s)$$

$$e''(S) = (1 + \Gamma)\frac{E}{s}$$

$$= \left[1 + \frac{(1 - CZs)}{(1 + CZs)}\right]\frac{E}{s}$$

$$= \left[2 - \frac{(2CZs)}{(1 + CZs)}\right]\frac{E}{s}$$

$$= \left[\frac{1}{s} - \frac{CZ}{(1 + CZs)}\right]\frac{2E}{s}$$

Taking the inverse transforms

$$e'' = 2\left[1 - \exp\left(-\frac{t}{CZ}\right)\right]E\,U(t)$$

From the expression for e'', it can be inferred that the steepness of the front is reduced and the wave rises slowly in an exponential manner. The capacitor initially acts as a short circuit and is charged through the line impedance Z. The voltage at the junction point finally rises to twice the magnitude of the incident wave.

Case (v): Transmission terminated by an inductance L:
In this case, $e = E\ U(t)$

$$Z_1 = Z, \text{ and } Z_2 = Ls$$

The coefficient of reflection $\Gamma = \dfrac{(Ls - Z)}{(Ls + Z)} = \dfrac{\left(s - \dfrac{Z}{L}\right)}{\left(s + \dfrac{Z}{L}\right)}$

Voltage of the reflected wave, $e' = \Gamma e$

∴ $$e'(s) = \dfrac{\left(s - \dfrac{Z}{L}\right)}{\left(s + \dfrac{Z}{L}\right)} \dfrac{E}{s}$$

$$= \left[1 - \dfrac{2\dfrac{Z}{L}}{s + \dfrac{Z}{L}}\right] \dfrac{E}{s}$$

$$= \left(-\dfrac{1}{s} + \dfrac{2}{s + \dfrac{Z}{L}}\right) E$$

Taking inverse transforms,

$$e' = -\left[1 - 2\exp\left(-\dfrac{Z}{L}t\right)\right] E\ U(t)$$

Voltage of the transformed transmitted wave, $e''(s) = (1 + \Gamma)\ e(s)$

∴ $$e''(s) = \left[1 + \dfrac{\left(s - \dfrac{Z}{L}\right)}{\left(s + \dfrac{Z}{L}\right)}\right] \dfrac{E}{s}$$

$$= \dfrac{2E}{\left(s + \dfrac{Z}{L}\right)}$$

∴ $$e'' = 2E\ \exp\left(-\dfrac{Z}{L}t\right)$$

The voltage across the inductor initially rises to double the value of the incident wave and decays exponentially. This is of importance when long lines are terminated with inductors or transformers on open circuit.

Case (vi): Line having a series inductor:

Let the surge impedance of the line be Z before and after the series inductor L. Considering the junction point after the inductor L, $e = E\ U(t)$, $Z_1 = (Z + Ls)$, and $Z_2 = Z$.

The coefficient of reflection
$$\Gamma = \frac{(Z_2 - Z_1)}{(Z_2 + Z_1)}$$

$$= -\frac{Ls}{2Z + Ls}$$

Voltage of the transformed reflected wave, $e'(s) = \Gamma\ e(s)$

$$\therefore \qquad e'(s) = -\frac{Ls}{2Z + Ls}\frac{E}{s}$$

$$= -\frac{EL}{2Z + Ls}$$

$$= -\frac{1}{\left(s + 2\dfrac{Z}{L}\right)}E$$

Taking the inverse transform,

$$e' = -E\ \exp\left(-\frac{2Z}{L}t\right)$$

$$e'' = (1 + \Gamma)e = \left[\frac{E}{s} - \frac{E}{\left(s + 2\dfrac{Z}{L}\right)}\right]$$

$$\therefore \qquad e'' = \left[1 - \exp\left(-\frac{2Z}{L}t\right)\right]E\ U(t)$$

As seen from the expression for e'', the steepness of the propagated wave through the inductor, i.e. the transmitted wave into the second portion of the line is reduced. The series inductor produced the same effect as that of a shunt capacitor on a transmission line.

Case (vii): Line terminated with a transformer
(taken as an L–C parallel combination):

$$e = E\ U(t)$$
$$Z_1 = Z$$
$$Z_2 = L\text{ and }C\text{ in parallel}$$

i.e.
$$Z_2 = \frac{\left(LS\dfrac{1}{Cs}\right)}{\left(LS+\dfrac{1}{Cs}\right)}$$

$$= \frac{s}{C}\left(\frac{1}{s^2+\dfrac{1}{LC}}\right)$$

∴ coefficient of reflection,

$$\Gamma = \frac{(Z_2+Z_1)}{(Z_2+Z_1)} = \frac{\left(\dfrac{s}{C(s^2+1/LC)}-Z\right)}{\left(\dfrac{s}{C(s^2+1/LC)}+Z\right)}$$

$$= -\frac{(s^2-s/CZ+1/LC)}{(s^2+s/CZ+1/LC)}$$

∴
$$1+\Gamma = \frac{\left(2\dfrac{s}{CZ}\right)}{\left(s^2+\dfrac{s}{CZ}+\dfrac{1}{LC}\right)}$$

Let $\dfrac{1}{CZ}=\alpha$ and $\dfrac{1}{LC}=\omega_0^2$ so that Γ can be written as

$$\Gamma = -\frac{(s^2-\alpha s+\omega_0^2)}{(s^2+\alpha s+\omega_0^2)}$$

and
$$1+\Gamma = \frac{2\alpha s}{(s^2+\alpha s+\omega_0^2)}$$

The reflected wave,
$$e'(s) = \Gamma e(s) = -\frac{(s^2-\alpha s+\omega_0^2)}{(s^2+\alpha s+\omega_0^2)}\left(\frac{E}{s}\right)$$

$$= -\left(+\frac{1}{s}-\frac{2\alpha}{(s^2+\alpha s+\omega_0^2)}\right)E$$

Taking the inverse transform,

$$e' = -\left\{+1\frac{2\alpha}{n-m}\Big[\exp(-mt)-\exp(-nt)\Big]\right\}E\,U(t)$$

$$\text{if } \omega_0^2 < \left(\frac{\alpha}{2}\right)^2$$

where n and m are roots of $(s^2+\alpha s+\omega_0^2)$.

Or $e' = -E\,U(t)\left[+1 - \dfrac{2\alpha}{\omega_0^2 - (\alpha/2)^2} \times \exp(-\alpha/2)\,t\right.$

$$\left. \times \sin\sqrt{\omega_0^2 - \left(\frac{\alpha}{2}\right)^2}\,t\,\right],\text{ if } \omega_0^2 > (\alpha/2)$$

The transmitted wave, $\quad e''(s) = (1 + \Gamma)\,e(s) = \dfrac{2\alpha E}{S^2 + \alpha\,s + \omega_0^2}\,\dfrac{E}{s}$

$$= \dfrac{2\alpha E}{S^2 + \alpha\,s + \omega_0^2}$$

Taking the inverse transform,

$$e'' = \frac{2\alpha}{n-m}\left[\exp(-mt) - \exp(-nt)\right] EU(t),\text{ if } \omega_0^2 < (\alpha/2)^2$$

and $\quad e'' = \left[\dfrac{2\alpha}{\omega_0^2 - (\alpha/2)^2} \exp(-\alpha/2t) \sin\sqrt{\omega_0^2 - (\alpha/2)^2}t\right] EU(t)$ if $\omega_0^2 > (\alpha/2)^2$

The transmitted wave reaching the transformer will be either a double exponential (standard impulse type) or a damped sinusoidal wave and the steepness of the wave front gets reduced.

The above analysis shows that a travelling wave is modified at the transition points, and the steepness of the wave front is reduced in certain cases. There can be doubling effect at the junction points such as an open ended line or an inductance termination. These also contribute for further overvoltages at the transition points in a transmission system.

8.2 OVERVOLTAGE DUE TO SWITCHING SURGES, SYSTEM FAULTS AND OTHER ABNORMAL CONDITIONS

8.2.1 Introduction

Till the time when the transmission voltages were about 220 kV and below, over voltages due to lightning were of very high order and over voltages generated inside the system were not of much consequence. In later years, with increase in transmission voltages, (400 kV and above) the overvoltages generated inside the system reached the same order of magnitude as those of lightning over voltages, or higher. Secondly, the overvoltages thus generated last for longer durations and therefore are severe and more dangerous to the system. Unlike the lightning overvoltages, the switching and other types of overvoltages depend on the normal voltage of the system and hence increase with increased system voltage. In insulation co-ordination, where the protective level of any particular kind of surge diverter is proportional to the maximum voltage, the insulation level and the cost of the equipment depends on the magnitudes of these overvoltages. In the EHV range, it is the switching surge and other types of overvoltages that determine the insulation level of the lines and other equipment and consequently, they also determine their dimensions and costs.

8.2.2 Origin of Switching Surges

The making and breaking of electric circuits with switchgear may result in abnormal overvoltages in power systems having large inductances and capacitances. The overvoltages may go as high as six times the normal power frequency voltage. In circuit breaking operation, switching surges with a high rate of rise of voltage may cause repeated restriking of the arc between the contacts of a circuit breaker, thereby causing destruction of the circuit breaker contacts. The switching surges may include high natural frequencies of the system, a damped normal frequency voltage component, or the restriking and recovery voltage of the system with successive reflected waves from terminations.

8.2.3 Characteristics of Switching Surges

The waveshapes of switching surges are quite different and may have origin from any of the following sources.

(i) De-energizing of transmission lines, cables, shunt capacitor, banks, etc.
(ii) Disconnection of unloaded transformers, reactors, etc.
(iii) Energization or reclosing of lines and reactive loads.
(iv) Sudden switching off of loads.
(v) Short circuits and fault clearances.
(vi) Resonance phenomenon like ferro-resonance, arcing grounds, etc. Typical waveshapes of the switching surges are given in Figs. 8.16a to (e).

From the figures of the switching surges it is clear that the overvoltages are irregular (oscillatory or unipolar) and can be of high frequency or power frequency with its harmonics. The relative magnitudes of the overvoltages may be about 2.4 p.u. in the case of transformer energizing and 1.4 to 2.0 p.u. in switching transmission lines.

8.2.3.1 Switching Overvoltages in EHV and UHV Systems

The insulation has the lowest strength for switching surges with regard to long air gaps. Further, switching overvoltages are of relatively higher magnitudes as compared to the lightning overvoltages for UHV systems. Overvoltages are generated in EHV systems when there is a sudden release of internal energy stored either in the electrostatic form (in the capacitance) or in the electromagnetic form (in the inductance). The different situations under which this happens are summarised as

(i) interruption of low inductive currents (current chopping) by high speed circuit breakers. This occurs when the transformers or reactors are switched off

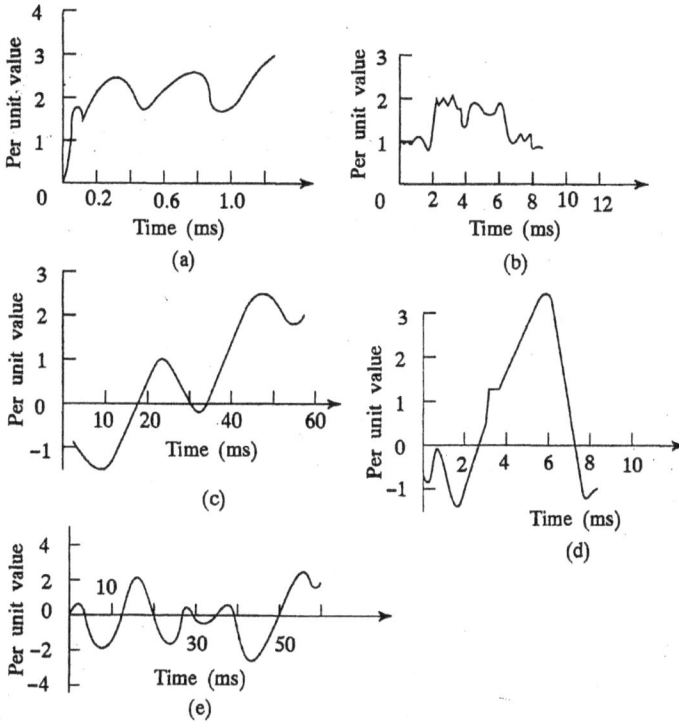

(a) Recovery voltage after fault clearing
(b) Fault initiation
(c) Overvoltage at the line end after fault clearing
(d) Energization of long transmission line
(e) Overvoltage at line end during (d)

Fig. 8.16   Typical waveshapes of switching surge voltages

(ii) interruption of small capacitive currents, such as switching off of unloaded lines etc.
(iii) ferro-resonance condition

This may occur when poles of a circuit breaker do not close simultaneously

(iv) energization of long EHV or UHV lines.

Transient overvoltages in the above cases can be of the order of 2.0 to 3.3 p.u. and will have magnitudes of the order of 1200 kV to 2000 kV on 750 kV systems. The duration of these overvoltages varies from 1 to 10 ms depending on the circuit parameters. It is seen that these are of comparable magnitude or are even higher than those that occur due to lightning. Sometimes the overvoltages may last for several cycles. The other situations of switching that give rise to switching overvoltages of shorter duration (0.5 to 5 ms) and lower magnitudes (2.0 to 2.5 p.u.) are:

(a) single pole closing òf circuit breaker
(b) interruption of fault current when the L-G or L-L fault is cleared
(c) resistance switching used in circuit breakers
(d) switching lines terminated by transformers
(e) series capacitor compensated lines
(f) sparking of the surge diverter located at the receiving end of the line to limit the lightning overvoltages

The overvoltages due to the above conditions are studied or calculated from

(a) mathematical modelling of a system using digital computers
(b) scale modelling using transient network analysers
(c) by conducting field tests to determine the expected maximum amplitude of the overvoltages and their duration at different points on the line. The main factors that are investigated in the above studies are
  (i) the effect of line parameters, series capacitors and shunt reactors on the magnitude and duration of the transients
  (ii) the damping factors needed to reduce the magnitude of overvoltages
  (iii) the effect of single pole closing, restriking and switching with series resistors or circuit breakers on the overvoltages, and
  (iv) the lightning arrester sparkover characteristics.

It is necessary in EHV and UHV systems to control the switching surges to a safe value of less than 2.5 p.u. or preferably to 2.0 p.u. or even less. The measures taken to control or reduce the overvoltages are

  (i) one step or multi-step energisation of lines by preinsertion of resistors,
  (ii) phase controlled closing of circuit breakers with proper sensors,
  (iii) drainage of trapped charges on long lines before the reclosing of the lines, and
  (iv) limiting the overvoltages by using surge diverters.

The first three methods, if used properly will limit the switching overvoltages to 1.5 to 2.0 p.u.

In Table 8.1, a summary of the extent of overvoltages that can be developed under various conditions of switching is given.

Table 8.1 Overvoltages due to Switching Operations Under Different Conditions

Maximum value of the system line-to-ground voltage = 1.0 p.u.

Sl. no.		Type of operation	Overvoltage (p.u.)
1		Switching an open ended line with:	
	(a)	infinite bus as source with trapped charges on line	4.1
	(b)	infinite bus as source without trapped charges	2.6

	(c)	de-energising an unfaulted line with a restrike in the circuit breaker	2.7
	(d)	de-energising an unfaulted line with a line to ground fault (about 270 km in length)	1.3
2	(a)	switching a 500 kV line through an auto-transformer, 220 kV/500 from the L.V. side	2.0
	(b)	switching a transformer terminated line	2.2
	(c)	series capacitor compensated line with 50% compensation	2.2
	(d)	series capacitor compensated line with shunt reactor compensation	2.6
3		High speed reclosing of line after fault clearance	3.6

8.2.4 Power Frequency Overvoltages In Power Systems

The power frequency overvoltages occur in large power systems and they are of much concern in EHV systems, i.e. systems of 400 kV and above. The main causes for power frequency and its harmonic overvoltages are

(a) sudden loss of loads,
(b) disconnection of inductive loads or connection of capacitive loads,
(c) Ferranti effect, unsymmetrical faults, and
(d) saturation in transformers, etc.

Overvoltages of power frequency harmonics and voltages with frequencies nearer to the operating frequency are caused during tap changing operations, by magnetic or ferro-resonance phenomenon in large power transformers, and by resonating over-voltages due to series capacitors with shunt reactors or transformers.

The duration of these overvoltages may be from one to two cycles to a few seconds depending on the overvoltage protection employed.

(a) Sudden Load Rejection

Sudden load rejection on large power systems causes the speeding up of generator prime movers. The speed governors and automatic voltage regulators will intervene to restore normal conditions. But initially both the frequency and voltage increase. The approximate voltage rise, neglecting losses, etc. may be taken as

$$v = \frac{f}{f_0} E' \left[\left(1 - \frac{f}{f_0} \right) \frac{x_s}{x_c} \right] \tag{8.38}$$

where x_s is the reactance of the generator ($\approx$ the sum of the transient reactances of the generator and the transformer), x_c is the capacitive reactance of the line at open end at increased frequency, E' the voltage generated before the over-speeding and load rejection, f is the instantaneous increased frequency, and f_0 is the normal frequency.

This increase in voltage may go to as high as 2.0 per unit (p.u.) value with 400 kV lines. The voltage at the sending end is affected by the line length, short circuit MVA at sending end bus, and reactive power generation of the line (due to line capacitive reactance and any shunt or series capacitors). Shunt reactors may reduce the voltage to 1.2 to 1.4 p.u.

(b) Ferranti Effect

Long uncompensated transmission lines exhibit voltage rise at the receiving end. The voltage rise at the receiving end V_2 is approximately given by

$$V_2 = \frac{V_1}{\cos \beta l} \tag{8.39}$$

where, V_1 = sending end voltage,

   l = length of the line,

   β = phase constant of the line

$$\approx \left[\frac{(R + j\omega L)(G + j\omega C)}{LC} \right]^{\frac{1}{2}}$$

   $\approx$ about 6° per 100 km line at 50 Hz frequency.

   $R, L, G,$ and C are as defined in Sec. 8.1.5, and

   ω = angular frequency for a line shown in Fig. 8.17.

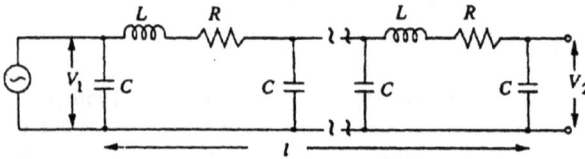

L, R and C — Inductance, resistance and capacitance
        per unit length of the line
   l — Length of the line

Fig. 8.17 Typical uncompensated long transmission line

Considering that the line capacitance is concentrated at the middle of the line, under open circuit conditions at the receiving end, the line charging current

$$I_C \approx j\omega CV_1 = \frac{V_1}{X_C} \tag{8.40}$$

and the voltage $V_2 \approx V_1 \left[1 - \frac{X_L}{2X_C} \right]$ \hfill (8.41)

where, X_L = line inductive reactance, and

   X_C = line capacitive reactance.

This approximation solution is shown in Fig. 8.18.

Equation (8.39) gives a better approximation when the line distributed parameters R, L, G, and C per unit length are known.

(c) Ground Faults and Their Effects

Single line to ground faults cause rise in voltages in other healthy phases. Usually, with solidly grounded systems, the increases in voltage (phase to ground value) will be less than the line-to-line voltage. With effectively grounded systems, i.e. with

$$\frac{X_0}{X_1} \leq 3.0 \text{ and } \frac{R_0}{X_1} \leq 1.0$$

(where, R_0 and X_0 are zero sequence resistance and reactance and X_1 is the positive sequence reactance of the system), the rise in voltage of the healthy phases does not usually exceed 1.4 per unit.

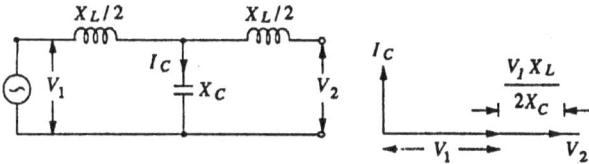

Transmission line approximation for line in Fig. 8.17.

Fig. 8.18 Vector (phase) diagram of an open circuited uncompensated line showing Ferranti effect

(d) Saturation Effects

When voltages above the rated value are applied to transformers, their magnetizing currents (no load currents also) increase rapidly and may be about the full rated current for 50% overvoltage. These magnetizing currents are not sinusoidal in nature but are of a peaky waveform. The third, fifth, and seventh harmonic contents may be 65%, 35%, and 25% of the exciting current of the fundamental frequency corresponding to an overvoltage of 1.2 p.u. For third and its multiple harmonics, zero sequence impedance values are effective, and delta connected windings suppress them. But the shunt connected capacitors and line capacitances can form resonant circuits and cause high third harmonic overvoltages. When such overvoltages are added, the voltage rise in the lines may be significant. For higher harmonics a series resonance between the transformer inductance and the line capacitance can occur which may produce even higher voltages. The analysis is involved and is given in Ref. [22].

8.2.5 Control of Overvoltages Due to Switching

The overvoltages due to switching and power frequency may be controlled by

> (a) energization of transmission lines in one or more steps by inserting resistances and withdrawing them afterwards,

      (b)   phase controlled closing of circuit breakers,

      (c)   drainage of trapped charges before reclosing,

      (d)   use of shunt reactors, and

      (e)   limiting switching surges by suitable surge diverters.

(a) Insertion of Resistors

It is normal and a common practice to insert resistances R in series with circuit breaker contacts when switching on but short circuiting them after a few cycles. This will reduce the transients occurring due to switching. The voltage step applied is first reduced to $Z_0/(R + Z_0)$ per unit where Z_0 is the surge impedance of the line. It is reflected from the far end unchanged and again reflected back from the near end with reflection factor $(R - Z_0)/(R + Z_0)$ per unit. If $R = Z_0$, there is no reflection from the far end. The applied step at the first instance is only 0.5 per unit.

    When the resistor is short circuited, a voltage step equal to the instantaneous voltage drop enters the line. If the resistor is kept for a duration larger than 5 m s (for 50 Hz sine wave = 1/4 cycle duration), it can be shown from successive reflections and transmissions, that the overvoltage may reach as high as 1.2 p.u. for a line length of 500 km. But for conventional opening of the breaker, the resistors have too high an ohmic value to be effective for resistance closing. Therefore, pre-insertion of suitable value resistors in practice is done to limit the overvoltage to less than 2.0 to 2.5 p.u. Normal time of insertion is 6 to 10 m s.

(b) Phase Controlled Switching

Overvoltages can be avoided by controlling the exact instances of the closing of the three phases separately. But this necessitates the use of complicated controlling equipment and therefore is not adopted.

(c) Drainage of Trapped Charge

When lines are suddenly switching off, "electric charge" may be left on capacitors and line conductors. This charge will normally leak through the leakage path of the insulators, etc. Conventional potential transformers (magnetic) may also help the drainage of the charge. An effective way to reduce the trapped charges during the lead time before reclosing is by temporary insertion of resistors to ground or in series with shunt reactors and removing before the closure of the switches.

(d) Shunt Reactors

Normally all EHV lines will have shunt reactors to limit the voltage rise due to the Ferranti effect. They also help in reducing surges caused due to sudden energizing. However, shunt reactors cannot drain the trapped charge but will give rise to oscillations with the capacitance of the system. Since the compensation given by the reactors will be less than 100%, the frequency of oscillation will be less than the power frequency and overvoltages produced may be as high as 1.2 p.u. Resistors in series with these reactors will suppress the oscillations and limit the overvoltages.

    Limiting the switching and power frequency overvoltages by using surge diverters is discussed in the next section.

8.2.6 Protection of Transmission Lines against Overvoltages

Protection of transmission lines against natural or lightning overvoltages and minimizing the lightning overvoltages are done by suitable line designs, providing guard and ground wires, and using surge diverters. Switching surges and power frequency overvoltages are accounted for by providing greater insulation levels and with proper insulation co-ordination. Hence, the above two protection schemes are dealt with separately in the next two sections.

Protection against Lightning Overvoltages and Switching Surges of short Duration

Overvoltages due to lightning strokes can be avoided or minimized in practice by
- (a) shielding the overhead lines by using ground wires above the phase wires,
- (b) using ground rods and counter-poise wires, and
- (c) including protective devices like expulsion gaps, protector tubes on the lines, and surge diverters at the line terminations and substations.

(a) Lightning Protection Using Shielded Wires or Ground Wires

Ground wire is a conductor run parallel to the main conductor of the transmission line supported on the same tower and earthed at every equally and regularly spaced towers. It is run above the main conductor of the line. The ground wire shields the transmission line conductor from induced charges, from clouds as well as from a lightning discharge. The arrangements of ground wires over the line conductor is shown in Fig. 8.19.

The mechanism by which the line is protected may be explained as follows. If a positively charged cloud is assumed to be above the line, it induces a negative charge on the portion below it, of the transmission line. With the ground wire present, both the ground wire and the line conductor get the induced charge. But the ground wire is earthed at regular intervals, and as such the induced charge is drained to the earth potential only; the potential difference between the ground wire and the cloud and that between the ground wire and the transmission line wire will be in the inverse ratio of their respective capacitances [assuming the cloud to be a perfect conductor and the atmospheric medium (air) a dielectric]. As the ground wire is nearer to the line wire, the induced charge on it will be much less and hence the potential rise will be quite small. The effective protection or shielding given by the ground wire depends on the height of the ground wire above the ground (h) and the protection or shielding angle θ_S (usually 30°) as shown in Fig. 8.19.

The shielding angle $\theta_S \approx 30°$ was considered adequate for tower heights of 30 m or less. The shielding wires may be one or more depending on the type of the towers used. But for EHV lines, the tower heights may be up to 50 m, and the lightning strokes sometimes occur directly to the line wires as shown in Fig. 8.19. The present trend in fixing the tower heights and shielding angles is by considering the "flashover rates" and failure probabilities.

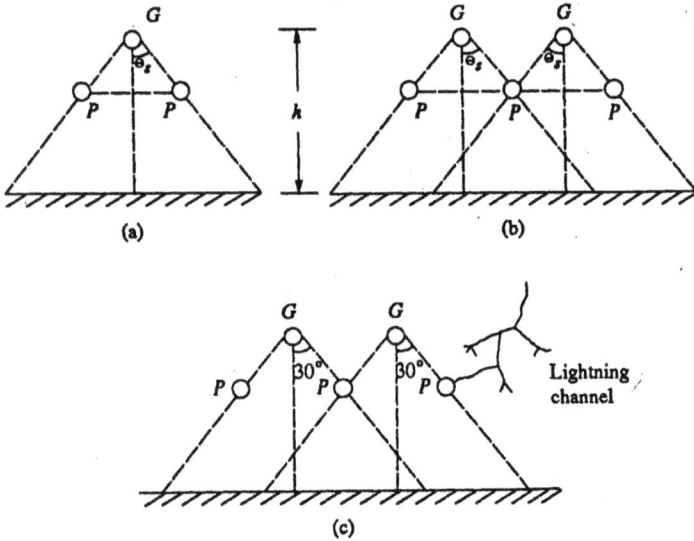

(a) (b)

(c)

G — Ground wire   h — Height of the ground wire
P-P — Phase wires   from the earth surface
θ_s — Shielding angle

Fig. 8.19 Shielding arrangement of overhead lines by ground wires

(b) Protection Using Ground Rods and Counter-Poise Wires

When a line is shielded, the lightning strikes either the tower or the ground wire. The path for drainage of the charge and lightning current is (a) through the tower frame to ground, (b) through the ground line in opposite directions from the point of striking. Thus the ground wire reduces the instantaneous potential to which the tower top rises considerably, as the current path is in three directions. The instantaneous potential to which tower top can rise is

$$V_T = \frac{I_0 Z_T}{\left(1 + \dfrac{Z_T}{Z_S}\right)} \qquad (8.42)$$

where,

Z_T = surge impedance of the tower, and

Z_S = surge impedance of the ground wire.

If the surge impedance of the tower, which is the effective tower footing resistance, is reduced, the surge voltage developed is also reduced considerably. This is accomplished by providing driven ground rods and counter-poise wires connected to tower legs at the tower foundation.

Ground rods are a number of rods about 15 mm diameter and 2.5 to 3 m long driven into the ground. In hard soils the rods may be much longer and can be driven to a depth of, say, 50 m. They are usually made of galvanized iron or copper bearing steel. The spacings of the rods, the number of rods, and the depth to which they are driven depend on the desired tower footing resistance. With 10 rods of 4 m long and spaced 5 m apart, connected to the legs of the tower, the dynamic or effective resistance may be reduced to 10 Ω.

The above effect is alternatively achieved by using counter-poise wires. Counter-poise wires are wires buried in the ground at a depth of 0.5 to 1.0 m, running parallel to the transmission line conductors and connected to the tower legs. These wires may be 50 to 100 m long. These are found to be more effective than driven rods and the surge impedance of the tower may be reduced to as low as 25 Ω. The depth does not materially affect the resistance of the counter-poise, and it is only necessary to bury it to a depth enough to prevent theft. It is desirable to use a larger number of parallel wires than a single wire. But it is difficult to lay counter-poise wires compared to ground or driven rods.

(c) Protective Devices

In regions where lightning strokes are intensive or heavy, the overhead lines within these zones are fitted with shunt protected devices. On the line itself two devices known as expulsion gaps and protector tubes are used. Line terminations, junctions of lines, and sub-stations are usually fitted with surge diverters.

(i) Expulsion gaps

Expulsion gap is a device which consists of a spark gap together with an arc quenching device which extinguishes the current arc when the gaps breakover due to over-voltages. A typical such arrangement is shown in Fig. 8.20a. This essentially consists of a rod gap in air in series with a second gap enclosed within a fibre tube. In the event of an overvoltage, both the spark gaps breakdown simultaneously. The current due to the overvoltage is limited only by the tower footing resistance and the surge imped-ance of the ground wires. The internal arc in the fibre tube due to lightning current vapourizes a small portion of the fibre material. The gas thus produced, being a mixture of water vapour and the decomposed fibre product, drive away the arc products and ionized air. When the follow-on power frequency current passes through zero value, the arc is extinguished and the path becomes open circuited. Meanwhile the insulation recovers its dielectric strength, and the normal conditions are estab-lished. The lightning and follow-up power frequency currents together can last for 2 to 3 half cycles only. Therefore, generally no disturbance in the network is produced. For 132 or 220 kV lines, the maximum current rating may be about 7,500 A.

(ii) Protector tubes

A protector tube is similar to the expulsion gap in, construction and principle. It also consists of a rod or spark gap in air formed by the line conductor and its high voltage terminal. It is mounted underneath the line conductor on a tower. The arrangement is shown in Fig. 8.20b. The hollow gap in the expulsion tube is replaced by a nonlinear element which offers a very high impedance at low currents but has low impedance for high or lightning currents. When an overvoltage occurs and the spark gap breaks

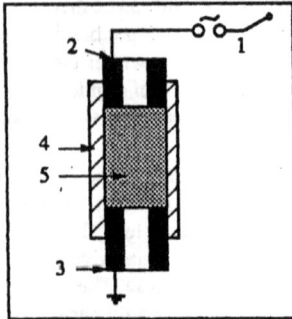

1. External series gap
2. Upper electrode
3. Ground electrode
4. Fibre tube
5. Hollow space

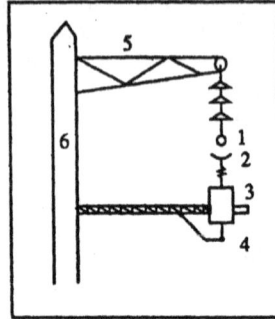

1. Line conductor on string insulator
2. Series gap
3. Protector tube
4. Ground connection
5. Cross arm
6. Tower body

Fig. 8.20a Expulsion gap   **Fig. 8.20b** Protector tube mounting

down, the current is limited both by its own resistance and the tower footing resistance. The overvoltage on the line is reduced to the voltage drop across the protector tube. After the surge current is diverted and discharged to the ground, the follow-on normal power frequency current will be limited by its high resistance. After the current zero of power frequency, the spark gap recovers the insulation strength quickly. Usually, the flashover voltage of the protector tube is less than that of the line insulation, and hence it can discharge the lightning overvoltage effectively.

(iii) Rod gaps

A much simpler and effective protective device is a rod-gap (refer Section 7.2.7 - rod-gaps, in Chapter 7). However, it does not meet the complete requirement. The sparkover voltage of a rod gap depends on the atmospheric conditions. A typical volt-time characteristic of a 67 cm-rod gap is shown in Fig. 8.21, with its protective margin. There is no current limiting device provided so as to limit the current after sparkover, and hence a series resistance is often used. Without a series resistance, the sparking current may be very high and the applied impulse voltage suddenly collapses to zero thus creating a steep step voltage, which sometimes proves to be very dangerous to the apparatus to be protected, such as transformer or the machine windings. Nevertheless, rod gaps do provide efficient protection where thunderstorm activity is less and the lines are protected by ground wires.

(iv) Surge diverters or lightning arresters

Surge diverters or lightning arresters are devices used at sub-stations and at line terminations to discharge the lightning overvoltages and short duration switching surges. These are usually mounted at the line end at the nearest point to the sub-station. They have a flashover voltage power than that of any other insulation or

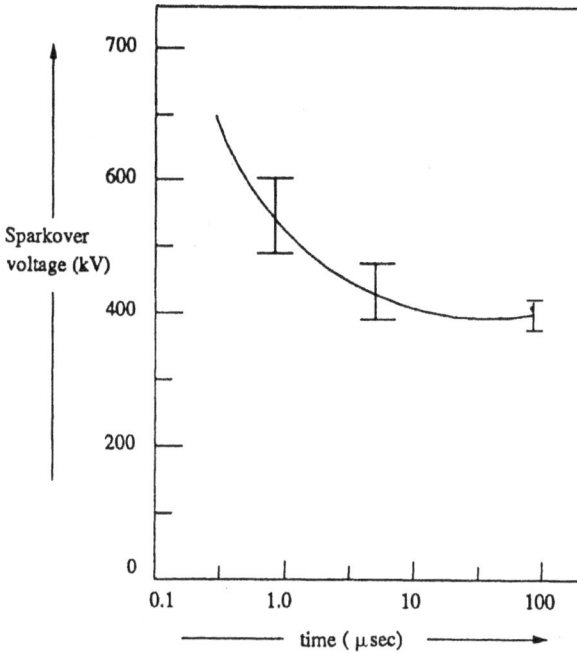

Fig. 8.21   Volt-time characteristic of a standard rod-rod gap

apparatus at the sub-station. These are capable of discharging 10 to 20 kA of long duration surges (8/20 μ s) and 100 to 250 kA of the short duration surge currents (1/5 μ s). The characteristics and detailed discussion of the surge diverters are presented in Sec. 8.3.1.

8.3   PRINCIPLES OF INSULATION COORDINATION ON HIGH VOLTAGE AND EXTRA HIGH VOLTAGE POWER SYSTEMS

Electric power supply should ensure reliability and continuity to the utility concerns. Hence, the power lines and sub-stations are to be operated and protected against overvoltages such that the number of failures are as few as possible. At the same time, the cost involved in the design, installation, and operation of the protective devices should not be too high. Hence, a gradation of system insulation and protective device operation is to be followed, keeping in view the importance of the various equipment involved.

Generally, sub-stations contain transformers, switchgear, and other valuable equipment with non-self restoring insulation, which have to be protected against failures and internal destruction. For other apparatus, which contain self-restoring insulation, like string insulators, they may be allowed to flashover in air. But the flashovers should be kept to a minimum so that the system disturbances are the least.

Hence, lightning and switching surge protection requires establishment of protective voltage levels called shunt protection levels, by means of protective devices like lightning arresters.

The lightning impulse withstand level known as the Basic Impulse Level (BIL) is established for each system nominal voltage for different apparatus. Various equipment and their component parts should have their BIL above the system protective level, by a suitable margin. This margin is usually determined with respect to air insulation by statistical methods. For non self-restoring insulation like the transformer insulation, the margin limit is fixed using conventional methods.

For system voltages below 400 kV, the switching surges are not of importance. If the BIL is chosen correctly, relative to the prevailing protective level, the equipment will have an adequate switching surge level. For higher system voltages, since the switching surges are of higher magnitude compared to the lightning overvoltages, switching surge magnitudes are of importance and the following criterion is to be adopted.

(i) The flashover voltage of a protective derive is chosen such that it will not operate for switching overvoltages and other power frequency and its harmonic overvoltages. But other long duration overvoltages, namely, sustained overvoltage due to faults and even the above-mentioned overvoltages may sometimes cause thermal overloading due to leakage currents. Therefore, the BIL has to be higher.

(ii) For EHV systems, it may be economical to use a protective device for limiting the overvoltages due to lightning as well as switching surges to a particular level. At present, there are surge diverters which operate for both types of overvoltages mentioned above. In such cases, it is preferable to assign to each protected equipment a Switching Impulse Level (SIL), so that there is a small margin above the controlled switching surge level, so that the surge diverters operate on switching surges, only when the controlling devices fail. Normally, only rod gaps and lightning arresters are used as protective devices for protection, and their characteristics are considered here.

The ideal requirements of a protective device connected in parallel or in shunt are:

(a) It should not usually flashover for power frequency overvoltages.
(b) The volt-time characteristics of the device must lie below the withstand voltage of the protected apparatus or insulation. The marginal difference between the above two should be adequate to allow for the effects of distance, polarity, atmospheric conditions, changes in the characteristics of the devices due to agencies, etc.
(c) It should be capable of discharging high energies contained in surges and recover insulation strength quickly.
(d) It should not allow power frequency follow-on current to flow.

The behaviour of shunt connected protective devices like rod gaps and surge diverters along with transformer insulation is given in Fig. 8.22.

In Fig. 8.22a the transformer insulation strength is given as a volt-time characteristic. Figure 8.20b gives the relative insulation strengths of the transformer (curve A), rod gaps (curves B and C), and that of a lightning arrester (curve D).

A lightning arrester protects the transformer insulation in the entire time region. The rod gap protects the transformer insulation, only if the rate of rise of surge is less than the critical slope (curve X). Thus, if the surge voltage rise is as shown by curve

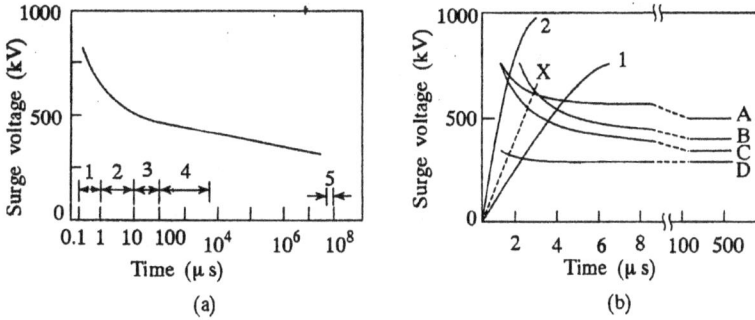

Typical transformer characteristic
1. For steep fronted lightning surge
2. Slow fronted lightning surge
3. Fast switching surge   Transformer — A
4. Slow switching surge   Rod gaps — B and C
5. Power frequency (1 minute withstand Surge diverter — D
   voltage)   Critical slope — X

Fig. 8.22 Volt-time characteristics of transformer rod gaps and surge diverters

1, rod gap flashes and protects the transformer. If the surge voltage rise follows curve 2, only the surge diverter can protect the transformer insulation.

Rod gaps are simple and cheap devices but do not meet all the requirements of a protective device. Moreover, their flashover characteristics depend on the atmospheric conditions, polarity of the wave, and waveshape. Also, it may give rise to very steep impulse waves on the transformer windings as chopped waves, because no current limiting resistance is used. Chopped impulse waves may lead to the destruction of the transformer turn to turn insulation. But still, rod gaps provide reasonable protection where lightning surge levels are low, and steep fronted surges are controlled by overhead ground wires.

8.3.1 Surge Diverters

These are non-linear resistors in series with spark gaps which act as fast switches. A typical surge diverter or lightning arrester is shown in Fig. 8.23 and its characteristics are given in Fig. 8.24. A number of non-linear resistor elements made of silicon carbide are stacked one over the other into two or three sections. They are usually separated by spark gaps (see Fig. 8.23). the entire assembly is housed in a porcelain water-tight housing. The volt-ampere characteristic of a resistance element is of the form

$$I = kV^a \tag{8.43}$$

where, I = discharge current,

V = applied voltage across the element, and

k and a are constants depending on the material and dimensions of the element.

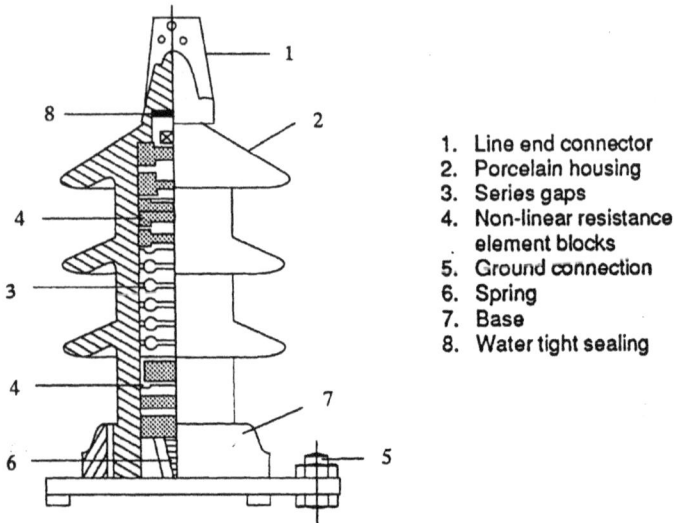

1. Line end connector
2. Porcelain housing
3. Series gaps
4. Non-linear resistance
   element blocks
5. Ground connection
6. Spring
7. Base
8. Water tight sealing

Fig. 8.23   Non-linear element surge diverter

The dynamic characteristic is shown in Fig. 8.24a.

When a surge voltage (V_i of Fig. 8.24b) is applied to the surge diverter, it breaks down giving the discharge current i_d and maintains a voltage V_d across it. Thus, it provides a protection to the apparatus to be protected above the protective level V_p (see Fig. 8.24b).

The lighter designs operate for smaller duration of currents, while the heavy duty surge diverters with assisted or active gaps are designed for high currents and long duration surges. The lighter design arresters can interrupt 100 to 300 A of power frequency follow-on current and about 5000 A of surge currents. If the current is to be more and has to be exceeded, the number of series elements has to be increased or some other method to limit the current has to be used. In heavy duty arresters, the gaps are so arranged that the arc burns in the magnetic field of the coils excited by power frequency follow-on currents. During lightning discharges, a high voltage is induced in the coil by the steep front of the surge, and sparking occurs in an auxiliary gap. For power frequency follow-on currents, the auxiliary gap is extinguished, as sufficient voltage will not be present across the auxiliary gap to maintain an arc. The main gap arcs occur in the magnetic field of the coils. The magnetic field, aided by the born shaped main gap electrodes, elongates the arc and quenches it rapidly. The follow-on current is limited by the voltage drop across the arc and the resistance element. During surge discharge the lightning protective level becomes low.

Sometimes, it is possible to limit the power frequency and other overvoltages after a certain number of cycles using surge diverters. The permissible voltage and duration depend on the thermal capacity of the diverter. The rated diverter voltage is normally chosen so that it is not less than the power frequency overvoltage expected (line to ground) at the point of installation, under any faulty or abnormal operating condition.

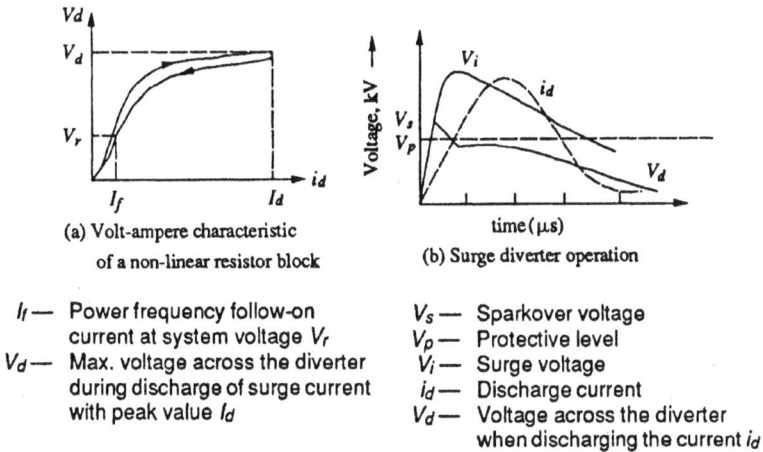

(a) Volt-ampere characteristic
of a non-linear resistor block

(b) Surge diverter operation

I_f — Power frequency follow-on current at system voltage V_r

V_d — Max. voltage across the diverter during discharge of surge current with peak value I_d

V_s — Sparkover voltage
V_p — Protective level
V_i — Surge voltage
i_d — Discharge current
V_d — Voltage across the diverter when discharging the current i_d

Fig. 8.24 Characteristics of a surge diverter

Typical characteristics of surge diverters in the voltage range 100 to 200 kV, 10 kA (heavy duty type) are given in Table 8.2.

Table 8.2 Characteristics of 100-200 kV Surge Diverters 10 kA and Heavy Duty Type (ref. 21)

	Characteristics	Per unit values (referred to the rated values of the diverter)
1.	Maximum 1.2/50 μs surge sparkover voltage	2.2 to 2.8
2.	Maximum front of wave sparkover voltage	2.9 to 3.1
3.	Maximum switching impulse sparkover voltage	2.3 to 3.0
4.	Maximum discharge voltage (V_d) for 8/20 μs current wave	
	5 kA	2.0 to 2.7
	10 kA	2.2 to 3.0
	20 kA	2.5 to 3.3

8.3.1.1 Surge Diverters for E.H.V. Systems

The selection of surge diverter voltage rating for EHV and UHV systems depends on
(i)   the rate of rise of voltage,
(ii) the type of system to be handled, i.e. whether effectively grounded or grounded through an impedance etc., and

 (*iii*) operating characteristic of the diverter.

The usual type of surge diverters used for the above purposes are

 (*i*) silicon carbide arresters with spark gaps,

 (*ii*) silicon carbide arresters with current limiting gaps, and

 (*iii*) the gapless metal oxide (zinc oxide) arresters.

The first two types of arresters have a *V-I* characteristic of the nature of $V = AI^n$, where *n* varies between 5 and 6 for the elements. The time to sparkover for the first type of arresters is around 1 to 2 μs and the voltage is limited to 2.0 p.u. of the power frequency voltage. The *V-I* characteristics of arresters with no spark gaps are not enough to limit the power frequency follow-on current, while the arresters with the spark gap provided will have a high limiting voltage. Further, arresters with spark gaps are not very well suited to limit the switching overvoltages. However, recent developments in solid state technology have led to the development of metal oxide non-linear resistors. With the use of these materials, the new class of surge arresters that can handle very small to very large current, with almost constant voltage across them, have been developed. One such arrester is the zinc oxide (ZnO) arrester which uses a base material of ZnO sintered into a different insulating medium such as BiO_3. The *V-I* characteristic of such a unit is of the form $V \propto I^n$ where $n = 0.02$ to 0.03. The *V-I* characteristic of silicon carbide and zinc oxide arresters are shown in Fig. 8.25 for comparison.

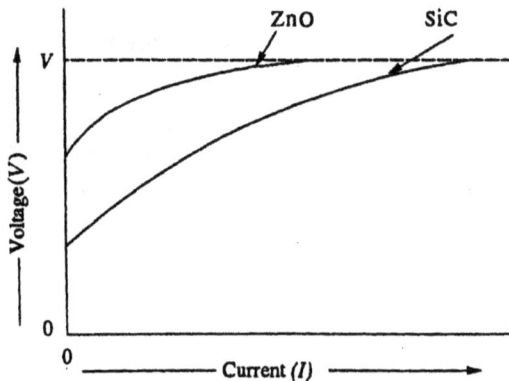

Fig. 8.25 Typical *V-I* characteristics of silicon carbide (SiC) and zinc oxide (ZnO) surge diverters

The advantages of zinc oxide arresters for EHV systems are

 (*i*) they are simple in construction,

 (*ii*) they have flat *V-I* characteristic over a wide current range, and

 (*iii*) the absence of a spark gap that produces steep voltage gradients when sparking occurs.

The main disadvantage of zinc oxide arresters is the continuous flow of power frequency current and the consequent power loss. Voltage grading system is not needed for each of the units of the zinc oxide arresters used in EHV systems. A typical

400 kV line arrester may be rated at 15 kA and may have a resistance of 100 Ohms at the peak current rating.

8.3.1.2 Protection of Lines with Surge Diverters

Since surge diverters are devices that provide low resistance paths for overvoltages through an alternate ground path, their operating characteristic and application is of importance. The spark gap inside the diverter acts as a fast acting switch while non-linear diverter elements provide the low impedance ground path. The arrester voltage at its terminal when connected to a·line of surge impedance Z to ground, is given as

$$V = [2(R + r)/(R + r + Z)]\, u(t) \qquad (8.44)$$

where, Z is the line surge impedance,

R is the resistance of the non-linear element,

r is the ground to earth resistance, and

$u(t)$ is the surge voltage.

The Thevinin equivalent circuit for the diverter is shown in Fig. 8.26.

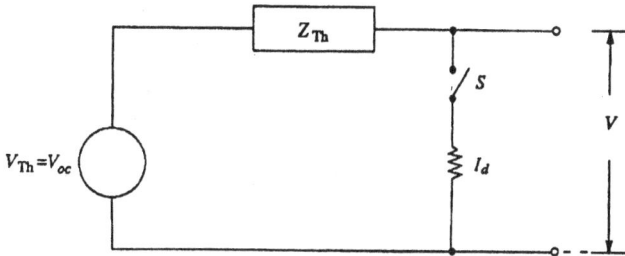

Fig. 8.26 Equivalent circuit of a surge diverter

The switch S is open for voltages less than the sparkover voltage of the surge diverter V_s, while it is closed for voltage magnitudes greater than V_s. The closing of the switch is represented by injecting a voltage cancellation wave having a negative amplitude equal to the potential difference between the voltage that appears when the switch is open V_{oc}, and the voltage developed across the impedance of the device after the switch is closed. Z_{Th} is the impedance of the system viewed from the terminals of the protective device.

8.3.2 Equipment Insulation Level and Insulation Coordination of Sub-stations

For steep fronted lightning waves at sub-stations and at different points on lines, the voltages at sub-stations may exceed the protective level depending on the distances involved and the diverter locations. Hence, it is necessary to decide the number of locations for diverters to optimize the overall cost. For high voltage sub-stations, it is usual to instal surge diverters between a transformer and its circuit breaker, in order to protect the transformer from current chopping and the overvoltage due to it. Further, nearness of the diverter to the transformer offers better protection. The basic insulation level is often determined by giving a margin of 30% to the protective level of surge diverter and selecting the next nearest standard BIL. The standard values of BIL for system voltages from 145 to 765 are given in Tables 8.3 and 8.4. When a surge diverter is used to give switching surge protection also, the margin allowed is only 15%. The insulation level for lines and other equipment is to be chosen separately. The selection of this level for lines depends on the atmospheric conditions, the lightning activity, insulation pollution present and the acceptable outrage or failure rate of the line.

The protective level of the sub-station insulation depends on the station location, the protective level of the diverter, and the line shielding used. The line insulation in the end spans near the sub-station is normally reduced to limit the lightning over-voltages reaching the sub-station. In a sub-station, the busbar insulation level is the

Table 8.3 Insulation Levels (BIL) for Various System Voltages (ref. 21)

Highest voltage of the system equipment	Impulse withstand voltage for standard impulse waves		Power frequency withstand voltage with respect to earth	
	Full insulation	Reduced insulation	Full insulation	Reduced insulation
kV (rms)	kV (peak)	kV (peak)	kV (rms)	kV (rms)
145	650	550	275	230
		450		185
245	1050	—	460	—
		900	—	395
		825	—	360
		750	—	325
362	—	1300	—	570
		1175	—	510
		1050	—	461
420	—	1675	—	740
		1550	—	680
		1425	—	630
		1300	—	570
525	—	1800	—	790
		1675	—	740
		1550	—	680
		1425	—	630
765	—	2400	—	1100
		2100	—	980
		1950	—	920
		1800	—	870

highest to ensure continuity of supply. The circuit breakers, isolators, instrument and relay transformers, etc. are given the next lower level. Since the power transformer is the costly and sensitive device, the insulation level for it is the lowest.

Table 8.4   Standard Insulation Levels for Equipment (> 300 kV) (ref. 21)

Highest voltage for equipment	Base for per unit voltage values	Rated switching impulse voltage		Ratio between lightning and switching withstand voltages	Rated lightning withstand voltage
U_m kV (rms)	$\left(U_m \dfrac{\sqrt{2}}{\sqrt{3}} \right)$ kV	pu	kV (peak)		kV (peak)
300	245	3.06	750	1.13	850
		3.45	850	1.27	950
362	296	2.86	850	1.12	950
				1.24	1050
		3.20	950	1.12	1050
				1.24	1175
420	343	2.76	950	1.12	1050
				1.24	1175
		3.06	1050	1.12	1175
				1.24	1300
				1.36	1425
525	429	2.45	1050	1.12	1175
				1.24	1300
				1.36	1425
		2.74	1175	1.12	1300
				1.21	1425
				1.32	1550
765	625	2.08	1300	1.10	1425
				1.19	1550
				1.38	1800
		2.28	1425	1.09	1550
				1.28	1800
				1.47	2100
		2.48	1550	1.16	1800
				1.26	1950
				1.55	2400

To illustrate the principles of insulation co-ordination, an example of a 132-kV sub-station is given below.

Nominal system voltage	: 132 kV
Highest system voltage	: 145 kV
Highest system voltage to ground	: $145 \times \dfrac{\sqrt{2}}{\sqrt{3}} = 119$ kV (peak)
Expected switching surge overvoltage (Table 8.1) 3.0 p.u.	: $3 \times 119 = 357$ kV (peak)

(a) Surge diverter

Rating	: 123 kV
Front of wave sparkover voltage	: 510 kV (peak)
V_d (discharge voltage at 10 kA, 8/20 μ s impulse current wave)	: 443 kV (peak)

(b) Transformer

Impulse withstand voltage	: 550 kV (peak)
Induced voltage (withstand) level	: 230 kV (rms)
Impulse protective margin	: $\dfrac{550 - 443}{443} \times 100 = 24\%$

(c) Switchgear

Impulse withstand voltage	: 650 kV (peak)
Bus insulation impulse withstand voltage	: 650 kV (peak)

In case a rod gap is used for the protection of a transformer, the rod gap with a negative sparkover voltage of 440 kV (59 cm gap) may be chosen to give the 25% margin. The protection is good for surges having a front time not shorter than 2 μ s. But the switching surge sparkover voltage which is about 380 kV is very near the maximum switching surge generated in the system and hence may cause many outrages. If a rod gap of 66 cm is used, the protection becomes doubtful, as the impulse sparkover voltage for 2 μ s front wave is 600 kV.

8.3.3 Insulation Levels at Sub-stations with Protective Zones

Magnitude and Shape of the Incoming Voltage Surges

Direct strokes to phase conductors near the station point are, very dangerous as they cause very high currents to flow through the surge diverter. The discharge voltages developed across the arrester elements are very high and the arresters may get destroyed. Therefore, the stations are completely shielded from direct strikes. The shielding is sometimes effected up to about 2 km on either side of the station. This protected zone gives the surges to originate only from outside regions. Usually the voltage wave at the station entrance is estimated by assuming a voltage magnitude at the beginning of the protected zone as equal to 1.3 times the negative critical flashover voltage of the line insulation. It is also important that the sloping off of the point of the surge is helpful for the effective performance of the surge diverter. Back flashovers are also important. There is a risk of a more severe surge occurring in the case of back flashovers. The probability of the occurrence of this depends on the rate of the back flashover of the protected zone and its length.

Equipment Insulation Level

For steep fronted travelling waves, the voltages at different points in the sub-station can exceed the protective level by amounts that depend on the distance from the diverter location, the steepness of the wave front and other electrical parameters. Hence, it is necessary to decide the number of locations at which surge diverters are to be located and their ratings. It is necessary to keep this number to a minimum. Also, care must be taken regarding switching overvoltages generated due to current chopping which may destroy the transformer or the equipment near the circuit breakers. The Basic Impulse Level (BIL) is often determined as simply 1.25 to 1.30 times the protective level offered by the surge diverter. Usually, the next higher BIL value from the standard values is chosen. This is quite sufficient for smaller stations and station ratings up to 220 kV. For bigger stations and stations of importance, the "distance effect" discussed in the next section is to be suitably allowed for when surge diverters are to be used for SIL also; a margin of 15 to 20% is normally allowed over the protective level. Distance effect is negligible for long fronted switching surges.

The Distance Effect

Usually the circuit breaker. The transformer and other equipment are placed at finite distances from the surge diverter and connected through a short distance overhead line or cable. When a surge arises, it suffers multiple reflections between each of the equipment which may give rise to over voltages of considerable magnitude (the travel time is usually less than a μs). It can be shown that when a surge diverter, a breaker and a transformer are in line, the voltage that can build up at a distance D from the surge diverter point is given as $V(D) = V_p + 2ST$, where V_p is the sparkover voltage/protective level, S is the steepness of the wave front, and T is the travel time = D/v. Here, v is the velocity of the wave travel, assuming that the line extends to a large distance such that no reflections come from the line end. The maximum value of $V(D)$ is attained when $2T = T_0$, the sparkover time of the diverter. The above simple expression shows that the transfer surge impedance is very high. The ratio of the transformer terminal voltage, V_T to that of the protective level V_p is a function of (T/T_0). For steep fronted waves, sometimes V_T may exceed even $2V_p$. It has been shown that in a 330 kV sub-station a 1.2/50 μs, 1500 kV incoming wave can give rise to a surge of peak value 1250 kV having rise time of 2 μs at the bus terminals (neglecting the transfer capacitance), whereas the protective level offered by the surge diverter at the transformer terminal is only 750 kV. The transformer terminals may get a surge voltage of 930 kV peak. This has been verified by computer calculations for a 1/50 μs wave, on a line connected to a 330 kV station.

8.3.4 Insulation Co-ordination in EHV and UHV Systems

The insulation design of EHV and UHV stations is based on the following principles
  (*i*) stations have transformers and other valuable equipment that have non-self restoring insulation, and

(*ii*) the protective levels for lightning surges and switching surges are almost equal and even overlap. If the basic impulse level for the equipment or the system is chosen, then this level cannot give protection against the switching impulses. Hence, a separate switching impulse level (SIL) has to be chosen. It is therefore, desirable to use protective devices for limiting both lightning and switching overvoltages. As such, the switching impulse insulation level above the controlled switching surge level has to be adopted so that the surge diverters operate only rarely on switching overvoltages when the controls of the control devices for switching voltages fail. A general guideline that can be adopted for different EHV and UHV systems for maximum switching surge levels are given in Table 8.5.

Table 8.5 Maximum Switching Surge Levels at Different Line Voltages

Highest system voltage (kV)	420	525	765	1150
Maximum switching surge level (kV) =	2.5	2.25	2.0	1.8 to 1.9
Highest system voltage multiplied by				

It is now necessary to allow a suitable margin in the insulation level above the maximum switching surge overvoltage and also permit a little risk for failure in the interest of economical adoption of insulation levels. Usually statistical methods are adopted based on a given risk of flashover which is calculated by combining the flashover voltage distribution function of the insulation structures with the over-voltage probability density function.

Let $P_0 (V_i) \, dV_i$ be the probability of a surge voltage occurring as an overvoltage between V_i and $(V_i + dV_i)$. Let the probability for flashover of the insulation be $P_0(V_i)$. Then, probability of both the above events occurring simultaneously in V_i and $(V_0 + V_i)$ will be given by $P_0(V_i)P_d(V_i)$. The risk of failure over the entire voltage range then becomes

$$\text{Risk of failure,} \quad R = \int_0^{V_i} P_0(V_i) \cdot P_d(V_i) \cdot dV_i \qquad (8.45)$$

If a number of insulation structures such as a string of insulators are subjected to a switching surge, then the failure is adjusted for appropriate probability, considering all the individual probabilities simultaneously.

The risk of failure is graphically represented in Fig. 8.27.

A simplified procedure to evaluate the risk of failure is given by the IEC which defines the safety factor as a ratio of the statistical overvoltage to that of statistical withstand voltage (T). The former voltage is the voltage likely to exceed 2% of all the overvoltages, while the latter is the 90% probability voltage for failure which is given as (CFO − 1.3 σ), where σ is the standard deviation of the overvoltage distribution function and CFO is the critical flashover voltage. A graph between the risk of failure and the statistical safety factor is given by the IEC and is shown in Fig. 8.28. It is evident from this figure that increasing the statistical safety factor will reduce the risk

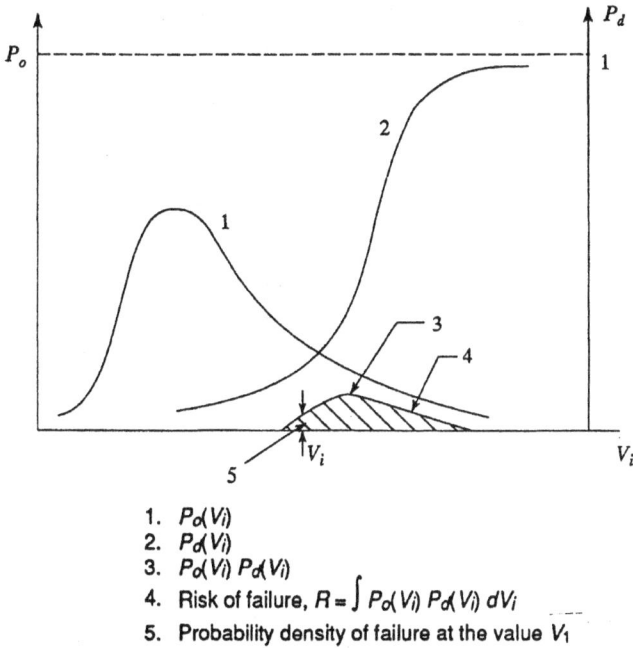

1. $P_o(V_i)$
2. $P_d(V_i)$
3. $P_o(V_i)\, P_d(V_i)$
4. Risk of failure, $R = \int P_o(V_i)\, P_d(V_i)\, dV_i$
5. Probability density of failure at the value V_1

Fig. 8.27 Risk of failure as a function of the probability of the occurrence of a surge voltage $[P_o(V_i)]$ and the probability of the insulation flashover $[P_o(V_i)]$

of failure but will cause an increase in the cost of insulation to be provided. For non-self restoring insulations like that of transformers, etc., the withstand voltage is expressed in terms of the breakdown voltage values.

The protective level provided by the protective devices like the surge diverters is established in the same manner as that for other apparatus, the difference being that the surge diverters must absorb the surge. The safety margin is arrived at by considering the risk factor R for the device used for the protection and the insulation structure to be protected, giving a safe margin.

In normal practice, the insulation level and the protective safety margin are arrived at by

(*i*) selecting the risk of failure R,

(*ii*) the statistical safety factor, γ, and

(*iii*) then fixing the withstand voltage and designing the insulation level of any equipment or apparatus corresponding to 90% or 95% of the withstand voltage thus fixed.

This type of approach may be understood as follows:

Let $P_o(V_i)$ represent the probability of occurrence of an overvoltage of magnitude V_{os} and has only 2% chance to cause the failure of the insulation. This voltage is known as statistical overvoltage. Let V_{ds} be the voltage that causes only 10%

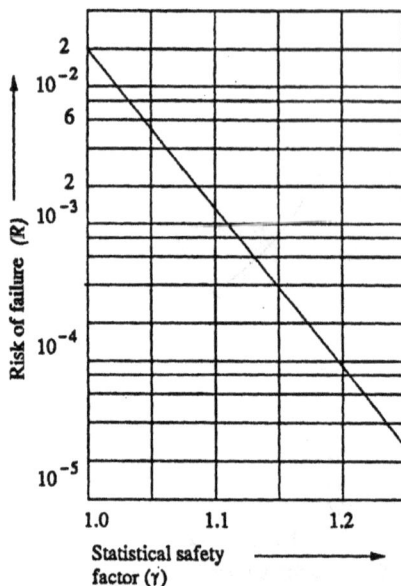

Fig. 8.28 Relation between the risk of failure (R) and the statistical safety factor (τ)

or less breakdowns or failures, when applied to the insulation system. Also, let $P_d(V_i)$ represent the statistical withstand voltage corresponding to the voltage V_{ds} for the given insulation system. These probabilities are shown in Fig. 8.29. The statistical safety factor, τ defined earlier, is given as the ratio V_{ds}/V_{os}. This statistical parameter τ, if increased, reduces the risk of failure R. This is illustrated in Fig. 8.29, where if the curve of the insulation flashover voltage $[P_d(V_i)]$ if shifted towards the right, will increase the safety factor and reduce the risk of failure. Hence, as has been stated earlier, proper selection of R and τ is to be done.

For proper insulation co-ordination, a certain margin of safety has to be provided by properly choosing the "protective level" for protective devices, such as spark gaps and surge diverters, and proper insulation level for the equipment and the apparatus. The correlation between the two is illustrated in Fig. 8.30.

In the figure, R_g gives the risk factor for the protective gap with $P_g(V_i)$ as its probability density function for failure and the overvoltage probability density function $P_o(V_i)$ occurring. The probability density function for insulation (to be protected) is given by $P_i(V_i)$ and R_i is its risk factor. The safety margin which is the difference between $P_g(V_i)$ and $P_i(V_i)$ is also shown. In reality, this computation is not simple and deviation in the occurrence of overvoltages greatly influences the risk factors and the safe margin usually gets reduced.

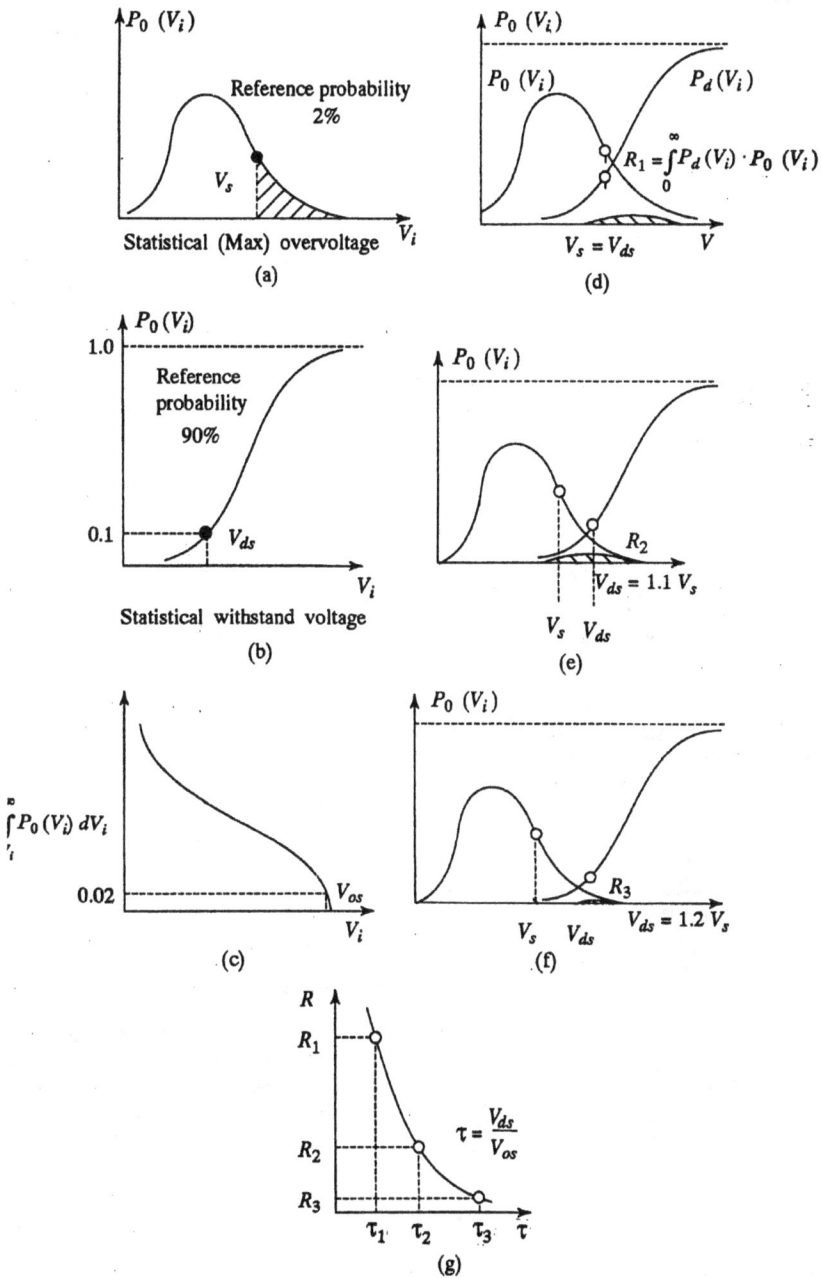

Fig. 8.29   The statistical safety factor ($\hat{\tau}$) and its relation to the risk of failure (R)

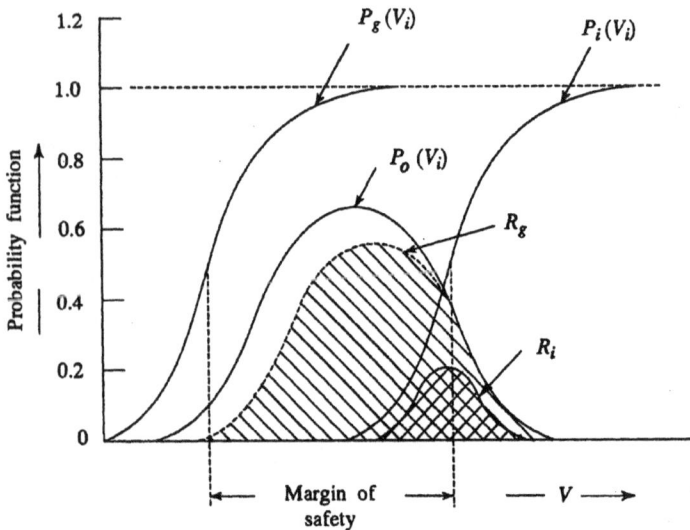

Fig. 8.30   Protective margin (margin of safety) and risk of failure provided
by a protective device

The insulation co-ordination and safety margins fixed for a typical 750 kV station
can be as follows:

surge diverter voltage rating	:	590 kV
maximum sealed off value of the power		
frequency harmonics	:	950 kV
lightning surge protective level	:	1450 kV
switching surge protective level	:	1200 kV
transformer or reactor BIL	:	1800 kV
transformer or reactor SIL		
circuit breaker and other apparatus SIL . . . :		1500 kV
when closed	:	1350 kV
when open	:	1500 kV
the maximum switching surge level		
taken as (2.0 p.u.)	:	1250 kV

If the 765 kV line uses a V type insulator string with 35 standard type disc insulators,
then the CFO for lightning surges will be about 1350 kV. These values overlap with
the SIL of the circuit breaker and other apparatus. Then, a suitable design for a
conductor to tower clearance and for the conductor to ground clearance have to be
given.

QUESTIONS

Q. 8.1   Explain the different theories of charge formation in clouds.

Q. 8.2   What are the mechanisms by which lightning strokes develop and induce overvoltages on overhead power lines ?

Q. 8.3   Give the mathematical models for lightning discharges and explain them.

Q. 8.4   What are the causes for switching and power frequency overvoltages ? How are they controlled in power systems ?

Q. 8.5   What are the different methods employed for lightning protection of overhead lines ?

Q. 8.6   Explain with suitable figures the principles and functioning of (a) expulsion gaps, (b) protector tubes.

Q. 8.7   What is a surge diverter ? Explain its function as a shunt protective device.

Q. 8.8   What is meant by insulation co-ordination ? How are the protective devices chosen for optimal insulation level in a power system ?

Q. 8.9   With suitable illustrations, explain how insulation level is chosen for various equipment in a 230/132 kV sub-station.

Q. 8.10 Write short notes on:

(a) Rod gaps used as protective devices.

(b) Ground wires for protection of overhead lines.

Q. 8.11 Derive the expressions for the voltage and current waves on long transmission lines and obtain the surge impedance of the line.

Q. 8.12 Define "surge impedance" of a line. Obtain the expressions for voltage and current waves at a junction or transition point.

Q. 8.13 Explain the terms "attenuation and distortion" of travelling waves propagating on overhead lines. What is the effect of corona on the transmission lines ?

Q. 8.14 A transmission line of surge impedance, Z_A equal to 500 Ω is connected through a cable of surge impedance 50 Ω to another line of surge impedance Z_B equal to 600 Ω. A travelling wave of 100 $u(t)$ kV travels from the 500 Ω line towards the 600 Ω line through a cable. Calculate the voltage at the junction of the 500 Ω line and the cable, after the first and second reflections.

Q. 8.15 A 500 kV, 2 μs rectangular wave travels on a line having a surge impedance of 350 Ω and approaches a termination with a capacitance C equal to 300 pF. Determine the magnitudes of the reflected and transmitted waves.

Q. 8.16 A 220 kV, 3-phase line has a horizontal configuration of conductors 5 m apart. The ground clearance is 15 m. Find the position and the number of ground wires required.

Q. 8.17 Explain the importance of switching overvoltages in EHV power systems. How is protection against overvoltages achieved ?

Q. 8.18 Explain the different aspects of insulation design and insulation co-ordination adopted for EHV systems.

WORKED EXAMPLES

Example 8.1: A 3-phase single circuit transmission line is 400 km long. If the line is rated for 220 kV and has the parameters, $R = 0.1$ ohms/km, $L = 1.26$ mH/km, $C = 0.009$ μF/km, and $G = 0$, find (a) the surge impedance and (b) the velocity of propagation neglecting the resistance of the line. If a surge of 150 kV and infinitely long tail strikes at one end of the line, what is the time taken for the surge to travel to the other end of the line ?

Solution : Velocity of propagation $= \dfrac{1}{\sqrt{LC}}$

$$= \frac{1}{\sqrt{1.26 \times 10^{-3} \times 0.009 \times 10^{-6}}}$$

$$= 3 \times 10^5 \text{ km/s}$$

$$\text{Surge impedance} = \sqrt{\frac{L}{C}}$$

$$= \sqrt{\frac{1.26 \times 10^{-3}}{9 \times 10^{-9}}}$$

$$= 374.2 \, \Omega$$

Time taken for the surge to travel to the other end is

$$= \frac{400}{3 \times 10^5}$$

$$= 1.33 \, \text{m s}$$

Example 8.2: A transmission line of surge impedance 500 Ω is connected to a cable of surge impedance 60 Ω at the other end. If a surge of 500 kV travels along the line to the junction point, find the voltage build-up at the junction ?

Solution:
$$e = 500 \, U(t) \, \text{kV}$$
$$Z_1 = 500 \, \Omega$$
$$Z_2 = 60 \, \Omega$$

Coefficient of reflection,

$$\Gamma = \frac{(Z_2 - Z_1)}{(Z_2 + Z_1)} = \frac{(500 - 60)}{(500 + 60)}$$

$$= 0.786$$

Magnitude of the transmitted wave to the cable

$$= (1 + \Gamma) \, e$$
$$= (1.786) \times 500$$
$$= 893 \, \text{kV}$$
$$= \text{junction voltage}$$

Example 8.3: An infinite rectangular wave on a line having a surge impedance of 500 Ω strikes a transmission line terminated with a capacitance of 0.004 μF. Calculate the extent to which the wave front is retarded ?

Solution: $e = E \, U(t)$, $Z_1 = 500 \, \Omega$, and $Z_2 = \dfrac{1}{Cs} = \dfrac{10^9}{4s}$

$$\Gamma = \frac{Z_2 - Z_1}{Z_2 + Z_1}$$

$$= \frac{\left(\dfrac{10^9}{4s} - 500\right)}{\left(\dfrac{10^9}{4s} + 500\right)}$$

$$= \frac{(10^9 - 2000s)}{(10^9 + 2000s)}$$

Hence,

$$(1 + \Gamma) = \frac{2 \times 10^9}{(10^9 + 2000 \, s)}$$

$$= \frac{10^6}{\left(\frac{10^6}{2} + s\right)}$$

$e''(s)$ voltage across capacitor $= (1 + \Gamma) \, e(s)$

$$= \frac{10^6}{\left(\frac{10^6}{2} + s\right)} \frac{E}{s}$$

Taking inverse transforms

$$e'' = 2 E \left[1 - \exp\left(-\frac{t}{2 \times 10^{-6}}\right) \right]$$

Taking the rise time to be $3CZ$, the wave is retarded by

$$3 \times 2 \times 10^{-6} = 6 \times 10^{-6} \, s$$

or by $6 \, \mu \, s$.

Example 8.4: A 10 MVA, 132 kV transformer is connected to the end of a transmission line of surge impedance 400 Ω. The transformer has an equivalent capacitance of 0.002 μ F and leakage inductance of 16 H. If a rectangular wave of 1000 kV travels through the line and strikes the transformer, find the surge voltage on the transformer ?

Solution: $e = 1000 \, U(t) \, kV$; $Z_1 = 400 \, \Omega$, Z_2 = combination of L and C given as,

$$\left[\frac{\frac{s}{C}}{\left(s^2 + \frac{1}{LC} \right)} \right]$$

$$\alpha = \frac{1}{CZ} = \frac{1}{0.002 \times 400 \times 10^{-6}}$$

$$= 1.25 \times 10^6$$

$$\omega_0 = \frac{1}{\sqrt{LC}} = \frac{1}{\sqrt{16 \times 0.002 \times 10^{-6}}}$$

or, $$\omega_0^2 = \frac{10^6}{0.032}$$

$$= 3.125 \times 10^7$$

$$(\alpha/2)^2 > \omega_0^2$$

Hence the voltage across the transformer $= e''(s)$

$$= (1 + \Gamma) \, e(s) = \frac{2\alpha E}{s^2 + \alpha s + \omega_0^2}$$

and the inverse transform is,

$$e''(s) = \left[\left(\frac{2a}{n-m}\right)\exp(-mt) - \exp(-nt)\right]E\,U(t)$$

where n and m are the roots of the denominator $s^2 + \alpha s + \omega_0^2 = 0$
i.e. $n = 1.25 \times 10^6$, and $m = 30$

$$e = 1000\,U(t)\text{ kV}$$

$\therefore$
$$e'' = + \frac{2.5 \times 10^6}{1.2497 \times 10^6}[\exp(-30t) - \exp(-1.25 \times 10^6 t)\,E\,U(t)$$

$$= 2000\,[\exp(-30t) - \exp(-1.25 \times 10^6 t)]$$

(The incident wave is modified to a double exponential wave when it reaches the transformer terminal.)

Example 8.5: An underground cable of inductance 0.189 mH/km and of capacitance 0.3 μF/km is connected to an overhead line having an inductance of 1.26 mH/km and capacitance of 0.009 μF/km. Calculate the transmitted and reflected voltage and current waves at the junction, if a surge of 200 kV travels to the junction, (*i*) along the cable, and (*ii*) along the overhead line.

Solution: Surge impedance of the cable $Z_1 = \sqrt{\dfrac{L_1}{C_1}}$

$$= \sqrt{\frac{0.189 \times 10^{-3}}{0.3 \times 10^{-6}}}$$

$$= 25.1\,\Omega$$

Surge impedance of the line $\qquad Z_2 = \sqrt{\dfrac{L_2}{C_2}}$

$$= \sqrt{\frac{1.26 \times 10^{-3}}{0.009 \times 10^{-6}}}$$

$$= 374.2\,\Omega$$

When the surge travels along the cable:

$$\Gamma = \frac{(374.2 - 25.1)}{(374.2 + 25.1)}$$

$$= 0.8742$$

The reflected wave, $\quad e' = \Gamma e = 0.8742 \times 200\text{ kV}$

$$= 174.84\text{ kV}$$

The transmitted wave, $\quad e'' = (1 + \Gamma)\,e$

$$= 1.8742 \times 200\text{ kV}$$

$$= 374.84\text{ kV}$$

The reflected current wave, $\quad I' = \dfrac{e'}{Z_1} = \dfrac{174.84 \times 10^3}{25.1}$

$$= 6.97\text{ kA}$$

The transmitted current wave, $I'' = \dfrac{e''}{Z_2} = \dfrac{374.84 \times 10^3}{374.20}$

$$= 1.002 \text{ kA}$$

When the wave travels along the line:

$$\Gamma = \frac{(25.1 - 374.2)}{(25.1 + 374.2)}$$

$$= -0.8742$$

$\therefore$ reflected wave $= e' = \Gamma e = -174.84$ kV

The transmitted wave $= e'' = (1 + \Gamma) e = (1 - 0.8742) \times 200$

$$= 25.16 \text{ kV}$$

The transmitted current wave $= I''$

$$= \frac{e''}{Z_2} = \frac{25.16}{25.1}$$

$$= 1.006 \text{ kA}$$

The reflected current wave $= I'$

$$= \frac{e'}{Z_1} = \frac{-174.84}{374.2}$$

$$= 0.467 \text{ kA or } 467 \text{ A}$$

Example 8.6: A long transmission line is energised by a unit step voltage 1.0 V at the sending end and is open circuited at the receiving end. Construct the Bewley lattice diagram and obtain the value of the voltage at the receiving end after a long time. Take the attenuation factor $\alpha = 0.8$.

Solution: Let the time of travel of the wave = 1 unit

At the receiving end

Reflection coefficient $\gamma = (\infty - Z)/(\infty + Z) = 1.0$

Transmission coefficient $= 1 + \gamma = 2.0$

At the sending end

Reflection coefficient $\gamma = (0 - Z)/(0 + Z) = -1.0$

Transmission coefficient $= 1 + \gamma = 0$

Since the source impedance $Z_s = 0$ and Z_2, the open receiving end impedance is ∞ (infinity), as shown in the lattice diagram of Fig. E8.6(b).

From the lattice diagram, the wave magnitudes are tabulated as shown below :

At the receiving end	At the sending end	Time_unit
1	0	0
1	1	α
$1 + \alpha^2$	2	2α
1	3	$2\alpha - \alpha^3$

$1 - \alpha^4$	4	$2\alpha - 2\alpha^3$
1	5	$2\alpha - 2\alpha^3 + \alpha^5$
$1 + \alpha^6$	6	$2\alpha - 2\alpha^3 + 2\alpha^5$

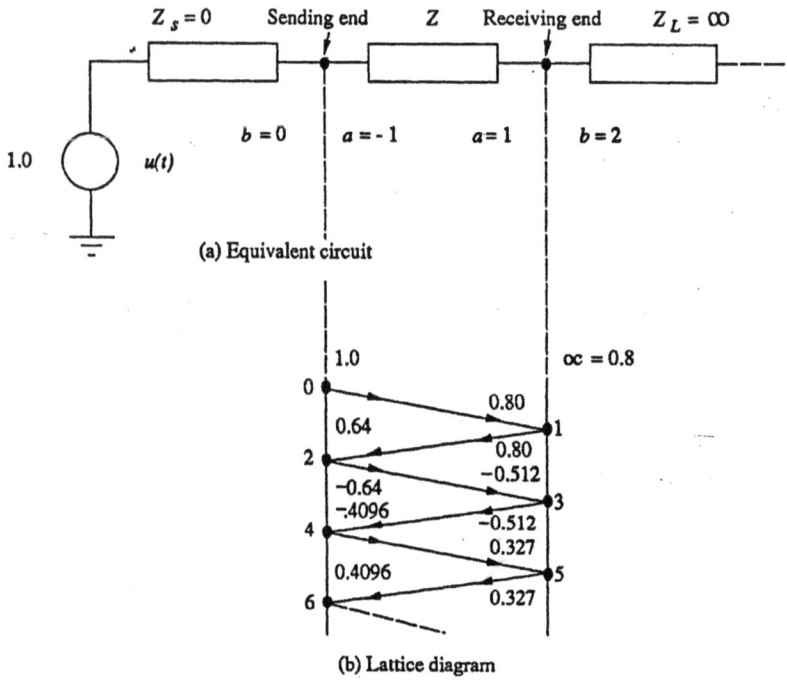

(a) Equivalent circuit

(b) Lattice diagram

Fig. E.8.6 Equivalent circuit and lattice diagram of a transmission line

The voltage at the receiving end after $4n$ units of time is

$$V = 2(\alpha - \alpha^3 + \alpha^5 - + \dots)\, u(t)$$
$$= 2\alpha\,[(1 - \alpha^{4(n+1)})/(1 + \alpha^2)]\, u(t)$$

Voltage at the receiving end after a long time (i.e. $t = \infty$) is $V_{u(t)} = [2\alpha/(1 + \alpha^2)]\, u(t)$. Substituting $\alpha = 0.8$, we get $V_\infty = 0.9756\, u(t)$.

Example 8.7: Determine the sparkover voltage and the diverter current when a surge diverter is connected at the end of a transmission line of surge impedance 400 Ω. Assume that a surge of 1000 kV (peak) strikes the diverter. The surge diverter characteristic for impulse currents may be taken as follows:

Surge current (kA)	1.5	3.0	5.0	10.0
Arrester voltage (kV)	264	308	336	360

Solution: (a) *Calculation of sparkover voltage with the line terminated*

Assuming that the line is terminated, the equivalent voltage is twice the surge voltage = 2000 kV. Neglecting the ground and diverter resistances, the maximum diverter current

$$= 2 \times 2000/400 = 5 \text{ kA}$$

Referring to Fig. E 8.7, the characteristic is cut by the line, $\tan^{-1}Z$ at $V_d = 330$ kV. Hence the voltage drop across the diverter is 330 kV and the current through the diverter 3.9 kA. For steep fronted current waves, the voltage drop due to lead inductance is taken to be 5% more. Hence, the sparkover voltage, $V_d = 350$ kV.

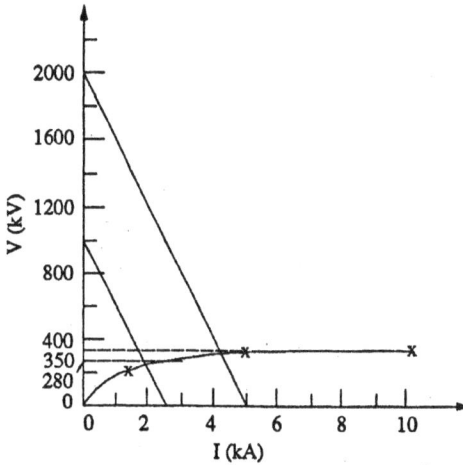

Fig. E.8.7   *V-I* characteristic of a surge diverter

(b) *Calculation of surge diverter sparkover voltage when the line is continuous*

If the line is continuous, the Thevinin equivalent voltage is

$$= \text{surge voltage} = 1000 \text{ kV}$$

Hence, the maximum diverter current = 1000/400 = 2.5 kA.

From Fig. 8.7, the sparkover voltage $\approx$ 280 kV and the diverter current = 1.8 kA. Adding 5% extra voltage for the steep-fronted current waves, the sparkover voltage, $V_d = 295$ kV.

Example 8.8: A transmission line has the following line constants $R = 0.1$ ohm/km, $L = 1.26$ mH/km, $C = 0.009$ μF/km, and $G = 0$. If the line is a 3-phase line and is charged from one end at a line voltage of 230 kV, find the rise in voltage at the other end, if the line length is 400 km.

Solution: Method I: Neglecting the resistance of the line,

$$V_2 = V_1 \left(1 - \frac{X_L}{2X_C}\right)$$

$$X_L = j\omega L = j(314)(1.26 \times 10^{-3} \times 400) = j \, 158.26 \, \Omega$$

$$X_C = -j/\omega C = -j\,(10^6/314 \times 0.009 \times 400) = -j\,884.6\,\Omega$$

$$\therefore \qquad V_2 = V_1\left[1 - \frac{j\,158.26}{2(-j\,884.6)}\right]$$

$$= V_1\,(1 + 0.0896)$$

The given line voltage = 230 kV

$$\therefore \qquad \text{phase voltage} = \frac{230}{\sqrt{3}} = 132.8\,\text{kV}$$

Hence, the rise in voltage $\quad = 0.0896 \times 132.8\,\text{kV}$

$$= 11.9\,\text{kV}$$

Method 2 : Considering all the parameters,

$$V_2 = \frac{V_1}{\cos\beta\,l}$$

where β is the phase constant of the line and l is the length of the line.

$$(\alpha + j\beta) = \text{transmission line constant}$$

$$= \sqrt{(R + j\omega L)\,(G + j\omega C)}$$

$$= [(0.1 + j314 \times 1.26 \times 10^{-3})(j \times 314 \times 0.009 \times 10^{-6})]^{1/2}$$

$$(\alpha + j\beta) = 400[(0.1 + j314 \times 1.26 \times 10^{-3})(j314 \times 0.009 \times 10^{-3})]^{1/}$$

$$= (0.1962 + j0.3822)\,\text{rad}:$$

$$\therefore \qquad \cos\beta l = 0.9278$$

$$V_1 = \frac{230}{\sqrt{3}} = 132.8\,\text{kV}$$

$$\therefore \qquad V_2 = \frac{V_1}{\cos\beta l}$$

$$= V_1 \times 1.0778$$

$$= 143.13\,\text{kV}$$

Hence, the rise in voltage is 10.33 kV.

Note: Even with resistance neglected, the rise in voltage is changed by 1.58 kV or ~ 1.2 %. Hence, in such calculations the effect of the line resistance is negligible.

REFERENCES

1. Bewley, L.V., *Travelling Waves on Transmission Systems*, Dover Publications Inc., New York (1963).
2. Lewis, W.W., *Protection of Transmission Lines and Systems Against Lightning*, Dover Publications, New York (1965).
3. *Transmission and Distribution Reference Book*, Westinghouse Electric Corporation and Oxford University Press, New Delhi (1962).
4. Marshall, J.L., *Lightning Protection*, John Wiley and Sons, New York (1973).
5. Diesendorf, W., *Insulation Coordination in H.V. Electric Power Systems*, Butterworths, London (1974).

6. Begamudre, R.D., E.H.V, *A.C. Transmission Engineering*, Wiley Eastern, New Delhi (1986).
7. Golde, R.H., *Lightning*, Vol. 1 and 2, Academic Press, London (1977).
8. R.M. Black and E.H. Raynolds, "Ionization and irradiation effects in high voltage dielectric materials", *Journal of Institute of Engineers*, London, 112, 1226 (1965).
9. Pitman, R.R., *et al.*, "Lightning investigations", *Tr. AIEE*, 50, 508 (1931).
10. Bell, E., *et al.*, "Lightning investigations on 220 kV systems", *Tr. AIEE*, 150, 1101 (1931).
11. *ERA Report*, 1941.
12. McEachron, K.B., "Lightning stroke on Empire State Building", *Tr. AIEE*, 60, 885 (1941).
13. Muller Hiller Brand *et al.*, "Lightning counter measurements", *Proc. IEE*, 112, 203 (1965).
14. Anderson, J.G., *EHV Transmission Reference Book*, Edison Electric Co, New York, 1968.
15. *CIGRE Report No.* 22, 139 (1972).
16. "AIEE Committee report", *Tr. AIEE*, Part 3, 69, 1187 (1950).
17. Berger, K., "Observations on lightning discharges", *J. Franklin Institute*, 283, 478 (1967).
18. Newman, S.E. *et al.*, "Insulation coordination in H.V. stations", *English Electric Journal*, 13, No. 3, 120 (1953).
19. Wilson, D.D., "EHV insulation coordination", *Electric World*, 165, 92 (1966).
20. "Insulation co-ordination", *IEC Technical Committee 28, Report No. 35* (1970).
21. *IEC Publication* on "Insulation coordination", No. 71, 5th edition, (1972).
22. Smith, D.C., *et al.*, "Dynamic overvoltages in 560 kV transmission lines", *Tr Electrical Engineering (Australia)*, EE4, No. 1 (1968).
23. Indian Standard on Insulation Coordination, *IS: 2165-1977*.
24. Paris, *et al.*, "Switching surge characteristics of large air gaps and long insulator strings", *Tr. IEEE PAS*, PAS 87, 947 (1968).
25. Udo, T., "Sparkover characteristics of long air gaps and insulator strings", *Tr. IEEE PAS*, PAS-83, 471 (1964).
26. Phelps J.D. *et al.*, "765 kV station insulation co-ordination" *Tr. IEEE PAS*, PAS-88, 1377 (1965).
27. Kuffel E. and Zaengl W., *"High Voltage Engineering Fundamentals"*, Pergamon Press, Oxford, England, 1984.

9
Non-Destructive Testing of Materials and Electrical Apparatus

9.1 INTRODUCTION

Electrical insulating materials are used in various forms to provide insulation for the apparatus. The insulating materials may be solid, liquid, gas, or even a combination of these such as paper impregnated with oil. These materials should possess good insulating properties over a wide range of operating parameters, such as a wide temperature range (0°C to 110°C) and a wide frequency range (d.c. to several MHz in the radio and high frequency ranges). Since it is difficult to test the quality of an insulating material after it forms part of an equipment, suitable tests must be done to ensure their quality in the said ranges of operation. Also, these tests are devised to ensure that the material is not destroyed as in the case of high voltage testing.

These tests are mainly done to assess the electrical properties, such as the resistivity (d.c.), the dielectric constant, and loss factor over a wide frequency range. In the high voltage apparatus, the quality of insulation is assessed by measuring the loss factor at high voltages and also by conducting partial discharge tests to detect any deterioration or faults in the internal insulation of the apparatus.

These tests may be conducted at a desired temperature or over a temperature range by keeping the test specimen in controlled temperature ovens. A knowledge of the variation of electrical properties over the operating range can be obtained from these tests and this will help the design engineer to take into account such variations in the design of electrical insulation for equipment.

9.2 MEASUREMENT OF d.c. RESISTIVITY

9.2.1 Specimens and Electrodes

The specimen shape and the electrode arrangement should be such that the resistivity can be easily calculated. For a solid specimen, the preferable shape is a flat plate with plane and parallel surfaces, usually circular. The specimens are normally in the form of discs of 5 to 10 cm diameter and 3 to 12 mm thickness.

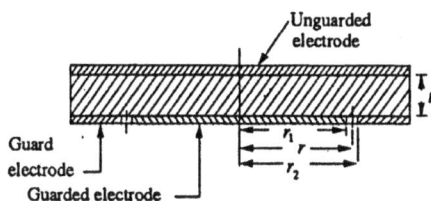

Fig. 9.1 The three electrode system

If the electrodes are arranged to be in contact with the surfaces of the specimen, the measured resistance will be usually greater due to the surface conductivity effects. Often, a three electrode arrangement shown in Fig. 9.1 is used. The electrode which completely covers the surface of the specimen is called the "unguarded" electrode and is connected to the high voltage terminal. The third electrode which surrounds the other measuring electrode is connected to a suitable terminal of the measuring circuit. The width of this "guard" electrode must be at least twice the thickness of the specimen, and the unguarded electrode must extend to the outer edge of the guard electrode. The gap between the guarded and guard electrodes should be as small as possible. The effective diameter of the guarded electrode is greater than the actual diameter and is given as follows.

Let r_1, r_2, and r be the radii of the guarded electrode, guard electrode including the gap, and the effective radius of the guarded electrode. Let the gap width = g and the specimen thickness = t. Then from references 5 and 6,

$$r = r_1 + \frac{g}{2} - \delta \qquad (9.1)$$

where,

$$\frac{\delta}{t} = \frac{2}{\pi} \ln \cosh\left(\frac{\pi g}{4t}\right)$$

9.2.2   Electrode Materials

For accurate measurements, the electrodes should have very good contact with the surface of the insulator specimen. Hence, it is necessary to use some type of thin metallic foil (usually of lead or aluminium of about 10 to 50 μm thickness), usually pressed on to the surface by a roller and made to stick by using a conducting adhesive like petroleum jelly or silicone grease. The electrodes are made simultaneously by cutting out a narrow strip by means of a compass provided with a narrow cutting edge. Sometimes conducting silver paint is also used for electrode deposition. Evaporated metal deposition on the specimen surface or use of mercury electrodes, applied by floating the specimen on a pool of mercury using confining rings with sharp edges for holding the metal, are also convenient.

9.2.3   Measuring Cells

The three terminal electrode system and the measuring cell used are shown in Fig. 9.2. The measuring cell is usually a shallow metalbox provided with insulating terminals. The box itself is connected to the guard electrode and is grounded if the guard terminal is grounded. The connecting lead for the guarded electrode is taken through a shielded wire. In case the unguarded electrode is grounded, the entire box is to be placed on insulated supports and is to be placed in a grounded shield to eliminate induced voltages, and the lead from the guard electrode is doubly shielded.

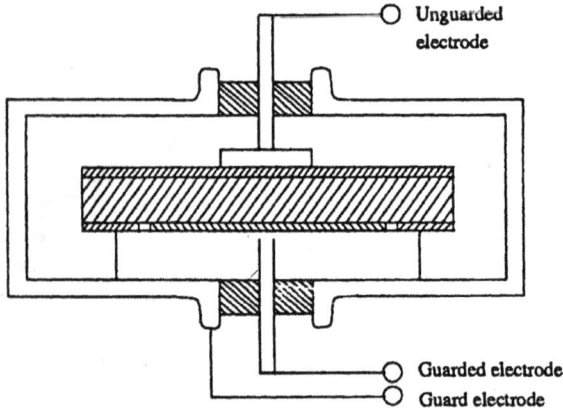

Fig. 9.2   The three terminal cell for study of solids

In the simple two terminal system, the measuring cell itself is the grounded support for the specimen and a small solid wire is connected to the high voltage terminal of the measuring circuit, as can be seen from Fig. 9.3. This is a simple and compact arrangement for quick measurements and requires less skill.

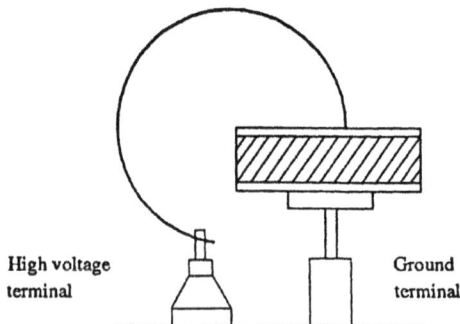

Fig. 9.3   Stiff wire connection for two terminal measurement

The arrangement used for the study of liquids is shown in Fig. 9.4. This consists of an outer cylindrical case and an inner cylinder with a cylindrical guard electrode. The opposing surfaces of the measuring electrodes should be carefully finished to give a polished surface, and a uniform spacing of about 0.25 mm is maintained. The insulation should be able to maintain the alignment of the electrode even at the highest temperatures used and should still allow easy disassembling and cleaning. Another

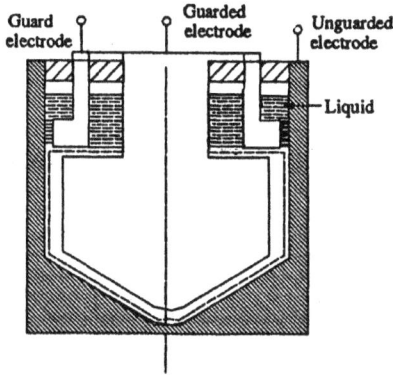

Fig. 9.4   Three terminal cell for study of liquids

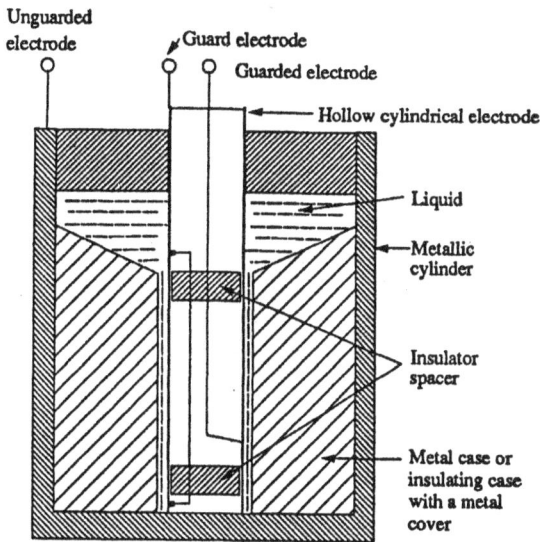

Fig. 9.5   Three terminal cell for the study of liquids

simple arrangement of the three electrode system for the study of liquids is shown in Fig. 9.5. This consists of a metallic cylindrical container with concentric hollow cylindrical electrodes as guard and guarded electrodes. The inner surface of the container electrode (unguarded) and the outer surfaces of the guard and unguarded electrodes should be carefully finished and a clearance of about 0.25 to 0.5 mm should be accurately maintained. The arrangement requires less liquid (usually only about 1 to 2 ml).

9.2.4 Measuring Circuits

A simple measuring circuit for the measurement of resistance is shown in Fig. 9.6. The galvanometer is first calibrated by using a standard resistance of 1 to 10 MΩ ($\pm$ 0.5% or $\pm$ 1%). If necessary, a standard universal shunt is used with the galvanometer. The deflection in cm per microampere of current is noted. The specimen (R_p) is inserted in the current as shown in Fig. 9.6, and maintaining the same supply voltage, the galvanometer current is observed by adjusting the universal shunt, if necessary. The galvanometer gives a maximum sensitivity of 10^{-9} A/cm deflection and a d.c. amplifier has to be used along with the galvanometer for higher sensitivities (up to 10^{-12} to 10^{-13} A/cm).

The resistance of the specimen is given by

$$R_p = \frac{V}{(D \times G)} \qquad (9.2)$$

where, D = deflection in cm (with specimen), and

G = galvanometer sensitivity

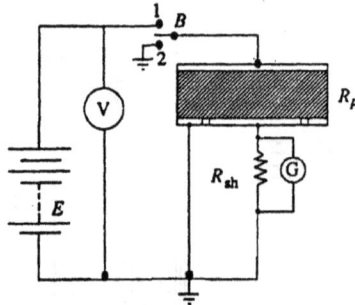

E — d.c. stabilized power supply, 500 to 2000 V
V — Voltmeter,
G — Galvanometer 10^{-8} to 10^{-9} A/cm deflection
R_{sh} — Universal shunt
R_p — Specimen

Fig. 9.6 d.c. galvanometer arrangement

$$\left(\frac{V}{R_s}\right) \times \left(\frac{1}{n}\right) \times \left(\frac{1}{D_s}\right)$$

in which, n = the universal shunt ratio,

D_s = deflection in cm with the standard resistance in position,

R_s = standard resistance used for calibration, and

V = the supply voltage.

It may be noted that the deflection with the specimen in the circuit, will change with time. The initial high deflections indicate the high charging current required by the dielectric specimen. For the purpose of calculations, the steady current obtained after a considerably long changing time should only be taken. Switch B is used to discharge the specimen after the measurements are complete and as well as for the initial discharging before the measurements are undertaken.

In Fig. 9.7 the d.c. galvanometer of Fig. 9.6 is replaced by a d.c. amplifier for resistivity measurements. Here, the d.c. amplifier is used as a null detector. A separate potentiometer circuit is used for obtaining a signal e equal and opposite to the voltage drop across the standard resistance R_s. At balance, the voltage drop across R_s is effectively made zero. Any a.c. voltage appearing across the standard resistance R_s is amplified only by the net gain of the amplifier which is small. Using a recording meter after the d.c. amplifier in this arrangement, the volt-ampere-time curves for long durations, of the order of several hours (days also) can be obtained. This type of information is essential to determine the relaxation time at low frequencies or with d.c.

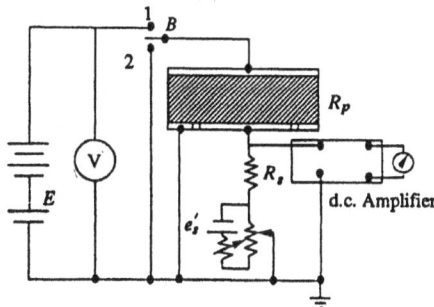

E — Stabilized power supply, 500 to 2000 V
V — Voltmeter
R_p — Specimen
R_s — Standard resistance ($\approx$ 1 M Ω)
e'_s — Auxiliary source

Fig. 9.7 d.c. amplifier circuit for resistivity measurements

Figure 9.8 gives the Wheatstone bridge network. One of the resistances (usually R_A) is made variable for balancing the bridge. At balance, R_p is given by

$$R_p = R_s \frac{R_A}{R_B} \qquad (9.3)$$

With the specimen arrangement used, the volume resistivity ρ can be given as

$$\rho = \frac{R\pi r^2}{t} \, \Omega \, m \qquad (9.4)$$

where, $R =$ measured resistance in ohms,

$t =$ thickness of the specimen in metres, and

$r =$ effective radius in metres.

Using this network, the volume resistivity ρ for liquid specimens with cylindrical electrodes as shown in Fig. 9.5, can be expressed as

$$\rho = \frac{2\pi R l}{\ln \dfrac{d_2}{d_1}} \, \Omega \, m \qquad (9.5)$$

where, $R =$ measured resistance in ohms,

$l =$ effective length of the guarded electrode in metres,

$d_2 =$ inner diameter of unguarded electrode, and

$d_1 =$ outer diameter of guarded electrode, both in metres.

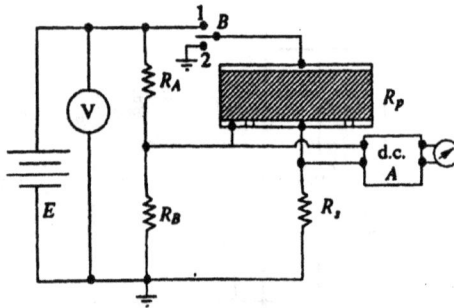

Fig. 9.8 Wheatstone bridge arrangement for resistivity measurements

9.2.5   Loss of Charge Method

Using the circuit shown in Fig. 9.6, the current-time characteristic of the discharge current can be obtained by placing the switch B in position 2. The capacitance of the specimen is discharged through its own volume resistance. The slope of the current-time characteristic gives the time constant $\tau = CR$, where C is the capacitance of the specimen. The volume resistance of the specimen R can thus be obtained from this relationship by measuring τ and C.

Alternatively, a capacitance previously charged (C_1) may be discharged through the specimen by inserting it in place of the battery E. By knowing the value of the capacitance C_1 which is very large compared to the specimen capacitance ($\approx 10\mu F$) and knowing the current-time characteristic of the discharge current, its time constant can be calculated. Hence, the resistance of the specimen is obtained from the relationship

$$R = \frac{C_1}{\tau}$$

where C_1 is the capacitance used and τ is the time constant.

In the deflection method described, a limit is reached when the insulation resistance is so high that the galvanometer deflection is not at all adequate. This happens when the resistance is around 10^{12} Ω or more. Under these conditions, the loss of charge method is convenient. It is also convenient to measure the initial voltage V_0 on the capacitor and the voltage V across it at any time t. Then the unknown resistance R can be calculated from the relationship

$$R = \frac{t}{C_1 \ln \frac{V_0}{V}} \tag{9.6}$$

It is convenient to use an electrometer amplifier or electrostatic voltmeter for the measurement of the voltage. The measured value of R includes the resistance of the voltage measuring device in parallel.

The accuracy of this method depends on the accuracies with which the values of the capacitance and the voltage are measured.

9.3 MEASUREMENT OF DIELECTRIC CONSTANT AND LOSS FACTOR

9.3.1 Introduction

Many insulating substances have dielectric constant greater than unity and have dielectric loss when subjected to a.c. voltages. These two quantities, namely, the dielectric constant and the loss depend on the magnitude of the voltage stress and on the frequency of the applied voltage. When a dielectric is used in an electrical equipment such as a cable or a capacitor, the variation of these quantities with frequency is of importance. The microscopic properties of the dielectric are described by combining the variation of the above two quantities into one "complex quantity" known as "complex permittivity" and determining them at various frequencies.

A capacitor connected to a sinusoidal voltage source $v = v_0 \exp (j\omega t)$ with an angular frequency $\omega = 2\pi f$ stores a charge $Q = C_0 v$ and draws a charging current $I_c = dQ/dt = j\omega C_0 v$. When the dielectric is vacuum, C_0 is the vacuum capacitance or geometric capacitance of the condenser, and the current leads the voltage v_c by 90°.

If the capacitor is filled with a dielectric of permittivity ε', the capacitance of the condenser is increased to $C = C_0 \varepsilon'/\varepsilon_0 = C_0 K'$ where K' is the relative dielectric constant of the material with respect to vacuum.

Under these conditions, if the same voltage v is applied, there will be a charging current I_c and loss component of the current I_l. I_l will be equal to Gv where G represents the conductance of the dielectric material. The total current $I = I_c + I_l = (j\omega C + G)V$. The current leads the voltage by an angle θ which is less than $90°$. The loss angle δ is equal to $(90 - \theta)°$. The phasor diagrams of an ideal capacitor and a capacitor with a lossy dielectric are shown in Figs. 9.9a and b.

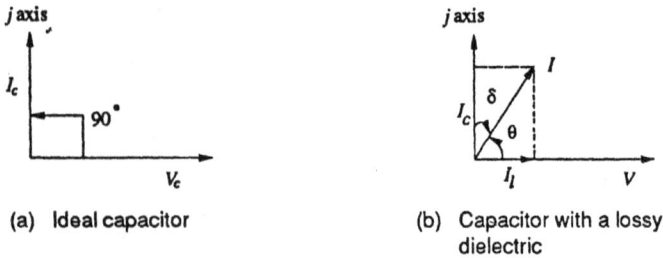

(a) Ideal capacitor

(b) Capacitor with a lossy dielectric

Fig. 9.9   Capacitor phasor diagrams

It would be premature to conclude that the dielectric material corresponds to an R-C parallel circuit in electrical behaviour. The frequency response of this circuit which can be expressed as the ratio of the loss current to the charging current, i.e. the loss tangent

$$\tan \delta = D = \frac{I_l}{I_c} = \frac{1}{\omega CR} \tag{9.7}$$

may not at all agree with the result actually observed, because the conductance need not be due to the migration of charges or charge carriers but may represent any other enery consuming process. Hence, it is customary to refer the existence of a loss current in addition to the charging current by introducing "complex permittivity"

$$s^* = \varepsilon' - j\varepsilon'' \tag{9.8}$$

so that current I may be written as

$$I = (j\omega\varepsilon' + \omega\varepsilon'')\frac{C_0}{\varepsilon_0} v \tag{9.9}$$

$$= j\omega C_0 K^* v$$

where $\qquad K^* = (\varepsilon' - j\varepsilon'')/\varepsilon_0 = K' - jK'' \tag{9.10}$

K^* is called the complex relative permittivity or complex dielectric constant, ε' and K' are called the permittivity, and relative permittivity and ε'' and K'' are called the loss factor and relative loss factor respectively.

The loss tangent

$$\tan \delta = \frac{\varepsilon''}{\varepsilon'} = \frac{K''}{K'} \tag{9.11}$$

The product of the angular frequency and ε'' is equivalent to the dielectric conductivity σ.

$$\sigma = \omega\varepsilon'' \qquad (9.12)$$

The dielectric conductivity sums up all the dissipative effects and may represent the actual conductivity as well as the energy loss associated with the frequency dependence (dispersion) of ε', i.e. the orientation of disposal in a dielectric.

In dielectric measurements, often, the geometrical capacitance and the capacitance of the system with a dielectric material are obtained. The ratio of the above two measurements gives the relative permittivity $\varepsilon'/\varepsilon_0 = K'$. This is sometimes referred to as the dielectric constant or ε_r.

9.3.2   Measurement Ranges

Lumped circuits are used in the measurements of the dielectric constant and tan δ over the frequency range from d.c. (0 Hz) to about 100 MHz. Neither of these quantities are determined directly. Measurements of capacitance using either a null method or a deflection method is employed over the low frequency range (0-10 Hz), where bridge methods are difficult. Normally, the shape of the current-time characteristic curve is determined from which these parameters are deduced. Sometimes, modified bridge methods are extended to this frequency range also. Over the medium frequency range (10 Hz to 10^6 Hz), bridge methods are employed. The commonly employed bridge networks are the four arm Schering bridge to operate from power frequency (50 Hz) to about 100 kHz and the "High Voltage Schering Bridge" for power frequency (50 Hz) when the effect of voltage stress on dielectric constant and tan δ is required. Over the high frequency range (100 kHz to 100 MHz), bridge circuits present problems in shielding and the errors of measurements become excessive. Micrometer electrodes using vernier capacitors (electrode separation being controlled and measured to micrometer (μm) accuracy have to be employed to avoid errors due to residual impedances. Using these methods, dissipation factor, tan δ, is usually obtained from the width of the resonant curve by varying either the frequency or the susceptibility of the circuit. At frequencies above 100 MHz, the lumped circuit approximation is not valid as the wavelength of the frequency approaches that of the specimen thickness. Microwave techniques like "the travelling wave" or "the standing wave" measurements are to be employed beyond the frequency limit of 10^8 Hz. The information obtained from the measurements of tan δ and the complex permittivity help in assessing the quality of the dielectric and the insulation system.

(i)   Variation and sudden change in tan δ value with applied voltage is an indication of the inception of partial discharge (P.D.). This is used to determine the inception level of internal discharges and losses due to P.D. in high voltage equipment.

(ii)   To study the variation of the dielectric properties with frequency. Much of the interest in this study is concerned with the frequency region where dispersion occurs, i.e. where the permittivity reduces with rise in the frequency.

9.3.3 Low Frequency Measurement Methods (0-10 Hz)

In testing equipment of large kVA ratings at high voltages and to study the properties of dielectrics at zero frequency (d.c.) or near to it, it is advantageous to carry out tests in the frequency range of zero to 10 Hz. Conventional Wheatstone type bridges are not suitable over this frequency range. The d.c. amplifier method described in Sec. 9.2.4 can be used in testing the dielectrics over this frequency range. The accuracy of the measurements will depend on the stability of the supply sources and the detectors.

Mole's Bridge

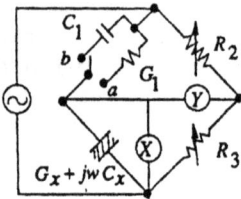

Fig. 9.10 Mole's bridge

The bridge that is usually used for low frequency measurements is the Mole's bridge whose schematic diagram is shown in Fig. 9.10. This is primarily devised to measure the dispersion in insulation of installed equipment and gives a reliable measure of the moisture content in the insulation. It is well known that dielectric dispersion increases rapidly with decrease in frequency and hence a more sensitive measurement is obtained at low frequencies. The bridge shown in Fig. 9.10 does not have the conventional balancing system. A cathode ray oscillograph (CRO) is used as the null indicator and the bridge is said to be balanced when the Y plate voltage V_y is in quadrature with X plate voltage V_x. Thus, with the switch in position a balance occurs when

$$\frac{V_x}{R_3}R_2 - \frac{V_x}{G_1}(G_x + j\omega C_x) = V_y = jkV_x$$

where

$$K = \frac{V_y}{V_x}$$

Hence,

$$G_x = \left(\frac{R_2}{R_3}\right)G_1 \qquad (9.13)$$

The bridge can be made direct reading in terms of G. Similarly, balance is obtained with the switch in position b when

$$V_x'\frac{R'_2}{R'_3} - V_x'\frac{(G_x + j\omega C_z)}{j\omega C_1} = V'_y = jK'V'_x$$

where

$$K' = \frac{V'_y}{V'_x} \text{ in the second case.}$$

Hence,

$$C_x = \left(\frac{R'_2}{R'_3}\right)C_1 \qquad (9.14)$$

R'_2 and R'_3 are new values of the ratio arms R_2 and R_3. By fixing the ratio of R_2 and R_3, the bridge may be made direct reading for C_x also in terms of C_1.

Usually, the bridge employes the ratio arms with a ratio range of 10 : 1 to 1 : 10, capacitance in the range 10 pF to 1.0 µF and conductance in the range 10^{-10} to 10^{-6} Siemens (mhos). Oscillators (the source voltages) operate in the frequency range of 0.05 Hz to 100 Hz.

9.3.4 Power Frequency Measurement Methods — High Voltage Schering Bridge

In the power frequency range (25 to 100 Hz) Schering bridge is a very versatile and sensitive bridge and is readily suitable for high voltage measurements. The stress dependence of K' or ε_r and tan δ can be readily obtained with this bridge.

The schematic diagram of the bridge is shown in Fig. 9.11. The lossy capacitor or capacitor with the dielectric between electrodes is represented as an imperfect capacitor of capacitance C_x together with a resistance r_x. The standard capacitor is shown as C_s which will usually have a capacitance of 50 to 500 µF. The variable arms are R_4 and $C_3 R_3$. Balance is obtained when

$$\frac{Z_1}{Z_2} = \frac{Z_4}{Z_3}$$

where,
$$Z_1 = r_x + \frac{1}{j\omega C_x}, \qquad Z_2 = \frac{1}{j\omega C_s}$$

$$Z_3 = \frac{R_3}{1 + j\omega C_3 R_3}, \text{ and } Z_4 = R_4$$

The balance equations are

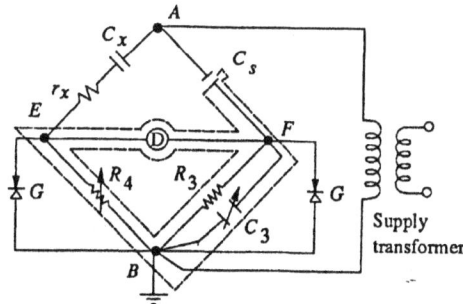

--- dotted line is the shielding arrangement. Shield is connected to
B, the ground

Fig. 9.11 Schematic diagram of a Schering bridge

$$C_x = \frac{R_3}{R_4} C_s; \text{ and } r_x \frac{C_3}{C_2} R_1 \tag{9.15}$$

The loss angle, $\tan \delta_x = \omega C_x R_x$

$$= \omega C_3 R_3 \tag{9.16}$$

Usually δ_x will be small at power frequencies for the common dielectrics so that

$$\cos \theta_x = \sin \delta_x = \delta_x = \tan \theta_x = \omega C_2 R_3 \tag{9.17}$$

The lossy capacitor which is made as an equivalent C_x in series with r_x can be represented as a parallel combination of C_x and R_x where the parallel combination R_x is found to be

$$R_x = \frac{1}{\omega^2 C_x^2 r_x} \tag{9.18}$$

with C_x having the same value.

The normal method of balancing is by fixing the value of R_3 and adjusting C_3 and R_4. If R_3 is chosen as $(1000/\pi)$ ohms for $\omega = 100 \pi$ and if C_3 is expressed in microfarads, then $\tan \delta = 0.1 C_3$ giving a direct reading of $\tan \delta$. R_4 will be a decade box with 5 to 6 decade dials. The maximum value of R_4 is limited to $10^4 \Omega$ and the lowest value will not be less than 0.01Ω. This range adequately takes care of the errors due to contact resistances as well as the stray capacitance effects across R_4 which are usually very small. It is important to see that the resistances are pure and not reactive and the standard capacitor has negligible $\tan \delta$ (air or gas filled capacitor is used).

The arrangement shown in Fig. 9.11 is suitable when the test specimen is not grounded. The standard condenser C_s is usually a three terminal condenser. The low voltage arms of the bridge (R_4 and $R_3 C_3$) and the detector are enclosed in grounded shielded boxes to avoid stray capacitances during the measurements. The detector is either a vibration galvanometer or in modern bridges a tuned electronic null detector of high sensitivity. The protective gaps G are so arranged that the low voltage arms are protected from high voltages in case the test objects fail. The values of the impedances of the low voltage arms are such that the voltage drop across EB or FB does not exceed 10 to 20 V. The arms will be usually rated for a maximum instantaneous voltage of 100 V.

For a very accurate measurement of the dissipation factor at power frequency, the stray and grounded capacitances should be eliminated and the indirect capacitive and inductive coupling of the arms are to be minimized to a level lower than the accuracy of the bridge arms. In this bridge the main source of error is the ground capacitance of the low voltage terminals of high voltage arms, i.e. the stray capacitances from E and F to ground. These are eliminated by shielding the low voltage arms using doubly shielded cables for connections and using the "Wagner earthing device". Sometimes, compensation for the stray capacitances is given by providing a parallel R-L circuit across R_4.

Schering Bridge Arrangement for Grounded Capacitors

For safety reasons and to define the shield potentials with respect to each other, one of the terminals of the bridge is earthed. This is usually the low voltage arm connection of the supply source (Terminal B of Fig. 9.11), because a low impedance arm with respect to the detector branch gives a high signal to noise ratio in measurements.

While testing grounded test objects like underground cables or bushings with flanges grounded to the tank of a transformer, one of the detector terminals (E or F) has to be grounded. In such cases, either an inverted bridge (Fig. 9.12) or a grounded detector arrangement (Fig. 9.13) has to be adopted. In the inverted bridge operation, the self-contained bridge is located inside a Faraday cage at the high voltage terminal and a standard capacitor is mounted on insulating supports. The variable arms are operated by insulated isolating rods. Sometimes, the operator himself will be inside the Faraday cage.

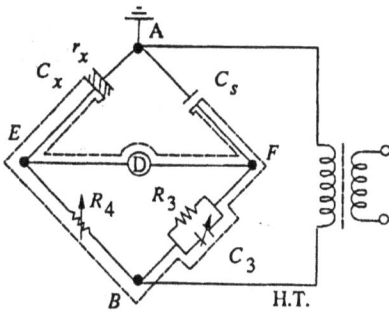

Fig. 9.12   Inverted Schering bridge   **Fig. 9.13**   Schering bridge for grounded
for grounded capacitors   objects (detector end grounded)

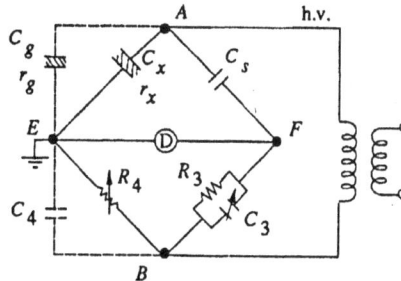

In the grounded detector arrangement, stray capacitances of the high voltage terminal (C_g) and of the source, leads, etc. come in parallel with the test object. Hence, the balancing is to be done in two steps. First, the test object is disconnected and the capacitance C_g and tan (δ_z) are measured. Then the test object is connected and a new balance is obtained. The second balance gives,

$$C'_x = C_x + C_g$$

and

$$\tan \delta'_x = \frac{C_x \tan \delta_x + C_g \tan \delta_g}{C_x + C_g} \qquad (9.19)$$

Hence, the actual capacitance and dissipation factor of the test object are

$$C_x = C'_x - C_g$$

and

$$\tan \delta_x = \frac{C'_x \tan \delta'_x - C_g \tan \delta_g}{C_x} \qquad (9.20)$$

The accuracy of the measurement is poor, if the ground capacitance is large compared to the test object capacitance.

Schering Bridge for High Charging Currents

High capacitance test objects, such as high voltage power cables and power factor correction capacitors take high charging currents which may exceed the rating of the variable resistance R_4 (Fig. 9.11). Further, the value of the resistance R_4 will become so low that the contact resistances and residual inductances can no longer be neglected. For these cases, the range is increased by connecting a shunting resistor R_s in parallel across R_4 as shown in Fig. 9.14. Hence, the balance conditions are modified as

$$C_x = C_2 R_4 \left[\frac{R_s + r + S + R_4}{R_s (R + \Delta S)} \right]$$

and
$$\tan \delta_x = \omega C_3 R_3 - \omega C_s R_3 \left[\frac{r + S - \Delta S}{R_4 + \Delta S} \right] \tag{9.21}$$

where,

R_s = shunt resistance,

r = a small resistance added in the first arm,

S = slide wire resistance

R_4 = the decade variable resistor, and

ΔS = portion of the slide wire resistance.

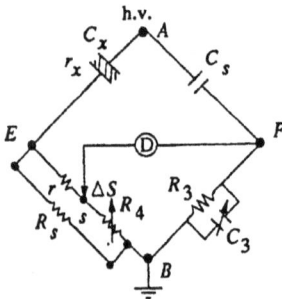

R_s — Shunt resistance
S — Slide wire
ΔS — Topped portion of slide wire
r — Fixed small resistance
R_4 — Decade resistance box

Fig. 9.14 Modified Schering bridge for large charging currents

Usually R_s, the shunt resistance is kept outside, and special arrangements are made for the correction of its inductive phase angle error, if necessary.

Schering Bridge for High Dissipation Factors

Extension of range of the dissipation factor can be achieved by connecting in parallel an additional large capacitance across C_3. But to avoid expensive and large-size decade capacitances the $C_3 R_3$ arm is modified as shown in Fig. 9.15a. R_3 is made as a slide wire along with a decade resistance, and C_3 is extended by connecting a single fixed capacitor. Usually the modification is limited to $\tan \delta \approx 1.0$.

For $\tan \delta$ values greater than 1.0 and in the range 1 to 10 or greater, the $C_3 R_3$ arm is no longer made a parallel combination. Instead, it is made a series combination of C_3 and R_3 where R_3 is variable. This makes the test object to be an equivalent of C_x in parallel with R_x, a better approximation for the capacitors and dielectrics. With this

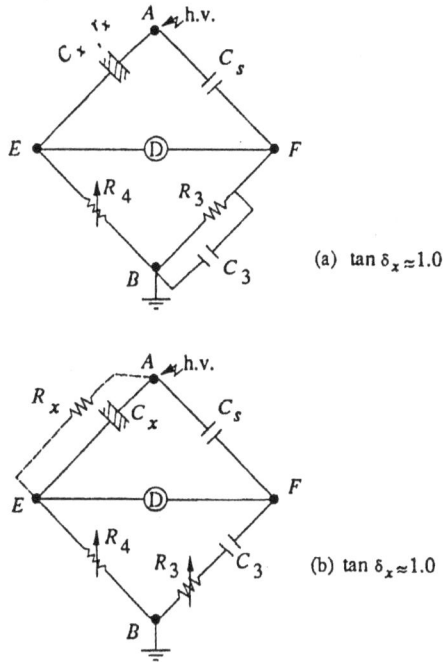

Fig. 9.15   Schering bridge for large dissipation factors

modification, tan δ range is made wider, from 10^{-4} to values greater than 10. The arrangement is shown in Fig. 9.15b.

9.3.5   Audio Frequency Measuring Techniques

The basic Schering bridge discussed in Sec. 9.3.4 may be used also in the audio frequency range for dielectric measurements. But it presents problems at higher frequencies, and the values of C_3 and R_3 have to be greatly altered. Sometimes, balancing also becomes difficult for low values of the test specimen capacitance, and the errors due to the stray capacitances across the arm R_4 may be appreciable. Therefore, in the case of small capacitance measurements (1 to 100 picofarads and less) substitution methods are used which, over this capacitance range may be more accurate. Also, in order to take care of all these effects modified Schering bridge shown in Fig. 9.16 is preferred.

Schering Bridge for Audio Frequency Range (50 Hz to 100 kHz)

The capacitance bridge used for dielectric measurements is shown in Fig. 9.16, where $C_A R_A$ and $C_B R_B$ are the ratio arm capacitances. Substitution method is preferred in this range for greater accuracy. The specimen is connected across the standard variable capacitance C_N. The bridge is balanced with and without the dielectric

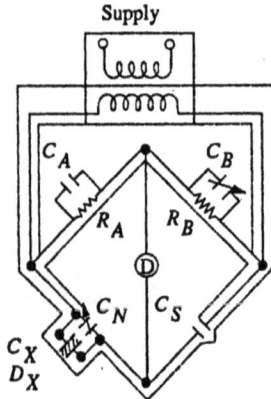

Fig. 9.16 Schering bridge for audio frequency measurements with shields and source transformer

specimen C_X. The specimen capacitance connected (C_X) is the difference of the two readings of the standard capacitor C_N.

$$C_X = \Delta C_N = C'_N - C_N \qquad (9.22)$$

The balance for the loss factor (resistive component balance) may be done in any one of the following ways:

(i) a small variable resistance in series with the standard capacitor C_N,

(ii) a variable resistor of high value in parallel with C_T, and

(iii) a variable capacitor in parallel with R_B (C_B variation).

Of these three, the third method is normally used, as C_B can be made a variable air condenser of high quality and the errors that arise within the resistances at high frequencies (skin effect, etc.) are avoided. The dissipation factor tan δ for the third method is given as,

$$\tan \delta = \frac{C'_N}{\Delta C} \Delta D \qquad (9.23)$$

where, $\Delta D = D - D' = \omega R_B (C_B - C'_B)$

Usually, the total capacitance C_N in the variable arm is made small to get a large variation in ΔD.

Whenever the specimen capacitance is greater than that of the maximum value of C_N, the substitution method is not possible. In such cases, direct measurement without C_N is done. The expressions, in the case of a direct measurement for C_X and tan δ are:

$$C_x = \frac{R_A}{R_B} C_N \qquad (9.24)$$

$$\tan \delta = \frac{f}{f_0} R_B C_B \qquad (9.25)$$

$$= \frac{f}{f_0} x \text{ (dial reading)}$$

where f_0 is the calibration frequency.

The range of this bridge extends from 1 pF to about 1000 pF when the substitution method is used, from 1000 pF to 100 μF when the direct method is used.

Earthing and Shielding

For two terminal measurements the bridge is grounded at its junction points. The supply transformer, detector, and all the components of the bridge are enclosed in earthed shields. For three terminal measurements, it is necessary to avoid stray capacitances, for accurate measurements. Hence, a guard circuit and earthing device (known as Wagner's earthing device) are used. The bridge is balanced once with the ratio arms and again with the earthing device, alternately, so that no change in balance occurs. This ensures the elimination of stray ground capacitances and coupling. The schematic arrangement is shown in Fig. 9.17. The arms, containing R_V and $R_W C_W$ are the Wagner earthing device arms. The bridge is first balanced with R_A, R_B, C_N, and C_T and later with R_y, R_w, C_N, and C_T. The detector is alternately connected to either the earthing device or to the bridge arms A and B. The balance is achieved, when at either position the detector indicates the same null indication.

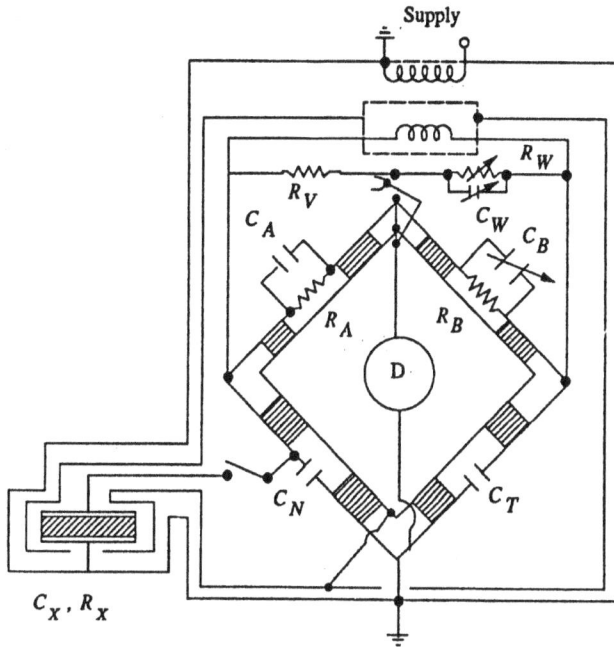

Fig. 9.17　Schering bridge for three terminal measurements with Wagner's earthing device

This bridge can be extended to frequencies as high as 500 kHz beyond which the lead lengths and the residual inductances in various arms become too high to achieve proper balance. Hence, for higher frequencies other methods have to be adopted.

Transformer Ratio Arm Bridges

It is a common practice to use the four arm Wheatstone bridge network for a.c. measurements. In high frequency measurements, the arms with high values of resistances lead to difficulties due to their residual inductances, capacitances, and skin effect. Also, shielding and grounding becomes difficult in large arms. Hence, at high frequencies the transformer ratio arm bridges which eliminate at least two arms are preferred. These bridges are also useful for the measurement of low value of capacitances accurately.

The ratio arm bridges can be either voltage ratio type or current ratio type; the former being used for high frequency low voltage applications.

The schematic diagram of a ratio arm bridge (voltage ratio) is given in Fig. 9.18. Assuming ideal transformer conditions, for a null indication of the detector,

$$\frac{V_s}{V_x} = \frac{N_s}{N_x} = \frac{C_x}{C_s}; \text{ and } \frac{R_x}{R_a} = \frac{N_x}{N_a} \tag{9.26}$$

where C_x and C_s are unknown and standard capacitances respectively, R_x and R_a are unknown and standard resistances, and N_x, N_a, and N_s are the corresponding turns of the transformer ratio windings.

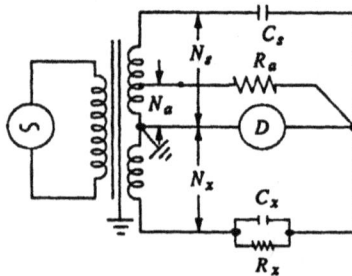

Fig. 9.18 Transformer voltage ratio arm bridge

In practical transformers, the voltage ratio slightly differs from the turns ratio due to the no load magnetizing current and is also affected by the load current. Therefore, the balance conditions shown above involve errors. The errors are classified as the ratio and loading errors and are determined separately and compensated for in the construction. A practical bridge constructed by General Radio Company (USA) has a useful range from a fraction of one pF to about 100 µF and covers a wide range of frequency from 100 Hz to 100 kHz, the accuracy being better than 0.5%.

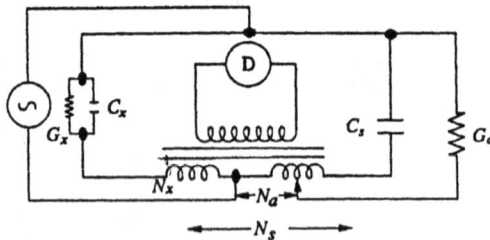

Fig. 9.19a Current comparator bridge

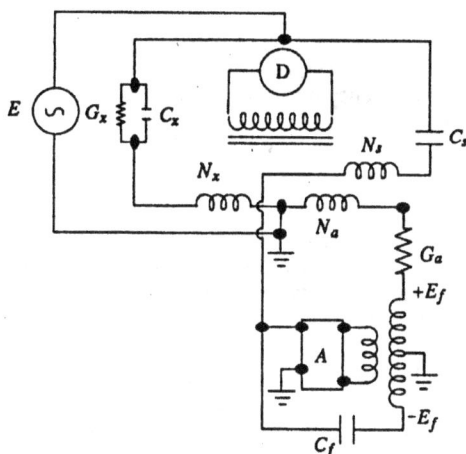

E_f — Proportional e.m.f.
A — Amplifier
C_f — Balancing capacitor

Fig. 9.19b Current comparator for high voltage application

For high voltage applications where sensitive measurements at fixed frequency (at 50 Hz) are required, the current comparator or the current ratio method (Fig. 9.19a) is used. This bridge has the advantage that full voltage is applied across the test capacitor but also has the drawback that a standard conductance has to be built for high voltages. It is difficult to construct a precision conductance suitable for high voltage operation. This disadvantage is overcome by generating a low voltage signal E_f proportional to and in phase with the supply voltage E as shown in the modified Fig. 9.19b. At balance, there is no voltage across the current comparator winding. If the gain of the amplifier (A) is high, that is,

$$E_f \approx \frac{C_s}{C_f} E$$

The balance equations of the bridge are

$$C_x = C_s \frac{N_s}{N_x}; \quad G_x = \frac{C_s}{C_f} \frac{N_a}{N_x} G_a$$

and
$$\tan \delta = \frac{G_x}{\omega C_x} = \frac{1}{\omega C_f} \frac{G_a N_a}{N_x} \tag{9.27}$$

where $C_x, C_s, N_s, N_a,$ and N_x are as defined in Eq. (9.26), G_x and G_a are unknown and standard conductances, and C_f is the balancing condenser.

9.3.6 Detectors in Dielectric Measurements

Since the measurements cover a large range of frequency, the detectors used in various ranges differ considerably.

For d.c. and low frequency measurements, d.c. galvanometer and d.c. amplifiers with a microammeter are used.

In power frequency a.c. measurements (50 Hz/60 Hz) vibration galvanometers serve as excellent tuned detectors as their bandwidth is quite narrow (± 1 to 2 Hz). But often tuned electronic amplifier null detectors are also used.

In the audio frequency range wide band or tuned null detectors are used with a sensitivity better than 10 μV. This consist of a filter, an attenuator, a multi-stage amplifier using a bridge rectifier with a microammeter. The bandwidth of the detector is from 50 Hz to 50 kHz and is protected from high input signals at off balance position.

Cathode ray oscilloscopes are also used as detectors, if too high a sensitivity is not required.

The selection of the detector depends on the type of bridge circuit used for measurement and the sensitivity required in the particular application.

9.4 PARTIAL DISCHARGE MEASUREMENTS

9.4.1 Introduction

Earlier the testing of insulators and other equipment was based on the insulation resistance measurements, dissipation factor measurements and breakdown tests. It was observed that the dissipation factor (tan δ) was voltage dependent and hence became a criterion for the monitoring of the high voltage insulation. In further investigations it was found that weak points in an insulation like voids, cracks, and other imperfections lead to internal or intermittent discharges in the insulation. These imperfections being small were not revealed in capacitance measurements but were revealed as power loss components in contributing for an increase in the dissipation factor. In modern terminology these are designated as "partial discharges" which in course of time reduce the strength of insulation leading to a total or partial failure or breakdown of the insulation.

If the sites of partial discharges can be located inside an equipment, like in a power cable or a transformer, it gives valuable information to the insulation engineer about the regions of greater stress and imperfections in the fabrication. Based on this information, the designs can be considerably improved.

Electrical insulation with imperfections or voids leading to partial discharges can be represented by an electrical equivalent circuit shown in Fig. 9.20. Consider a capacitor with a void inside the insulation (C_a). The capacitance of the void is represented by a capacitor in series with the rest of the insulation capacitance (C_b). The remaining void-free material is represented by the capacitance C_c. When the voltage across the capacitor is raised, a critical value is reached across the capacitor C_a and a discharge occurs through the capacitor, i.e. it becomes short circuited. This is represented by the closure of the switch.

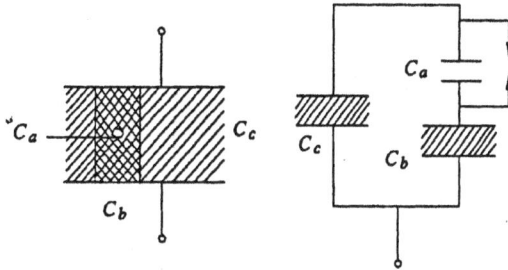

C_a — Capacitance of the void acting as a spark gap
C_b — Capacitance of the remaining series insulation with the void
C_c — Remaining part of the discharge free insulation of the test object

Fig. 9.20 Insulating device with a void C_a and its simplified electrical equivalent circuit

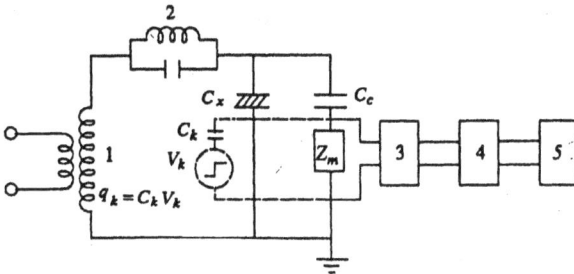

1 — H.V. testing transformer
2 — Filter
3 — Band pass filter
4 — Amplifier
5 — Display unit (CRO or pulse counter or multi-channel analyser unit)

C_x — Sample or test piece
C_c — Coupling condenser
Z_m — Detector impedance
V_k — Calibrating pulse
C_k — Calibrating capacitor
q_k — Calibrator charge

Fig. 9.21 Straight discharge detection circuit

Generally, $C_a \ll C_b \ll C_c$. A charge Δq_a which was present in the capacitor C_a flows through C_b and C_c giving rise to a voltage pulse across the capacitor C_c. A measure of the voltage pulse across the capacitor gives the amount of discharge quality. But this measurement is difficult in practice, and an apparent charge measurement across a detecting impedance is usually made.

Partial Discharge Phenomenon

Definition of terms normally used: The following terminology is often used in partial discharge detection and as such their definitions are of importance:

Electrical discharge: The movement of electrical charges through an insulating (dielectric) medium, initiated by electron avalanches.

Partial discharge: An electrical discharge that only partially bridges the dielectric or insulating medium between two conductors. Examples are: internal discharges, surface discharges and corona discharges. Internal discharges are discharges in cavities or voids which lie inside the volume of the dielectric or at the edges of conducting inclusions in a solid or liquid insulating media.
   Surface discharges are discharges from the conductor into a gas or a liquid medium and form on the surface of the solid insulation not covered by the conductor.
   Corona is a discharge in a gas or a liquid insulation around the conductors that are away or remote from the solid insulation.

Discharge inception voltage is the lowest voltage at which discharges of specified magnitude will recur when an increasing a.c. voltage is applied.

Discharge extinction voltage is the lowest voltage at which discharges of specified magnitude will recur when an applied a.c. voltage, which is more than the inception voltage, is reduced.

Discharge magnitude is the quantity of charge, as measured at the terminals of a sample due to a single discharge.

Discharge energy is the energy dissipated by a single discharge.

Average current is the average value of the discharge current during a cycle due to a single or multiple discharges. I_a, the average current over an interval T can be expressed as

$$I_a = \sum_{r=1}^{m} |q_r|$$

when, q is the apparent charge in rth discharge.

Quadratic rate is the average value of the square of the discharge magnitudes. D, the quadratic rate is given as

$$D = \frac{1}{T} \left| \sum q_r^2 \right|$$

Discharge detector is a device or an instrument used for either detecting and/or measuring the discharges.

Sensitivity is the magnitude of the smallest individual discharge that can be measured under particular test conditions.

Resolution is the minimum interval between two discharges which can be measured without the magnitude of one discharge affecting the other.

Referring to the equivalent circuit shown in Figure 9.20, when a discharge takes place, the charge transfer across the cavity C_a reduces the voltage across it from V_a to $\pm V$ and the discharge will be complete. If the a.c. voltage applied is further increased, the voltages across the cavity will again increase, till V_a reaches the discharge inception voltage u and a new discharge occurs. It has been shown (Kreuger) that the discharges remain stationary if displayed on an oscilloscope having a time base whose frequency is the same as that of the supply voltage. The discharge pattern is shown in Fig. 4.6 (refer to Chapter 4, section 4.5.3). It may be observed that the current pulses occur at the peaks and zeros of the voltage waveform. As the voltage applied increases, the number of discharges will also increase and the discharge pattern will become irregular.

9.4.2 Discharge Detection Using Straight Detectors

The circuit arrangement shown in Fig. 9.21 gives a simplified circuit for detecting "partial discharges". The high voltage transformer shown is free from internal discharges. A resonant filter is used to prevent any pulses starting from the capacitance of the windings and bushings of the transformer. C_x is the test object, C_c is the coupling capacitor, and Z_m is a detection impedance. The signal developed across the impedance Z_m is passed through a band pass filter and amplifier and displayed on a CRO or counted by a pulse counter multi-channel analyzer unit.

In Figure 9.22, the discharge pattern displayed on the CRO screen of a partial discharge detector with an elliptical display is shown. The sinusoidal voltage and the corresponding ellipse pattern of the discharge are shown in Fig. 9.22a and a single corona pulse in a point-plane spark gap geometry is shown in Figs. 9.22b and c. When the voltage applied is greater than that of the critical inception voltage, multiple pulses appear (see Fig. 9.22c), and all the pulses are of equal magnitude. A typical discharge pattern in cavities inside the insulation is shown in Fig. 9.22d. This pattern of discharge appears on the quadrants of the ellipse which correspond to the test voltage rising from zero to the maximum, either positively or negatively. The discharges usually start near the peaks of the test voltage but spread towards the zero value as the test voltage is increased beyond the inception level. The number and magnitude of the discharges on both the positive and negative cycles are approximately the same. A typical discharge pattern from a void bounded on one side by the insulation and the other side by a conductor is shown in Fig. 9.22c. This pattern of discharge is common in insulated cables (like polyethylene and XLPE cables) when the discharge is made up of a large number of pulses of small magnitude on the positive cycle and a much smaller number of large magnitude pulses on the negative half-cycle.

In the narrow band detection scheme Z_m is a parallel $L - C$ circuit tuned to 500 kHz. The bandpass filter has a bandwidth of about ± 10 kHz. The pulses after amplification are displayed in an elliptical time base of a CRO, and the resolution for the pulses is about 35 per quadrant.

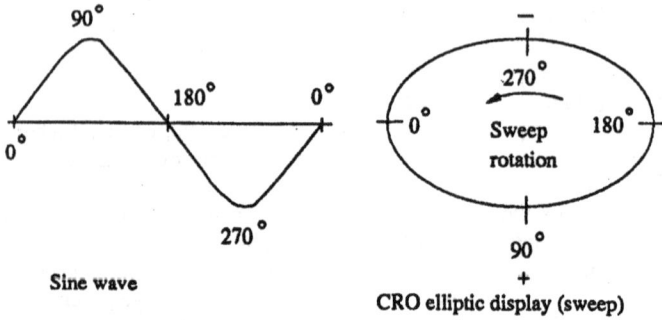

(a) Elliptic sweep display

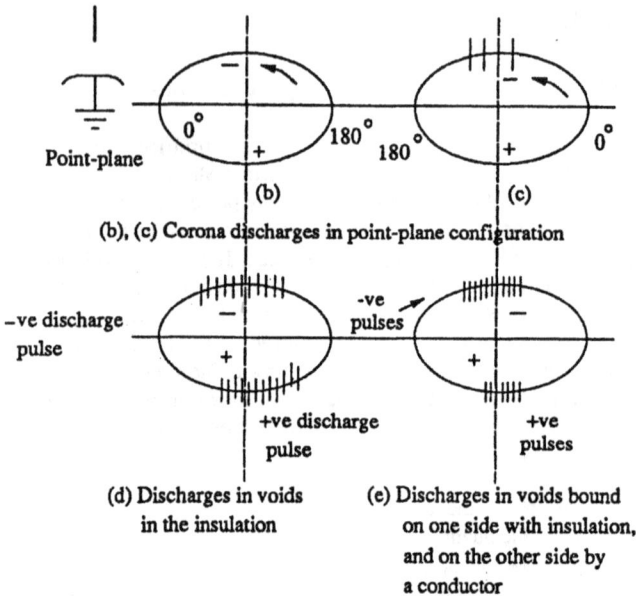

(b), (c) Corona discharges in point-plane configuration

(d) Discharges in voids
in the insulation

(e) Discharges in voids bound
on one side with insulation,
and on the other side by
a conductor

Fig. 9.22 Partial discharge patterns

In the wide band detection scheme Z_m is an R-C network connected to a double tuned transformer. The bandwidth is about 250 kHz with centre frequency between 150 to 200 kHz. A wide band amplifier is used, and the signal is displayed on the CRO as in the previous case. The resolution is about 200 pulses per quadrant.

With tuned narrow band detectors, the discharges can be detected with a sensitivity less than one pico coulomb for a test piece capacitance of 100 pF. Test pieces with capacitances in the range 100 pF to 0.1 pF can be tested. With the wide band detector samples up to 250 µF capacitance can be tested. Sensitivity of the measurement varies from 0.005 pico coulomb at 6 pF sample capacitance to about 15 pC at a sample capacitance of 250 µF.

9.4.3 Balanced Detection Method

In the straight detection method, the external disturbances are not fully rejected. The filter used to block the noise sources may not be effective. The Schering bridge employed for the tan δ measurement is sometimes used. In this method, the test object is not grounded. A modification to the Schering bridge detector is the differential discharge detector, given by Kreuger. Both the schemes are given in Fig. 9.23. The bridges are tuned and balanced at 50 Hz. A filter is used across the detector terminals to block the 50 Hz components present. Signals in the range from 5 to 50 kHz are allowed to pass through the filter and amplified. The CRO gives the display of the pulse pattern. Any external interference from outside is balanced out, and only internally (test piece) generated pulses are detected. In the modified scheme, another test sample called dummy sample is used in the place of the standard condenser. The capacitance and tan δ of the dummy sample are made approximately equal, but need not be equal. The disadvantage is that if two discharges occur in both the samples simultaneously, they cancel out, but this is very rare. The main advantage of the second method is its capacity for better rejection of external noise and use of the wide frequency band with better resolution of the individual pulses.

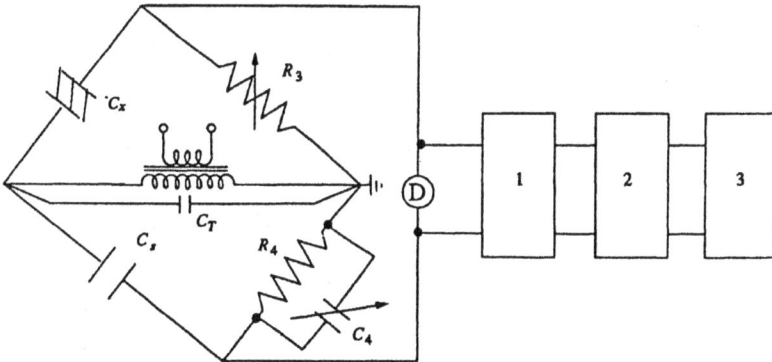

D — Power frequency detector
1 — Filter (band pass)
2 — Amplifier
3 — Display unit — CRO or counter

(a) Balanced detector using Schering bridge

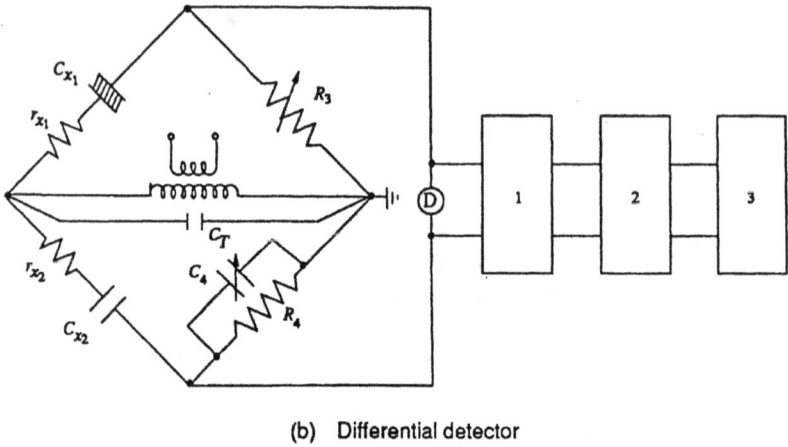

(b) Differential detector

Fig. 9.23 Balanced discharge detector schemes

9.4.4 Calibration of Discharge Detectors

Partial discharge detectors are connected across a measuring impedance Z_m as shown in Fig. 9.21 and the signal measured across this impedance is read by the detector. The signal voltage developed across Z_m depends on the circuit parameters C_x and C_c and also on the internal circuitry of the instrument (blocks 3, 4, 5 shown in Fig. 9.21).

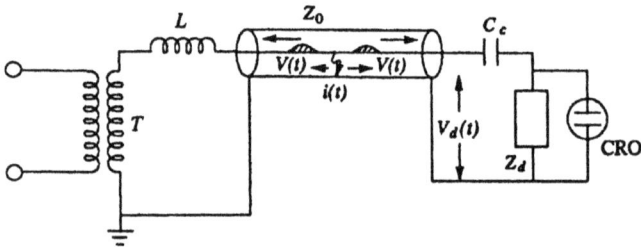

T — Transformer
Z_0 — Surge impedance of cable under test
C_c — Coupling condenser
Z_d — Detector impedance
$V_d(t)$ — Voltage signal across the detector impedance, Z_d
L — Inductor to isolate the short-circuiting influence of the transformer secondary on the termination

Fig. 9.24 Partial discharge testing arrangement for distributed parameter equipment (cables, transformer windings, etc.)

Hence, the measuring instrument or detector is calibrated by injecting a pulse having a charge of known magnitude into the detecting system. For this purpose, a square wave generator and a calibrating capacitor (C_k) are usually used. The magnitude of the charge injected is $q_k = C_k V_k$, where V_k is the magnitude of the voltage pulse. The rise time of the pulse is about 0.1 μ sec, and the pulse width varies from 10 to 20 μ sec. With suitable attenuation, the output voltage of the pulse generator can be varied from a minimum output of about 10 μV to a maximum value of 100 V in steps. The value of C_k, usually used, lies between 1000 and 2500 pF. If the calibrating pulse is directly injected at the H.V. terminal of the test object, the magnitude of the calibration pulse will be $C_k \cdot V_k$. If the pulse is injected across the measuring impedance (as shown in Fig. 9.21), then the calibrating pulse magnitude should be multiplied by $(C_x + C_c)/C_c$.

Another method of calibration is to use a secondary standard, consisting of a point-hemisphere electrode system of specified dimensions (refer to I.E. publication 270, 1968 (reference no. 8)).

This method is more accurate and is easily reproducible. With an over voltage of 10-20% applied above the discharge inception voltage, the arrangement gives discharges which are used for calibration purposes.

9.4.5 Discharge Detection in Power Cables

Power cables of great lengths may be considered to be specimens with distributed capacitance rather than a lumped capacitance. The same feature is true for windings of the transformers and rotating machines of high voltage ratings. In such cases, the location of discharge is done using the travelling wave technique, and the circuit for the same is shown in Fig. 9.24. The breakdown or discharge in a cavity at location F causes a charge accumulation at the fault point F. This creates two travelling waves which travel in opposite directions towards the cable ends. The voltage wave is given by

$$v_i = \frac{Z_0\, i(t)}{2}$$

If C_c is neglected in comparison with the detection impedance Z_d, the voltage signal across the detection impedance becomes

$$v_d(t) = \frac{Z_d}{(Z_d + Z_0)}\, Z_0\, i(t) \qquad (9.28)$$

If $\qquad\qquad Z_d = Z_0, \quad v_d(t) = \dfrac{Z_0\, i(t)}{2} \qquad\qquad (9.29)$

If the partial discharge site is located exactly in the middle of the cable, the travelling wave moving to the left is reflected without any polarity change at the open end and then arrives at the detection impedance, Z_d, separated from the direct pulse by a time equal to the cable transit time τ. If the site is at any other point, the time difference between the original pulse and its reflection may be any value from 0 to 2τ. Usually, the supply transformer end is not strictly an open circuit because of the shunt capacitance of the windings of the transformer. Therefore, the travelling-wave is reflected from the transformer end with the surge impedance of the transformer as its termination. To avoid this reflection an inductor is inserted at the transformer end. By knowing the transit time of the cable τ and the actual time difference between the two pulses, the fault position can be located (also see Sec. 10.3.4 in Chapter 10).

9.4.6 Discharge Magnitude and Discharge Energy

From the equivalent circuit shown in Fig. 9.20, when a discharge takes place in the void, the instantaneous charge transferred will be given by the product of the equivalent capacitance as seen from the terminals of the cavity, C_a, and the change in the voltage across the cavity δV_a.

The equivalent capacitance, $C = C_a + C_b C_c/(C_b + C_c)$

Therefore, $$q_a = [C_a + C_b C_c/(C_a + C_c)]\, \delta \cdot V_a \tag{9.30}$$

For all cavities in the insulation, $C_c \gg C_b$

Hence, $$q_a \simeq (C_a + C_b) \cdot \delta V_a \tag{9.31}$$

The charge q_a cannot be measured directly. The change in the voltage across the sample and hence C_c can be measured.

Thus, δV_c, the change in the voltage across the sample is given by

$$\delta V_c = \left[C_b/(C_b + C_c) \right] \cdot \delta V_a \tag{9.32}$$

Therefore, the apparent charge transfer, q_c, which is the product of δ_{V_c} and the equivalent capacitance across the terminals, C_c can be written as

$$q_c = [C_c + C_a C_b/(C_a + C_b)]\, \delta V_c = (C_b + C_c)\, \delta V_c \tag{9.33}$$

Substituting for V_c from equation (9.31) and (9.32), we get

$$q_c = [C_b/(C_a + C_b)]\, q_a = C_b \cdot \delta V_a \tag{9.34}$$

q_c is called the apparent charge or the discharge magnitude, as already defined in Sec. 9.4.1.

If $\pm u$ is the voltage across the cavity before the discharge occurs, i.e. the voltage required for the breakdown of the gas in the cavity, and $\pm v$ is the voltage across the cavity while the discharge takes place, then, the discharge voltage,

$$V_a = u - v.$$

The energy transfer from the cavity is the difference in the stored energy before and after the discharge. Hence, w, energy, transferred $= (1/2)C_a u^2 - (1/2) C_a v^2$

$$= (1/2)\, C_a\, (u + v)\, (u - v)$$
$$= (1/2)\, C_a\, (u + v) \cdot \delta V_a$$

If the residual voltage v is neglected, then

$$w = (1/2)\, C_a\, u \cdot \delta V_a \tag{9.35}$$

u is related to the peak value of discharge inception voltage, V_i, and is given by

$$u = [C_b/(C_a + C_b)] \cdot V_{i(peak)}$$

δ substituting for u in Eq. (9.35), we get

$$w = (1/2)C_a \cdot \delta V_a \cdot [C_b/(C_a + C_b)] \cdot V_{i\text{(peak)}}$$
$$\approx (1/2)C_b \cdot \delta V_a \cdot V_{i\text{(peak)}}$$

from Eq. (9.34), we have

$$W = (1/2)q_c V_{i\text{(peak)}}$$
$$= (1/\sqrt{2})q_c V_{i\text{(rms)}}$$
$$= 0.71 \, q_a V_i \tag{9.36}$$

Here, W is a measure of the discharge energy.

Even a small discharge whose magnitude is 10 pc, at an inception voltage of 7.1 kV (rms), will release an energy of 5×10^8 J. Since, this energy is released in a small duration of about 10 ns (10^{-8} s), the power is quantitatively about 5 watts which can cause heating and damage to the insulation.

The observable effects due to partial discharge are: (i) increase in the loss component of the charging current taken by the capacitive sample, (ii) increased dielectric power loss, (iii) radiation of energy in the form of sound, light or ultraviolet radiation, (iv) changes in pressure, and (v) chemical changes in the dielectric material. Some of these effects are useful in locating the sites of partial discharges. Further details about this can be obtained from references 7, 9 and 11.

QUESTIONS

Q.9.1   How is a lossy dielectric represented? Explain how an ideal capacitor in parallel with a resistor can represent a lossy dielectric over a wide frequency range?

Q.9.2   Define "complex permittivity". What are the factors that govern the quantities "relative permittivity" and "loss factor"?

Q.9.3   Explain how the volume resistivity of a solid dielectric is determined.

Q.9.4   What is the three electrode arrangement used in dielectric measurements? Explain with sketches the electrode arrangements for (a) solid specimen, (b) liquid specimen.

Q.9.5   Describe Mole's arrangement for measuring high dissipation factors in the low frequency range.

Q.9.6   Explain the high voltage Schering bridge for the tan δ and capacitance measurement of insulators or bushings.

Q.9.7   Explain the modifications to be made to the Schering bridge for the following situations:

(a) high dissipation factor test objects,

(b) high capacitance test objects, and

(c) one end of the test object to be grounded.

Q.9.8   Why are the earthing and shielding arrangements needed in the Schering bridge measurements?

Q.9.9   What is "Wagner's earthing device" and how is it used for eliminating stray capacitances?

Q.9.10 Why is the substitution method used for measuring low capacitances?

Q.9.11 Explain the transformer ratio arm bridge for audio frequency range measurements. Discuss its merits and demerits over other methods.

Q.9.12 Why is the micrometer electrode system needed for radio frequency ranges? Give a sketch of the micrometer electrode system for solid and liquid specimens.

Q.9.13 Discuss the type of detectors used in dielectric measurements in the following cases:

(a) d.c. measurements,

    (b) 50 Hz measurements,

    (c) audio frequency measurements, and

    (d) radio frequency measurements.

Q.9.14 Briefly explain how partial discharges in an insulation system or equipment can be detected and displayed.

Q.9.15 Briefly explain the methods used for calibrating the partial discharge detectors.

Q.9.16 What are partial discharges and how are they detected under power frequency operating conditions?

Q.9.17 Discuss the method of balanced detection for locating partial discharges in electrical equipment.

Q.9.18 Describe how a fault in a long cable can be detected and located using partial discharge technique.

Q.9.19 What are "broad band" and "narrow band" detectors? What is the sensitivity in each of the above detectors?

WORKED EXAMPLES

Example 9.1: The volume resistivity of a bakelite piece was determined by using standard circular electrodes, a sensitive galvanometer, and a stabilized power supply. When the applied voltage was 1000 V, the galvanometer deflection with the specimen was 3.2 cm. When a standard resistance of $R_S = 10$ MΩ is used for calibration, the deflection was 33.30 cm with a universal shunt ratio of 3,000. The diameter of the electrodes is 10 cm, and the thickness of the specimen is 2 mm. Find the volume resistivity.

Solution: Let

$$G = \text{galvanometer sensitivity,}$$
$$R_S = \text{standard resistance,}$$
$$n = \text{shunt ratio,}$$
$$D_S = \text{deflection in cm, and}$$
$$V = \text{applied voltage.}$$

Then,

$$G = \frac{V}{R_S} \frac{1}{n} \frac{1}{D_S}$$

$$= \frac{1000}{10^7} \frac{1}{3000} \frac{1}{33.3}$$

$$\approx 10^{-9} \text{ A/cm}$$

The resistance R of the specimen $= \dfrac{V}{D \times G}$

$$= \frac{1000}{3.2 \times 10^{-9}} = 3.125 \times 10^{11} \ \Omega$$

Also, $R = \dfrac{\rho l}{\pi r^2}$ or the volume resistivity

$$\rho = \frac{\pi r^2 R}{t} = \frac{\pi (10/2)^2 \, 3.125 \times 10^{11}}{2 \times 10^{-1}}$$

$$= 1.227 \times 10^{14} \, \Omega \, cm$$

Hence, the volume resistivity $= 1.227 \times 10^{14} \, \Omega \, cm$

Example 9.2: The resistivity of the specimen referred to in Ex. 9.1 was determined using the loss of charge method. A 0.1 μF, 1000 V standard condenser was charged to 1000 V and was discharged through the specimen. If the time taken for the voltage to fall from 1000 V to 500 V was 30 min 20 s, find the resistivity of the specimen.

Solution : $t = 30 \, min \, 20 \, s = 1820 \, s$

Resistance, $R = \dfrac{t}{C \ln \dfrac{V_n}{V}} = \dfrac{1820}{0.1 \times 10^{-6} \times \ln \dfrac{1000}{500}}$

$$= 2.627 \times 10^{10} \, \Omega$$

∴ volume resistivity, $\rho = \dfrac{\pi r^2 R}{\text{thickness}}$

$$= \frac{\pi \left(\dfrac{10}{2}\right)^2 2.627 \times 10^{11}}{0.2}$$

$$= 1.031 \times 10^{13} \, \Omega \, cm$$

(*Note*: In this method, the leakage resistance of the standard condenser insulation and the voltmeter impedance came in parallel with the specimen. Hence, the volume obtained is much lower.)

Example 9.3: The capacitance and loss angle of the above specimen were measured using the same electrode set-up. The capacitance and tan δ with the specimen are 147 pF and 0.0012 respectively. The air capacitance of the electrode system was 35 pF. What is the dielectric constant and complex permittivity of bakelite ?

Solution: The dielectric constant, $\varepsilon_r = \dfrac{147}{35} = 4.2$

Complex permittivity, $\varepsilon^* = \varepsilon' - j\varepsilon''$

$$= \varepsilon_0 (K' - jK'')$$

$$\tan \delta = \frac{K''}{K'} = 0.0012, \text{ and } K' = \varepsilon_r = 4.2$$

$$\therefore K'' = 0.0012 \, K'$$

$$= 0.00504$$

$$\therefore \varepsilon^* = \varepsilon_0 (4.2 - j \, 0.00504)$$

Hence, the complex permittivity, $\varepsilon^* = (3.71 - j \, 0.0445) \times 10^{-11} \, F/m$

Example 9.4: A Schering bridge was used to measure the capacitance and loss angle of a h.v. bushing. At balance, the observations were: the value of the standard condenser = 100 pF, R_3 = 3180 Ω, C_3 = 0.00125 μF and R_4 = 636 Ω. What are the values of capacitance and tan δ of the bushing ?

Solution: The unknown capacitance $C_X = \dfrac{R_3}{R_4} C_S$

$$= \frac{3180}{636} \times 100$$

$$= 500 \text{ pF}$$

$$\tan \delta = \omega \, C_X R_X = 314 \times 3180 \times 0.00125 \times 10^{-5}$$

$$= 0.00125$$

Example 9.5: An audio frequency Schering bridge was used to determine the dielectric constant and tan δ of transformer oil at 1 kHz. The observations obtained were as follows.

(Method employed: Substitution method)

(i) With standard condenser and leads, the capacitance, C_1 = 504 pF the dissipation factor, D_1 = 0.0003.

(ii) With standard condenser in parallel with the empty test cell, capacitance C_2 = 525 pF, and dissipation factor D_2 = 0.00031.

(iii) With the standard condenser in parallel with the test cell and oil, capacitance C_3 = 550 pF and dissipation factor D_3 = 0.00075.

Find the dielectric constant and tan δ of the transformer oil?

Solution: Capacitance of the test cell = 525 – 504 = 21 pF
Capacitance of the test cell + oil

$$= 550 - 504 = 46 \text{ pF}$$

∴ ε_r =   dielectric constant of oil

$$= \frac{46}{21} = 2.19$$

ΔD of empty cell   =   0.00001
ΔD of oil   =   0.00075 – 0.00031 = 0.00044
∴   tan δ of oil   =   0.00044

Example 9.6: The dielectric constant and dissipation factor of a perspex sheet were measured at 10 MHz using the susceptibility variation technique. The observation obtained were as follows:

(i) With specimen clamped into the circuit, geometrical capacitance

$$C_a = 7.8 \text{ pF}$$

Calibration capacitance for specimen spacing

$$C_s = 6.2 \text{ pF}$$

(ii) Calibration capacitance C of micrometer electrode to obtain resonance

$$= 21.6 \text{ pF}$$

(iii) Capacitance of detuning micrometers for half power points

$$C_1 = 3.6 \text{ pF and } C_2 = 3.71 \text{ pF}$$

Total capacitance including detuning capacitors $= C_c$

$$= 13.0 \text{ pF}$$

Find the relative permittivity and loss factor.

Solution: Specimen capacitance $= C_x = C' - C_s + C_a$

$$= 21.6 - 6.2 + 7.8$$

$$= 23.2 \text{ pF}$$

Geometrical capacitance, $\quad C_a = 7.8 \text{ pF}$

$\therefore$ relative permittivity, $\quad \varepsilon_r = \dfrac{23.2}{7.8} = 2.97$

The loss factor, $\quad \tan \delta = \dfrac{C}{2(C' + C_e)}$

$$= \dfrac{(3.71 - 3.6)}{2(21.6 + 13.0)}$$

$$= \dfrac{0.11}{(2 \times 34.6)}$$

$$= 1.589 \times 10^{-3}$$

REFERENCES

1. Schwab, A., *High Voltage Measurement Technique*, M.I.T. Press (1972).
2. Harris, F.H., *Electrical Measurements*, John Wiley, New York (1966).
3. Bhimani, B.V., "Low frequency a.c. bridges", *Tr. A.I.E.E.*, 80, Part 3, 155, (1961).
4. Mole, G., "Design and performance of a portable a.c. discharge detector", *ERA Research Report*, V/T, p. 116 (1953).
5. Von Hippel, A., *Dielectrics and Applications*, M.I.T. Press (1961).
6. *ASTM Standards on* "Electrical insulating materials", *D 150-50T*, 1961.
7. Kreuger, F.H., *Discharge Detection in High Voltage Equipment*, Haywood, London (1964).
8. "Partial Discharge Measurements", *I.E.C. specification No. 270* (1968).

10

High Voltage Testing of Electrical Apparatus

It is essential to ensure that the electrical equipment is capable of withstanding the overvoltages that are met with in service. The overvoltages may be either due to natural causes like lightning or system originated ones such as switching or power frequency transient voltages. Hence, testing for overvoltages is necessary.

10.1 TESTING OF INSULATORS AND BUSHINGS

In this section, different overvoltage tests done on insulators and bushings are discussed. The overvoltage tests are classified into two groups: (i) power frequency voltage tests; and (ii) impulse voltage tests. These tests together ensure the overvoltage withstand capability of an apparatus. Before discussing the actual tests, to know the general terminology of the technical terms used is essential.

10.1.1 Definitions

In test codes and standard specifications, certain technical terms are used to specify and define conditions or procedures. Hence, commonly used technical terms are defined here before the actual testing techniques are discussed.

(a) Disruptive Discharge Voltage

This is defined as the voltage which produces the loss of dielectric strength of an insulation. It is that voltage at which the electrical stress in the insulation causes a failure which includes the collapse of voltage and passage of current. In solids, this causes a permanent loss of strength, and in liquids or gases only temporary loss may be caused. When a discharge takes place between two electrodes in a gas or a liquid or over a solid surface in air, it is called flashover. If the discharge occurs through a solid insulation it is called puncture.

(b) Withstand Voltage

The voltage which has to be applied to a test object under specified conditions in a withstand test is called the withstand voltage [as per IS: 731 and IS: 2099-1963].

(c) Fifty Per Cent Flashover Voltage

This is the voltage which has a probability of 50% flashover, when applied to a test object. This is normally applied in impulse tests in which the loss of insulation strength is temporary.

(d) Hundred Per Cent Flashover Voltage

The voltage that causes a flashover at each of its applications under specified conditions when applied to test objects is specified as hundred per cent flashover voltage.

(e) Creepage Distance

It is the shortest distance on the contour of the external surface of the insulator unit or between two metal fittings on the insulator.

(f) a.c. Test Voltages

Alternating test voltages of power frequency should have a frequency range of 40 to 60 Hz and should be approximately sinusoidal. The deviation allowed from the standard sine curve is about 7%. The deviation is checked by measuring instantaneous values over specified intervals and computing the rms value, the average value, and the form factor.

(g) Impulse Voltages

Impulse voltages are characterized by polarity, peak value, time to front (t_f), and time to half the peak value after the peak (t_t). The time to front is defined as 1.67 times to time between 30% and 90% of the peak value in the rising portion of the wave. According to *IS: 2071 (1973)*, standard impulse is defined as one with $t_f = 1.2 \, \mu s$, $t_t = 50 \, \mu s$ (called 1/50 μs wave). The tolerances allowed are ±3% on the peak value, ±30% in the front time (t_f), and ±20% in the tail time (t_t).

(h) Reference Atmospheric Conditions

The electrical characteristics of the insulators and other apparatus are normally referred to the reference atmospheric conditions. According to the Indian Standard Specifications, they are:

Temperature	: 27°C
Pressure	: 1013 millibars (or 760 torr)
Absolute humidity	: 17 gm/m^3

Since it is not always possible to do tests under these reference conditions, correction factors have to be applied. In some cases, the following test conditions are also used as reference (British Standard Specifications) conditions.

Temperature	: 20°C
Pressure	: 1013 millibars (760 Torr)

Absolute humidity   : 11 g/m³ (65% relative humidity at 20°C)

The flashover voltage of the test object is given by

$$V_s = V_a \times \frac{h}{d}$$

where, V_a = voltage under actual test conditions,
                 V_s = voltage under reference atmospheric conditions,
                 h = humidity correction factor, and
                 d =   air density correction factor.

The air density correction factor is given by,

$$d = \frac{0.289b}{273+t} \text{ for 20°C}$$

$$\text{or,} \quad \frac{0.296b}{273+t} \text{ for 27°C}$$

where, b =   atmospheric pressure in millibars, and

                 t =   atmospheric temperature, °C.

Humidity correction factor h is obtained from the temperatures of a wet and dry bulb thermometer, by obtaining the absolute humidity and then computing h from the absolute humidity. These are graphically given in Figs. 10.1 and 10.2.

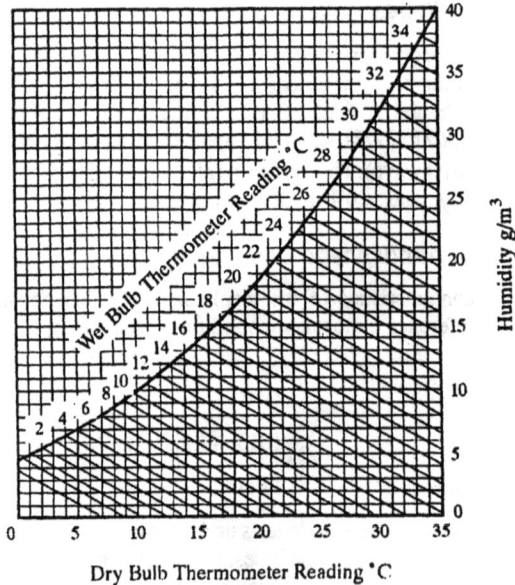

Fig. 10.1   Computation of absolute humidity from wet and dry bulb thermometer temperatures

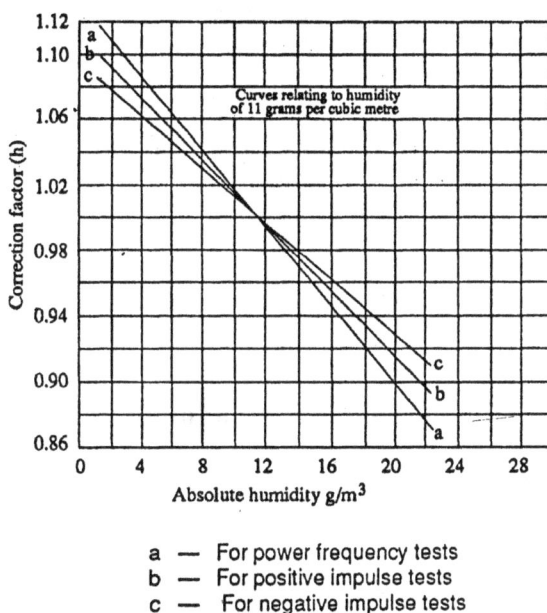

a — For power frequency tests
b — For positive impulse tests
c — For negative impulse tests

Fig. 10.2   Humidity correction factor

10.1.2 Tests on Insulators

The tests that are normally conducted are usually subdivided as (i) type tests, and (ii) the routine tests. Type tests are intended to prove or check the design features and the quality. The routine tests are intended to check the quality of the individual test piece. Type tests are done on samples when new designs or design changes are introduced, whereas the routine tests are done to ensure the reliability of the individual test objects and quality and consistency of the materials used in their manufacture.

High voltage tests include (i) the power frequency tests, and (ii) impulse tests. All the insulators are tested for both categories of test.

Power Frequency Tests

(a) Dry and Wet Flashover Tests   In these tests the a.c. voltage of power frequency is applied across the insulator and increased at a uniform rate of about 2 per cent per second of 75% of the estimated test voltage, to such a value that a breakdown occurs along the surface of the insulator. If the test is conducted under normal conditions without any rain or precipitation, it is called "dry flashover test". If the test is done under conditions of rain, it is called "wet flashover test". In general, wet tests are not intended to reproduce the actual operating conditions, but only to provide a criterion based on experience that a satisfactory service operation will be obtained. The test object is subjected to a spray of water of given conductivity by means of

nozzles. The spray is arranged such that the water drops fall approximately at an inclination of 45° to the vertical. The test object is sprayed for at least one minute before the voltage application, and the spray is continued during the voltage application. The characteristics of the spray are

   (i) Precipitation rate   : $3 \pm 10\%$ (mm/min)
   (ii) direction   : 45° to the vertical
   (iii) conductivity of water   : 100 micro Siemens $\pm$ 10%
   (iv) water temperature   : ambient $\pm 15°C$

The International Electrotechnical Commission (IEC) in its recent document (No. 42 of 1972) has revised the test procedure and precipitation conditions as follows. Average precipitation rate:

$$\text{vertical component} = 1 \text{ to } 1.5 \text{ mm/min}$$

$$\text{horizontal component} = 1 \text{ to } 1.5 \text{ mm/min}$$

limits for individual measurements   = 0.5 to 2.0 mm/min.

temperature of collected water   = ambient temperature $\pm 15°$ C.

and the conductivity of water corrected to 20°C = 100 ± 15 micro Siemens

Specifications are being modified for application of 15 positive and 15 negative impulses (reference 14). Two in each set are allowed to flashover. If more than two flashovers occur in each set, then the insulator is deemed to have failed the test. This procedure is satisfactorily equivalent to the one mentioned above.

(b)   Wet and Dry Withstand Tests (One Minute)   In these tests, the voltage specified in the relevant specification is applied under dry or wet conditions for a period of one minute with an insulator mounted as in service conditions. The test piece should withstand the specified voltage.

Impulse Tests

(a)   Impulse Withstand Voltage Test   This test is done by applying standard impulse voltage of specified value under dry conditions with both positive and negative polarities of the wave. If five consecutive waves do not cause a flashover or puncture, the insulator is deemed to have passed the test. If two applications cause flashover, the object is deemed to have failed. If there is only one failure, additional ten applications of the voltage wave are made. If the test object has withstood the subsequent applications, it is said to have passed the test.

(b)   Impulse Flashover Test   The test is done as above with the specified voltage. Usually, the probability of failure is determined for 40% and 60% failure values or 20% and 80% failure values, since it is difficult to adjust the test voltage for the exact 50% flashover values. The average value of the upper and the lower limits is taken. The insulator surface should not be damaged by these tests, but slight marking on its surface or chipping off of the cement is allowed.

(c) Pollution Testing Because of the problem of pollution of outdoor electrical insulation and consequent problems of the maintenance of electrical power systems, pollution testing is gaining importance. The normal types of pollution are (i) dust, micro-organisms, bird secretions, flies, etc., (ii) industrial pollution like smoke, petroleum vapours, dust, and other deposits, (iii) coastal pollution in which corrosive and hygroscopic salt layers are deposited on the insulator surfaces, (iv) desert pollution in which sand storms cause deposition of sand and dust layers, (v) ice and fog deposits at high altitudes and in polar countries. These pollutions cause corrosion, non-uniform gradients along the insulator strings and surface of insulators and also cause deterioration of the material. Also, pollution causes partial discharges and radio interference. Hence, pollution testing is important for extra high voltage systems.

At present there is no standard pollution test available. The popular test that is normally done is the salt fog test. In this test, the maximum normal withstand voltage is applied on the insulator and then artificial salt fog is created around the insulator by jets of salt water and compressed air. If the flashover occurs within one hour, the test is repeated with fog of lower salinity, otherwise, with a fog of higher salinity. The maximum salinity at which the insulator withstands three out of four tests without flashover is taken as the representative figure. Much work is yet to be done to standardize the test procedures.

10.1.3 Testing of Bushings

Power Frequency Tests

(a) Power Factor—Voltage Test In this test, the bushing is set up as in service or immersed in oil. It is connected such that the line conductor goes to the high voltage side and the tank or earth portion goes to the detector side of the high voltage Schering bridge. Voltage is applied up to the line value in increasing steps and then reduced. The capacitance and power factor (or tan δ) are recorded at each step. The character-istic of power factor or tan δ versus applied voltage is drawn. This is a normal routine test but sometimes may be conducted on percentage basis.

(b) Internal or Partial Discharge Test This test is intended to find the deterio-ration or failure due to internal discharges caused in the composite insulation of the bushing. This is done by using internal or partial discharge arrangement (see Sec. 9.4). The voltage versus discharge magnitude as well as the quadratic rate gives an excellent record of the performance of the bushing in service. This is now a routine test for high voltage bushings.

(c) Momentary Withstand Test at Power Frequency This is done as per the Indian Standard Specifications, IS: 2099, applied to bushings. The test voltage is specified in the specifications. The bushing has to withstand without flashover or puncture for a minimum time (~ 30s) to measure the voltage. At present this test is replaced by the impulse withstand test.

(d) One Minute Wet Withstand Test at Power Frequency The most common and routine tests used for all electrical apparatuses are the one minute wet, and dry

voltage withstand tests. In wet test, voltage specified is applied to the bushing mounted as in service with the rain arrangement as described earlier. A properly designed bushing has to withstand the voltage without flashover for one minute. This test really does not give any information for its satisfactory performance in service, while impulse and partial discharge tests give more information.

(e) Visible Discharge Test at Power Frequency This test is intended for determining whether the bushing is likely to give radio interference in service, when the voltage specified in IS: 2099 is applied. No discharge other than that from the arcing horns or grading rings should be visible to the observers in a dark room. The test arrangement is the same as that of the withstand test, but the test is conducted in a dark room.

Impulse Voltage Tests

(a) Full Wave Withstand Test The bushing is tested for either polarity voltages as per the specifications. Five consecutive full waves of standard waveform are applied, and, if two of them cause flashover, the bushing is said to have failed in the test. If only one flashover occurs, ten additional applications are done. The bushing is considered to have passed the test if no flashover occurs in subsequent applications.

(b) Chopped Wave Withstand and Switching Surge Tests The chopped wave test is sometimes done for high voltage bushings (220 kV and 400 kV and above). Switching surge flashover test of specified value is now-a-days included for high voltage bushings. The tests are carried out similar to full wave withstand tests.

Thermal Tests

(a) Temperature Rise and Thermal Stability Tests The purpose of these tests is to ensure that the bushing in service for long does not have an excessive temperature rise and also does not go into the "thermal runaway" condition of the insulation used.

Temperature rise test is carried out in free air with an ambient temperature below 40°C at a rated power frequency (50 Hz) a.c. current. The steady temperature rise above the ambient air temperature at any part of the bushing should not exceed 45°C. The test is carried out for such a long time till the temperature is substantially constant, i.e. the increase in temperature rate is less than 1°C/hr. Sometimes, the bushings have to be operated along with transformers, of which the temperature reached may exceed 80°C. This temperature is high enough to produce large dielectric losses and thermal instability. For high voltage bushings this is particularly important, and hence the thermal stability test is done for bushings rated for 132 kV and above. The test is carried out with the bushing immersed in oil at a maximum temperature as in service, and the voltage applied is 86% of the nominal system voltage. This is approximately $\sqrt{2}$ times the working voltage of the bushing and hence the dielectric losses are about double the normal value. The additional losses account for the conductor ohmic losses. It has been considered unnecessary to specify the thermal stability test for oil-impregnated paper bushings of low ratings; but for the large high voltage bushings (1600 A, 400 kV transformer bushings, etc.), the losses in the conductor may be high enough to outweigh the dielectric losses.

It may be pointed out here, that the thermal stability tests are type tests. But in the case of large sized high voltage bushings, it may be necessary to make them routine tests.

10.2 TESTING OF ISOLATORS AND CIRCUIT BREAKERS

10.2.1 Introduction

In this section, the testing of isolators and circuit breakers is covered, giving common characteristics for both. While these characteristics are directly relevant to the testing of circuit breakers, they are not much relevant as far as the testing of isolators are concerned since isolators are not used for interrupting high currents. At best, they interrupt small currents of the order of 0.5 A (for rated voltages of 420 kV and below) which may be the capacitive currents of bushings, busbars etc. In fact, the definition of an Isolator or a Disconnector as per IS: 9921 (Part I) - 1981 is as follows:

An isolator or a disconnector is a mechanical switching device, which provides in the open position, an isolating distance in accordance with special requirements. An isolator is capable of opening and closing a circuit when either negligible current is broken or made or when no significant change in the voltage across the terminals of each of the poles of the isolator occurs. It is also capable of carrying currents under normal circuit conditions, and carrying for a specified time, currents under abnormal conditions such as those of a short circuit.

Thus, most of the discussion here refers to the testing of circuit breakers.

Testing of circuit breakers is intended to evaluate (a) the constructional and operational characteristics, and (b) the electrical charactristics of the circuit which the switch or the breaker has to interrupt or make. The different characteristics of a circuit breaker or a switch may be summarized as per the following groups.

(i) (a) The electrical characteristics which determine the arcing voltage, the current chopping characteristics, the residual current, the rate of decrease of conductance of the arc space and the plasma, and the shunting effects in interruption.

(b) Other physical characteristics including the media in which the arc is extinguished, the pressure developed or impressed at the point of interruption, the speed of the contact travel, the number of breaks, the size of the arcing chamber, and the materials and configuration of the circuit interruption.

(ii) The charateristics of the circuit include the degree of electrical loading, the normally generated or applied voltage, the type of fault in the system which the breaker has to clear, the time of interruption, the time constant, the natural frequency and the power factor of the circuit, the rate of rise of recovery voltage, the restriking voltage, the decrease in the a.c. component of the short circuit current, and the degree of asymmetry and the d.c. component of the short circuit current.

To assess the above factors, the main tests conducted on the circuit breakers and isolator switches are (*i*) the dielectric tests or overvoltage tests,

    (*ii*) the temperature rise tests,

    (*iii*) the mechanical tests, and

    (*iv*) the short circuit tests

Dielectric tests consist of overvoltage withstand tests of power frequency, lightning and switching impulse voltages. Tests are done for both internal and external insulation with the switch or circuit breaker in both the open and closed positions. In the open position, the test voltage levels are 15% higher than the test voltages used when the breaker is in closed position. As such there is always the possibility of line to ground flashover. To avoid this, the circuit breaker is mounted on insulators above the ground, and hence the insulation level of the body of the circuit breaker is raised.

The impulse tests with the lightning impulse wave of standard shape are done in a similar manner as in the case of insulators. In addition, the switching surge tests with switching overvoltages are done on circuit breakers and isolators to assess their performance under overvoltages due to switching operations.

Temperature rise and mechanical tests are tube tests on circuit breakers and are done according to the specifications (reference 17).

10.2.2 Short Circuit Tests

The most important tests carried out on circuit breakers are short circuit tests, since these tests assess the primary performance of these devices, i.e. their ability to safely interrupt the fault currents. These tests consists of determining the making and breaking capacities at various load currents and rated voltages. In the case of isolators, the short circuit tests are conducted only with the limited purpose to determine their capacity to carry the rated short circuit current for a given duration; and no breaking or making current test is done.

    The different methods of conducting short circuit tests are

(i) Direct Tests

    (a) using a short circuit generator as the source

    (b) using the power utility system or network as the source.

(ii) Synthetic Tests

(a) Direct Testing in the Networks or in the Fields Circuit breakers are sometimes tested for their ability to make or break the circuit under normal load conditions or under short circuit conditions in the network itself. This is done during period of limited energy consumption or when the electrical energy is diverted to other sections of the network which are not connected to the circuit under test. The advantages of field tests are:

    (*i*) The circuit breaker is tested under actual conditions like those that occur in a given network.

(*ii*) Special occasions like breaking of charging currents of long lines, very short line faults, interruption of small inductive currents, etc. can be tested by direct testing only.

(*iii*) to assess the thermal and dynamics effects of short circuit currents, to study applications of safety devices, and to revise the performance test procedures, etc.

The disadvantages are:

(*i*) The circuit breaker can be tested at only a given rated voltage and network capacity.

(*ii*) The necessity to interrupt the normal services and to test only at light load conditions.

(*iii*) Extra inconvenience and expenses in installation of controlling and measuring equipment in the field.

(b) Direct Testing in Short Circuit Test Laboratories In order to test the circuit breakers at different voltages and at different short circuit currents, short circuit laboratories are provided. The schematic layout of a short circuit testing laboratory is given in Fig. 10.3. It consists of a short circuit generator in association with a master circuit breaker, resistors, reactors and measuring devices. A make switch initiates the short circuit and the master circuit breaker isolates the test device from the source at the end of a predetermined time set on a test sequence controller. Also, the master circuit breaker can be tripped if the test device fails to operate properly. Short circuit generators with induction motors as prime movers are also available.

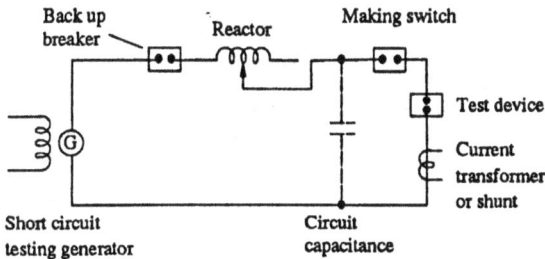

Fig. 10.3 Schematic diagram showing basic elements of a short circuit testing laboratory

(c) Synthetic Testing of Circuit Breakers Due to very high interrupting capacities of circuit breakers, it is not economical to have a single source to provide the required short circuit and the rated voltage. Hence, the effect of a short circuit is obtained as regards to the intensity of the current and the recovery voltage as a combination of the effects of two sources, one of which supplies the a.c. current and the other the high voltage.

In the initial period of the short circuit test, the a.c. current source supplies the heavy current at a low voltage, and then the recovery voltage is simulated by a source

~of comparatively high voltage of small current capacity. A schematic diagram of a synthetic testing station is shown in Fig. 10.4.

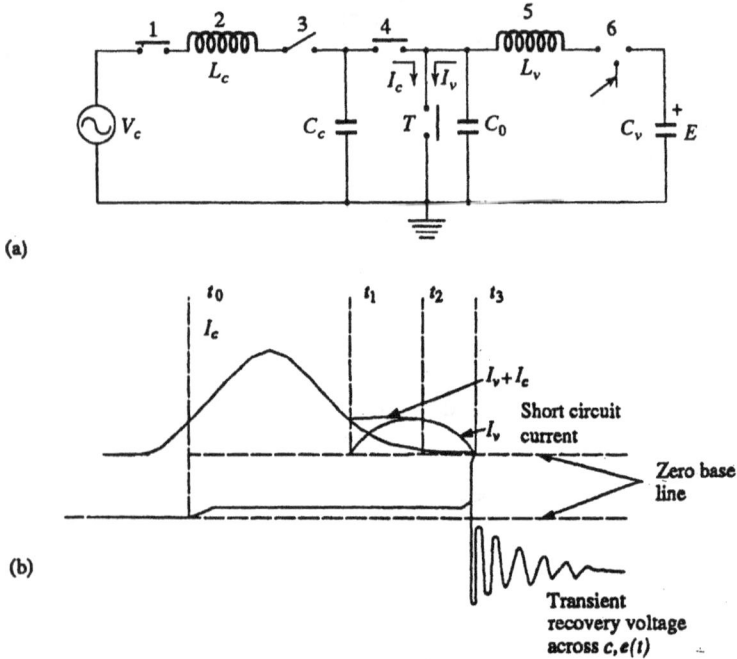

(a)

(b)

V_c– Low voltage, high current generator; L_c– Current controlling inductance (2); 1 – Master breaker; 3 – Main switch; (T – Circuit breaker) under test; 4 – Auxiliary breaker; (L_v – Voltage waveform) controlling choke (5); 6 – Trigger gap; (C_v – Capacitor charged to) E_v to give necessary recovery voltage; (C_0 – Capacitor to control) the frequency of the transient recovery voltage; t_0 – Opening of auxiliary circuit breaker (4); t_1 – Trigger gap 6 is fired; t_2 – Auxiliary circuit breaker clears and interrupts I_c; t_3 – I_v becomes zero

Fig. 10.4   (a) Schematic diagram of synthetic testing of circuit breakers

(b) Current and recovery voltage waveforms across the test circuit breaker

With the auxiliary breaker (3) and the test breaker (T) closed, the closing of the making switch (1) causes the current to flow in the test circuit breaker. At some instant say t_0, the test circuit breaker (T) begins to operate and the master circuit breaker (1) becomes ready to clear the generator circuit. At some times t_1, just before the zero of the generator current, the trigger gap (6) closes and the higher frequency current from the discharging capacitor C_v also flows through the arc. At time t_2, when the generator current is zero, the circuit breaker (1) clears that circuit, leaving only the current from C_v which has the required rate of change of current at its zero flowing in the test circuit breaker. At the zero of this current, full test voltage will be available. The closing of gap (6) would be a little earlier in time than shown in Fig.10.4, but it has been drawn as shown for clarity at current zeros. It is important to see that the high-current source is disconnected and a high-voltage source applied with absolute precision (by means of an auxiliary circuit breaker) at the instant of circuit breaking.

(d) Composite Testing In this method, the breaker is first tested for its rated breaking capacity at a reduced voltage and afterwards for rated voltage at a low current. This method does not give a proper estimate of the breaker performance.

(e) Unit Testing When large circuit breakers of very high voltage rating (220 kV and above) are to be tested and where more than one break is provided per pole, the breaker is tested for one break at its rated current and the estimated voltage. In actual practice, the conditions of arc in each gap may not be identical and the voltage distribution along several breaks may be uneven. Hence, certain uncertainty prevails in the testing of one break.

(f) Testing Procedure The circuit breakers are tested for their (i) breaking capacity B, and (ii) making capacity M. The circuit breaker, after the calibration of the short circuit generator, is tested for the following duty cycle.
   (1) *B-3-B-3-B* at 10% of the rated symmetrical breaking capacity
   (2) *B-3-B-3-B* at 30% of the rated symmetrical breaking capacity
   (3) *B-3-B-3-B* at 60% of the rated symmetrical breaking capacity
   (4) *B-3-MB-3MB-MB* at 100% breaking capacity with the recovery voltage not less than 95% of the rated service voltage.
The power factor in these tests is generally between 0.15 and 0.3. The numral 3 in the above duty cycle indicates the time interval in minutes between the tests.

(g) Asymmetrical Tests One test cycle is repeated for the asymmetrical breaking capacity in which the d.c. component at the instant of contact separation is not less than 50% of the a.c. component.

10.3 TESTING OF CABLES

Cables are very important electrical apparatus for transmission of electrical energy by underground means. They are also very important means for transmitting voltage signals at high voltages. For power engineers, large power transmission cables are of importance, and hence testing of power cables only is considered here. Of the different electrical and other tests prescribed, the following are important to ensure that cables withstand the most severe conditions that are likely to arise in service.
   Different tests on cables may be classified into
   (*i*) mechanical tests like bending test, dripping and drainage test, and fire resistance and corrosion tests,
   (*ii*) thermal duty tests,
   (*iii*) dielectric power factor tests,

(*iv*) power frequency withstand voltage tests,
(*v*) impulse withstand voltage tests,
(*vi*) partial discharge tests, and
(*vii*) life expectancy tests.
Here only the electrical tests are described, i.e. tests (*iii*) to (*vii*).

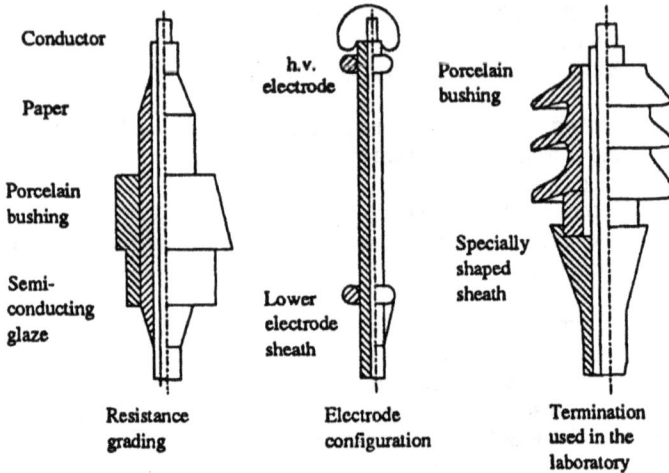

Fig. 10.5 Cable and terminals

10.3.1 Preparation of the Cable Samples

For overvoltage and withstand tests, samples have to be carefully prepared and terminated; otherwise, excessive leakage or end flashovers may occur during testing. The normal length of the cable sample used varies from about 50 cm to 10 m. The terminations are usually made by shielding the end conductor with stress shields or terminations to relieve the ends from excessive high electrical stresses. A few terminations are shown in Fig. 10.5. During power factor tests, the cable ends are provided with shields so that the surface leakage current is avoided from the measuring circuits.

10.3.2 Dielectric Power Factor Test

The dielectric power factor test is done using the high voltage Schering bridge (see Section 9.3.4). The power factor or dissipation factor tan δ is measured at 0.5, 1.0. 1.66, and 2.0 times the rated voltage (phase to ground) of the cable. The maximum value of the power factor and the difference in power factor between the rated voltage and 1.66 times the rated voltage, as well as, between the rated voltage and two times the rated voltage are specified. Sometimes, difficulty is felt in supplying the charging voltamperes of the cable from the available source. In such cases, a choke is used or a suitably rated transformer winding is used in series with the cable to form a resonant circuit. This improves the power factor and raises the test voltage between the cable core and the sheath to the required value, when a source of high voltage and high capacity is used. The Schering bridge has to be given protection against overvoltages, in case breakdown occurs in the cables.

10.3.3 High Voltage Tests on Cables

Cables are tested for withstand voltages using the power frequency a.c., d.c., and impulse voltages. At the time of manufacture, the entire cable is passed through a high voltage test at the rated voltage to check the continuity of the cable. As a routine test, the cable is tested applying an a.c. voltage of 2.5 times the rated value for 10 min. No damage to the cable insulation should occur. Type tests are done on cable samples using both high voltage d.c. and impulse voltages. The d.c. test consists of applying 1.8 times the rated d.c. voltage of negative polarity for 30 min., and the cable system is said to be fit, if it withstands the test. For impulse tests, impulse voltage of the prescribed magnitude as per specifications is applied, and the cable has to withstand five applications without any damage. Usually, after the impulse test, the power frequency dielectric power factor test is done to ensure that no failure occurred during the impulse test.

10.3.4 Partial Discharges

(a) Discharge Measurement

Partial discharge measurements and the discharge locations are important for cables, since the life of the insulation at a given voltage stress depends on the internal discharges. Also, the weakness of the insulation or faults can be detected with the help of these tests; the portion of the cable if weak may be removed, if necessary. The general arrangement for partial discharge tests is the same as described in Sec.9.4.

The equivalent circuit of the cable for discharges is shown in Fig. 10.6, and the cable connection to the discharge detector through the coupling condenser is shown in Figs. 10.7a and b. If the detector is connected through a coupling capacitor to one end of the cable as in Fig. 10.7a, it will receive the transient travelling wave directly from the cavity towards the nearer end, and after a short time, a second travelling wave pulse reflected from the far end is observed. Thus, the detected response is the combination of the above two transient pulses. But, if the connections are made as in Fig. 107b, no severe reflection is involved except as a second order effect of negligible magnitude. Now two transients will arrive at both the ends of the cable, and the superposition of the two pulses is detected. This can be obtained by adding the responses of the two transients. The superpositions of the two responses may give rise to a serious error in the measurement of the discharge magnitude. The magnitude of the possible error may be determined mainly by the shape of the response of the discharge detector.

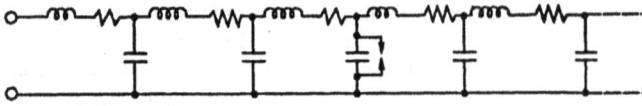

Fig. 10.6 Equivalent circuit of the cable for discharges

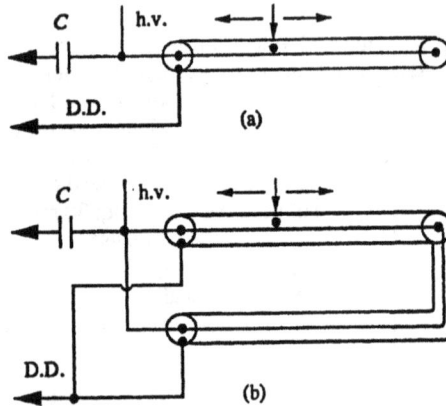

Fig. 10.7 Discharge detector connection to long length of cable
D.D.—Discharge detector

(b) Location of Discharges

The voltage dip caused by a discharge at a fault or a void is propagated as a travelling wave along the cable. This wave is detected as a voltage pulse across the terminals of the cable ends. By measuring the time duration between the pulses, the distance at which the discharge is taking place from the cable end can be determined. The shapes of the voltage pulses depend on the nature of the discharges. Typical waveshapes are given in Fig. 10.8. The detection circuits for the pulses are shown in Fig. 10.9, and the attenuation of the travelling wave in cables is given in Fig. 10.10. Usually, the pulses detected across the resistor are distorted after passing through the amplifier of the discharge detector.

(c) Scanning Method

In order to scan the entire cable length for voids or imperfections in manufacture, the bare core of the cable is passed through a high electric field and the discharge location is done. The core of the material is passed through a tube of insulating material filled with distilled water. Four electrodes in the form of rings are mounted at both ends of tube as well as at the middle, such that they have electrical contact with the water. The middle electrodes are energized with a high voltage, and the other two electrodes and

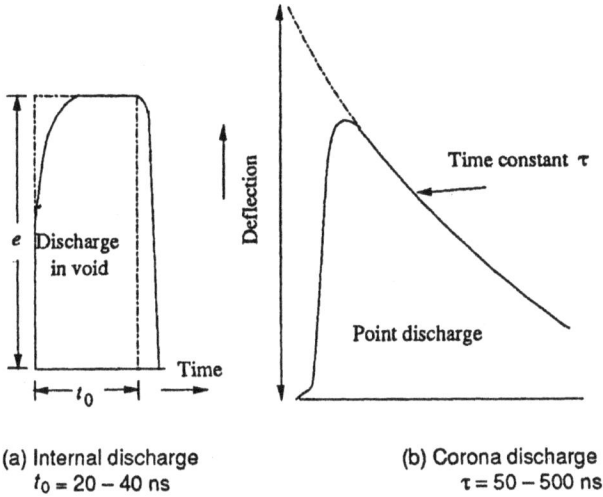

(a) Internal discharge
$t_0 = 20 - 40$ ns

(b) Corona discharge
$\tau = 50 - 500$ ns

------- Hypothetical waveshape
_____ Waveshape observed with oscilloscope

Fig. 10.8 Typical waveshapes of pulses at the cable ends

cable conductor are grounded. If a discharge occurs in the portion between the middle electrodes, as the cable is passed between the middle electrodes' portion, the discharge is detected and is located at that length of cable.

This test is very convenient for isolating the defective insulation at the factory site. The manufactured cable, before being rolled on to its former, can be conveniently passed through the test apparatus. "The defective part" can be isolated and cut off from the cable reel before it is sent from the factory.

(d) Life Tests

Life tests are intended for reliability studies in service. In order to determine the expected life to the cable under normal stress, accelerated life tests using increased voltages are performed on actual cable lengths. It is established that the relation between the maximum electrical stress E_m and the life of the cable insulation in hours t approximately follows the relationship

$$E_m = Kt^{-(1/n)}$$

where,

$K =$ constant which depends on the field conditions and the material, and

$n =$ life index depending on the material.

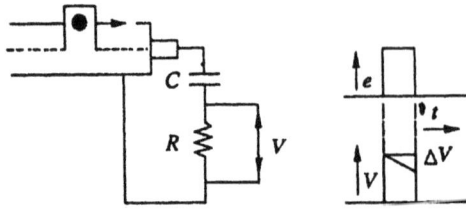

(a)   Resistor and capacitor

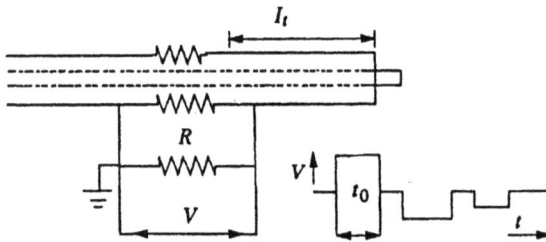

(b)   Without capacitor

Fig. 10.9   Detection circuits for long cables

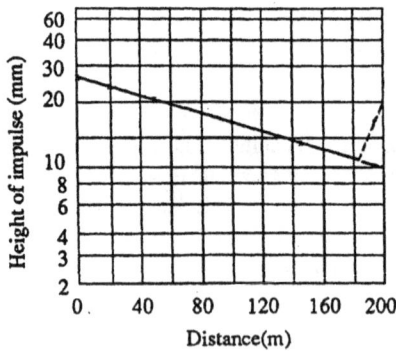

Fig. 10.10   Attenuation of travelling waves

By conducting long duration life tests at increased stress (1 hr to about 1000 hr) the expected life at the rated stress may be determined.

10.4 TESTING OF TRANSFORMERS

Transformers are very important and costly apparatus in power systems. Great care has to be exercised to see that the transformers are not damaged due to transient overvoltages of either lightning or power frequency. Hence, overvoltage tests become very important in the testing of transformers. Here, only the overvoltage tests are discussed, and other routine tests like the temperature rise tests, short circuit tests, etc. are not included and can be found in the relevent specifications.

(a) Induced Overvoltage Test

Transformers are tested for overvoltages by exciting the secondary of the transformer from a high frequency a.c. source (100 to 400 Hz) to about twice the rated voltage. This reduces the core saturation and also limits the charging current necessary in large power transformers. The insulation withstand strength can also be checked.

(b) Partial Discharge Tests

Partial discharge tests on the windings are done to assess the discharge magnitudes and the radio interference levels (see also Sec. 10.6). The transformer is connected in a manner similar to any other equipment (see Sec. 9.4) and the discharge measurements are made. The location of the fault or void is sometimes done by using the travelling wave technique similar to that for cables. So far, no method has been standardized as to where the discharge is to be measured. Multi-terminal partial discharge measurements are recommended. Under the application of power frequency voltage, the discharge magnitudes greater than 10^4 pico coulomb are considered to be severe, and the transformer insulation should be such that the discharge magnitude will be far below this value.

10.4.1 Impulse Testing of Transformers

The purpose of the impulse tests is to determine the ability of the insulation of the transformers to withstand the transient voltages due to lightning, etc. Since the transients are impulses of short rise time, the voltage distribution along the transformer winding will not be uniform. The equivalent circuit of a transfomer winding for impulses is shown in Fig. 10.11. If an impulse wave is applied to such a. network (shown in Fig. 10.11) the voltage distribution along the element will be uneven, and oscillations will be set in producing voltages much higher than the applied voltage.

Impulse testing of transformers is done using both the full wave and the chopped wave of the standard impulse, produced by a rod gap with a chopping time of 3 to 6 µs. To prevent large overvoltages being induced in the windings not under test, they are short circuited and connected to ground. But the short circuiting reduces the impedance of the transformer and hence poses problems in adjusting the standard waveshape of the impulse generators. It also reduces the sensitivity of detection.

(a) Procedure for Impulse Testing

The schematic diagram of the transformer connection for impulse testing is shown in Fig. 10.12, and the waveshapes of the full and chopped waves are shown in Fig. 10.13. In transformer testing it is essential to record the

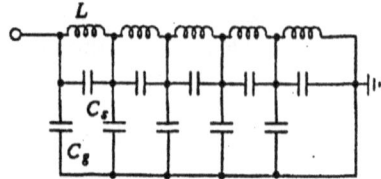

L — Inductance (series)
C_s — Series capacitance
C_g — Shunt capacitance to ground

Fig. 10.11 Equivalent circuit of transformer winding for impulses

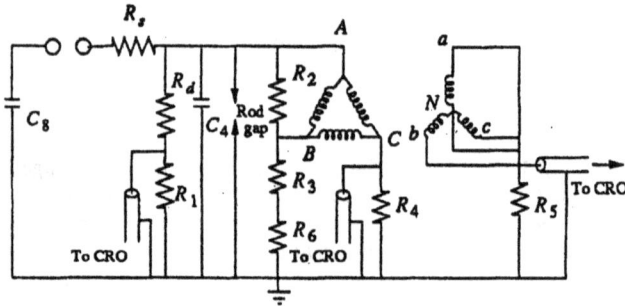

Fig. 10.12 Arrangement of transformer for impulse testing

waveforms of the applied voltage and current through the windings under test. Sometimes, the transferred voltage in the secondary and the neutral current are also recorded.

Impulse testing is done in the following sequence:

(i) applying impulse voltage of magnitude 75% of the Basic Impulse Level (BIL) of the transformer under test,

(ii) one full wave ovltage of 100% BIL,

(iii) two chopped waves of 100% BIL,

(iv) one full wave of 100% BIL, and

(v) one full wave of 75% BIL.

It is very important to see that the grounding is proper and the windings not under test are suitably terminated.

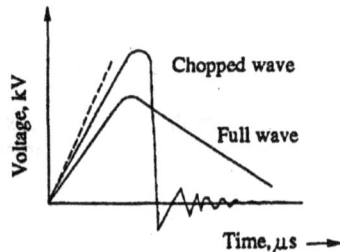

Fig. 10.13 Full wave and chopped wave

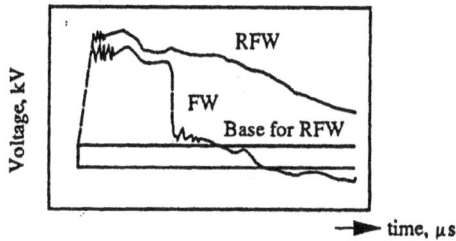

(a)   Failure from the line lead to ground through oil

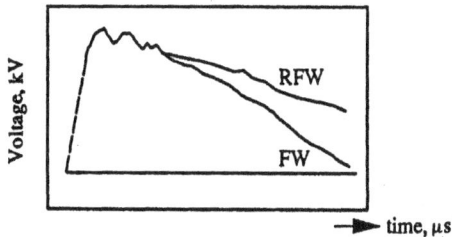

(b)   8.5% of winding failed

Fig. 10.14 Voltage oscillograms of transformer winding with a fault
RFW—Reduced full wave
FW—Full wave

(b) Detection and Location of Fault During Impulse Testing

The fault in a transformer insulation is located in impulse tests by any one of the
following methods.

General observations : The fault can be located by general observations like noise
in the tank or smoke or bubbles in breather.

Voltage oscillogram method : Fault or failure appears as a partial or complete
collapse of the applied voltage wave. Figure 10.14 gives the typical waveform. The
sensitivity of this method is low and does not detect faults which occur on less than
5% of the winding.

Neutral current method : In the neutral current method, a record of the impulse
current flowing through a resistive shunt between the neutral and ground point is used
for detecting the fault. The neutral current oscillogram consists of a high frequency
oscillation, a low frequency disturbance, and a current rise due to reflections from the
ground end of the windings. When a fault occurs such as arcing between the turns or
from turn to the ground, a train of high frequency pulses similar to that in the front of
the impulse current wave are observed in the oscillogram and the waveshape changes.

If the fault is local, like a partial discharge, only high frequency oscillations are observed without a change of waveshape. The sensitivity of the method decreases, if other windings not under test are grounded.

Transferred surge current method : In this method, the voltage across a resistive shunt connected between the low voltage winding and the ground is used for fault location. A short high frequency discharge oscillation is capacitively transferred at the event of failure and is recorded. Hence, faults at a further distance from the neutral are also clearly located. The waveshape is distorted depending on the location and type of fault, and hence can be more clearly detected.

After the location of the fault, the type of fault can be observed by dismantling the winding and looking for charred insulation or melted parts on the copper winding. This is successful in the case of major faults. Local faults or partial discharges are self healing and escape observation.

10.5 TESTING OF SURGE DIVERTERS

10.5.1 Introduction

In modern practices, surge diverters or lightning arresters are the most reliable apparatus to project the power system against transient voltages due to lightning and switching surges. They are invariably used from distribution voltages (400 V) to highest system transmission voltages of 765 kV or above. Hence, testing them precisely in standard laboratories with standard test procedures are of great importance in modern power system practice.

A surge diverter has to be a non-conductor for operating power frequency voltages. It should behave as a short circuit for transient overvoltages of impulse character, discharge the heavy current, and recover its insulation without allowing the follow-up of the power frequency current.

In Table 10.1, the impulse current ratings of the surge diverters in relation to their voltage are given, and the testing is usually done at these current ratings.

Table 10.1 Surge Diverter Voltage and Current Ratings

Diverter class	Diverter rating	Impulse current rating (8/20 μ s) (Amperes)	High current rating (4/10 μ s) (Amperes)	Long duration rating—duration is given in μ s (Amperes)
A	Low voltage (230 V to 600 V)	1500 2500	10,000 25,000	50 (500 μ s)
B	Distribution voltages (400 V to 33 kV)	5000	65,000	75 (1000 μ s)
C	Station type lightning arresters (11 kV and above)	10,000	100,000	150 (2000 μ s)

10.5.2 Tests on Surge Diverters

(i) Power Frequency Sparkover Test

This is routine test. The test is conducted using a series resistance to limit the current in case a sparkover occurs. The arrester has to withstand at least 1.5 times the rated value of the voltage for five successive applications. The test is generally done also under dry and wet conditions.

(ii) Hundred per cent Standard Impulse Sparkover Test

This test is conducted to ensure that the diverter operates positively when over-voltages of impulse nature occur. The impulse generator is adjusted to give the standard impulse voltage of a preset magnitude specified in the specifications. The arrester has to sparkover every time in each of the ten successive applications. The test is done with both positive and negative polarity waveforms. Sometimes, the test is done by starting at a voltage level that does not give flashover at all, and is repeated in increasing steps of voltage till hundred per cent flashover occurs. The magnitude of the voltage at which hundred per cent flashover occurs is the required sparkover voltage.

(iii) Front of Wave Sparkover Test

In order to ensure that the surge diverter flashes over for very steep fronted waves of high peaks, this test is conducted using an overvoltage having a rate of rise of 100 kV/ μs, per 12 kV of the rating. The estimated maximum steepness of the waves are specified in standards and specifications. The test is done by conducting hundred per cent sparkover voltage test for increasing magnitudes of the standard impulse wave. The time to sparkover is measured. The volt-time characteristic of the diverter is plotted, and the intersection of the V–t characteristic and the line with slope of the virtual steepness of the front gives the front of a wave sparkover voltage.

(iv) Residual Voltage Test

This test is conducted on pro-rated diverters of ratings in the range 3 to 12 kV only. The voltage developed across the Non-Linear Resistor units (NLR) during the flow of surge currents through the arrester is called the 'residual voltage'. A pro-rated arrester is a complete, suitably housed section of an arrester including series gaps and non-linear series resistors in the same proportion as in the complete arrester.) Standard impulse currents of the rated magnitudes are applied, and the voltage developed across the diverter is recorded using a suitable voltage divider and a CRO. The magnitudes of the currents are approximately 0.5, 1.0, and 2.0 times the rated currents. From the oscillogram, a graph is drawn between the current magnitudes and the voltage developed across the diverter pro-rated unit. From the graph, the residual voltage corresponding to the exact rated current is obtained.

Let $V_1 =$ rating of the complete unit,

$V_2 =$ rating of the pro-rated unit tested,

$$V_{R_1} = \text{residual voltage of the complete unit, and}$$
$$V_{R_2} = \text{residual voltage of the pro-rated unit.}$$

Then, it is assumed that

$$\frac{V_1}{V_2} = \frac{V_{R1}}{V_{R2}}$$

Let V_{RM} be the maximum permissible residual voltage for the complete unit. The ratio $V_{RM}/V_1 = r$, is defined as a multiplying factor of the rating for the residual voltage test, which depends on V_1. The "diverter" is said to pass the test, if

$$V_{R_2} < rV_2$$

10.5.3 High Current Impulse Test on Surge Diverters

This test is also done on pro-rated diverter units in the range of 3 to 12 kV. A high current impulse wave of 4/10 μs of peak value mentioned in the specifications is applied to a spare unit of identical characteristics. Two such applications are done on the units under test, allowing sufficient time for the cooling of the unit to the room temperature. The unit is said to pass the test, if

    (i) the power frequency sparkover voltage before and after the test does not differ by more than 10%,

    (ii) the voltage and current waveforms of the diverter do not differ significantly in the two applications, and

    (iii) the non-linear resistance elements in the diverter do not show any sign of puncture or external flashover.

(a) Long Duration Impulse Current Test

This test is also done on pro-rated units of 3 to 12 kV. The circuit used for generating a rectangular impulse wave consists of an artificial transmission line with lumped inductances and capacitances. The duration of the current pulse t is given by $2(n-1)$ $\sqrt{LC}$, where n is the number of stages or sections used, and L and C are the inductance and capacitance of each unit. Rectangular wave is generated, if the surge impedance of the diverter is equal to $\sqrt{L/C}$ at the test current. As per the specifications, 20 applications are made with specified current in five groups. The interval between the successive applications is about 1 min. It is usual to record the waveforms in the first two and the last two applications of the current wave. The diverter is said to have passed the test, if

    (i) the power frequency sparkover voltage before and after the application of the current wave does not differ by 10%,

    (ii) the voltage across the diverter at the first and the last application does not differ by more than 8%, and

    (iii) there is no sign of puncture or other damage.

(b) Operating Duty Cycle Test

This test is conducted on pro-rated units of diverters and gives better closeness to actual conditions. The diverter is kept energized at its rated power frequency supply voltage. The rated impulse current wave is applied first at a phase angle of about 30° from the a.c. voltage zero. If the power frequency follow-on current is not established, the angle at which current wave is applied is advanced in steps of 10° up to 90° or the peak position of the supply voltage wave till the follow-on current is established. In the course of application of the current wave, if the power frequency voltage is reduced during the flow of current, it can be compensated up to a maximum of 10% of the overvoltage. During the follow-on current period, the peak voltage across the diverter should be less than or equal to the rated peak voltage. Twenty applications of the impulse current at the selected points on the voltage wave are made in four groups. The time interval between each application is about 1 min, and between successive groups it is about half an hour. The arrester is said to have passed the test, if

(*i*) the average power frequency sparkover voltage before and after the test does not differ by more than 10%.

(*ii*) the residual voltage at the rated current does not vary by more than 10%,

(*iii*) the follow-on power frequency current is interrupted each time, and

(*iv*) no significant change, signs of flashover, or puncture occurs to the pro-rated unit.

(c) *Other Tests*

The other tests that are normally conducted on surge diverters are

(*i*) mechanical tests like porosity test, temperature cycle tests, and others,

(*ii*) pressure relief test,

(*iii*) the voltage withstand test on the insulator housing of the diverter,

(*iv*) the switching surge flashover test, and

(*v*) the pollution tests.

These tests are usually done on diverters used on Extra High Voltage (EHV) systems.

10.6 RADIO INTERFERENCE MEASUREMENTS

10.6.1 Introduction

Many electrical apparatuses like transformers, line conductors, rotating machines, etc. produce unwanted electrical signals in the radio and high frequency (television band, microwave bands, etc.) ranges. These signals arise due to corona discharges in air, internal or partial discharges in the insulation, sparking at commutators and brush gear in rotating machines, etc. It is important to see that the noise voltages generated in the radio and other transmission bands are limited to acceptable levels, and hence the radio interference voltage measurements are of importance. It has been found that the surface conditions of the overhead conductors subjected to high voltage stresses and varying atmospheric conditions greatly influence the magnitude of the noise voltage produced. In case of solid insulators, the bonding between the porcelain and the metal pin, the binding of high voltage conductor and the insulator surface, and the surface pollution were found to be the sources of this noise.

10.6.2 Measurements of Radio Interference Voltage

The noise generated in the radio frequency band as a result of corona or partial discharges in high voltage power apparatus may be measured

(i) by the radio frequency line to ground voltage known as the radio influence voltage or RIV, and

(ii) as an interfering field by means of an antenna known as the radiated radio interference voltage or RI.

Normally, the tests and measurements done in the laboratories are RIV measurements, whereas field investigations with portable radio receivers are RI measurements.

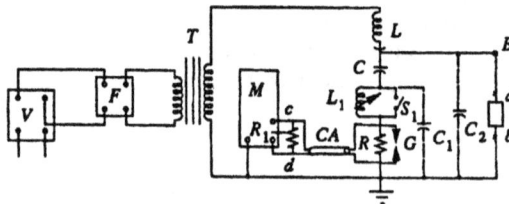

F — Voltage control unit; V — Voltmeter; T — High voltage transformer; L — Radio frequency choke; C — Coupling condenser; R_1 — Meter input impedance; M — Radio noise meter; a-b — Test apparatus; CA — Coaxial cable; G — Protective gap; S_1 — Shorting switch, C_1, C_2 — Stray capacitances; L_1 — Tuning choke; R — Measuring impedance

Fig. 10.15 Schematic diagram of circuit for the measurement of RIV of a high voltage appratus in 150 kHz to 30 MHz frequency range

A radio noise meter used in the laboratory consists of a portable radio receiver with a local oscillator, a radio frequency amplifier, a mixer, an intermediate frequency amplifier, and a detector similar to that of a standard radio receiver and operates in the frequency range 150 kHz to 30 MHz. In addition, the radio noise meter has multi-input circuits to accommodate a number of pick-up devices, attenuators, calibrators, and output circuits containing special detectors and meters. The detector circuit consists of a diode detector in series with a series resistance R_s, charging a parallel R-C circuit. The detector circuit is provided with a measuring device to measure either (a) the average value, (b) the peak value, or (c) quasi-peak value (the quasi-peak value of the impulse noise is equal to the rms value of the sine wave at the centre frequency of the pass band which produces the same deflection in the meter scale as that of the impulse). The voltmeter provided at the end of the detector has an input impedance of 50 to 75 Ω.

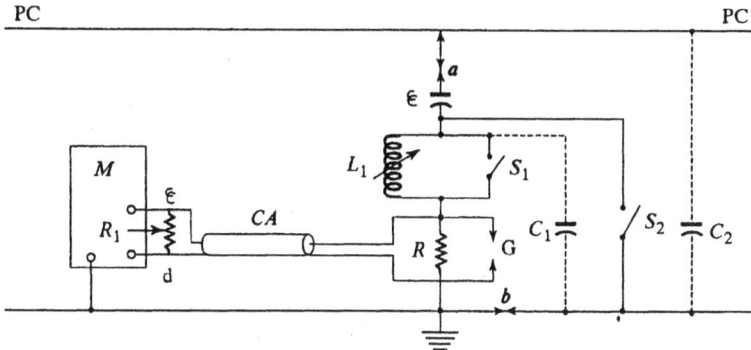

ϵ— Coupling condenser; M — Radio noise meter; a-b — Test apparatus; CA — Coaxial cable; O — Protective gap; S_1 — Shorting switch; C_1, C_2 — Stray capacitances; L_1 — Tuning choke; R — Measuring impedance; R_1 — Meter input impedance; P.C. — Power system conductor

Fig. 10.16 Circuit for measurement of RIV from the conductors of an energized system

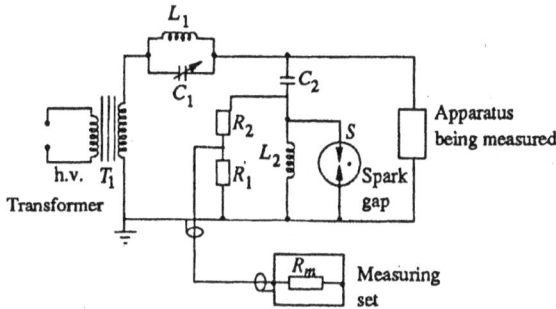

T_1 — High voltage transformers; L_1, C_1 — Filter; C_2 — Coupling condenser; L_2 — Measuring choke; R_1, R_2 — Potential divider; S — Protective spark gap; R_m — Input impedance of the meter

Fig. 10.17 Circuit for RIV measurement as given by British standards

10.6.3 Test Circuits for the Measurements

The schematic circuits used for RIV measurements are shown in Figs. 10.15, 10.16 and 10.17. The RIV meter is first calibrated as per standards. The important components of the circuits are:

(i) The radio frequency choke to limit the loss of the RIV voltage and to conduct energy from the sample. The choke itself should be free from noise, and its impedance should be less than 1500 Ω.

(*ii*) The coupling capacitor C (< 0.001 μ F); it should be free from noise in the operating range and the resistance of R should be equal to 800 Ω. The value indicated by the meter gives the conducted radio noise from the test sample.

(*iii*) Coaxial cable (CA): A coaxial cable of characteristic impedance 185 Ω shall be connected between the resistance R and the radio noise meter.

When the radio noise meter measurements are stated, the information regarding the specifications of the meter used, the frequency range of measurements, the band pass characteristics, and the open circuit and the detector characteristics have to be mentioned.

Now-a-days, for transmission systems of 400 kV and above, radio noise voltages are of importance, and corrective measures are to be adopted for various apparatus and hardware to minimize the radio and television band noise.

QUESTIONS

Q.10.1 Explain the terms (*a*) withstand voltage, (*b*) flashover voltage, (*c*) 50% flashover voltage, and (*d*) wet and dry power frequency tests as referred to high voltage testing.

Q.10.2 What are the different power frequency tests done on insulators? Mention the procedure for testing.

Q.10.3 What is the significance of impulse tests? Briefly explain the impulse testing of insulators.

Q.10.4 What are the significances of power factor tests and partial discharge tests on bushings? How are they conducted in the laboratory?

Q.10.5 Mention the different electrical tests done on isolators and circuit breakers.

Q.10.6 Why is synthetic testing advantageous over the other testing methods for short circuit tests? Give the layout for synthetic testing.

Q.10.7 Explain the partial discharge tests on high voltage cables. How is a fault in the insulation located in this test?

Q.10.8 Explain the method of impulse testing of high voltage transformers. What is the procedure adopted for locating the failure?

Q.10.9 What is an operating duty cycle test on a surge diverter? Why is it more significant than other tests?

Q.10.10 Explain the importance of RIV measurements for EHV power apparatus.

Q.10.11 Explain, with a schematic diagram, one method of measuring RIV of transmission line hardware.

REFERENCES

1. Allston, L.L. *High Voltage Technology*, Oxford University Press (1968).
2. *J and P. Switchgear Book*, Newnes-Butterworths, London, 7th edn. (1972).
3. *Reference Book on Transmission and Distribution*, Westinghouse and Oxford University Press, New Delhi (1962).
4. Kreuger, F.H., *Discharge Detection in H.V. Equipment*, Haywood London (1964).
5. C.H.A. Ely and Lambeth, P.J., "Artificial pollution test for H.V. outdoor insulators," *Proc. I.E.E.* **111**, 991 (1964).
6. Nandagopal, M.R. and Gopalakrishna, H.V., "Saltfog pollution test", *Tr. I.E.E.E. — Electrical Insulation, EI* -8, 33 (1973).

7. Nagabhushana, G.R. and Gopalakrishna, H.V., "Lightning arrester duties and test standards", *Electrical India*, 9, 23, Sept (1969).
8. Clark, C.H.W., "Radio interference from H.V. insulators", *Electrical Review*, 16, 491 (1959).
9. C.I.S.P.R. Publication 1, "R.I.V. measuring apparatus for the frequency range 150 kHz to 30 MHz" (1961).
10. Boulet, L., *et al.*, "Environmental studies of R.I. from conductors and hardware" Paper No. 408, *C.I.G.R.E.* (1966).
11. "R.I.V. from H.V. transmission lines", *ERA Bibliography/IB-202A* (1935-1960).
12. "Methods of h.v. testing", *IS: 2071-1976, IEEE Std.* 4-1978.
13. "Methods of impulse testing", *IS: 2070-1962*.
14. "Insulators testing", *IS : 2099-1986*.
15. "Testing of bushings", *IS: 2099-1986*
16. "Impulse testing of transformers", *IS: 2026--1981*.
17. "Testing of circuit breakers and Isolators", *IS: 116-1962, BS: 3659, BS: 3078, ASTA: 15, 17, 22 and 27, IS: 2516-1980*.
18. "Testing of power cables", *IS: 5959-1970* (Parts I and II), *IS: 1554* Parts I and II - 1964, 1970, *IS:7098* Parts I and II - 1973, and *IS 3070* Part I - 1965, 1974.
19. "Lighning arresters and surge diverters", *IS: 4004-1967, IS: 4850-1968, BS: 2914-1957, IEC Publication No. 99-1*, Part I-1970.
20. "Partial discharges and P.D. measurements", *SI: 6209-1971, IEC No. 270-1968*.
21. "R.I.V. Measurements", *BS: 5049-1973, NEMA Specification No. 107-1964*.

11

Design, Planning and Layout of High Voltage Laboratories

11.1 INTRODUCTION

Industrial and economic development in the present world demands the use of more and more electrical energy which has to be transported over long distances in large quantities. Transportation of large amounts of power needs extra high voltage transmission lines. Elsewhere in the world, transmission lines of 760 kV have come into operation, and transmission lines of ratings of 1000 kV or more are coming into operation in the USA and USSR. Extensive studies are being made in different countries on the possible use of complex extra high voltage d.c. systems of ± 400 kV and above.

This very fast development of power systems should be followed by system studies on equipment and service conditions which they have to fulfil. These conditions will also determine the values for test voltages of a.c. power frequency, impulse, or d.c., under specific conditions.

In India, at present the transmission voltage has reached a level of 400 kV with a few Electricity Boards adopting this voltage for their long distance transmission. It is proposed to have a National grid at a transmission voltage of 400 kV or even higher. In another decade 400 kV lines of about 10,000 km length will be in operation.

High voltage laboratories are an essential requirement for making acceptance tests for the equipment that go into operation in the extra high voltage transmission systems. In addition, they are also used in the development work on equipment for conducting research, and for planning to ensure economical and reliable extra high voltage transmission systems. Here a brief review of the planning and layout of testing laboratories and some problems and limitations of the test techniques are presented.

11.2 TEST FACILITIES PROVIDED IN HIGH VOLTAGE LABORATORIES

A high voltage laboratory is expected to carry out withstand and/or flashover tests at high voltages on the following transmission system equipment:

(i) Transformers

 (*ii*) Lightning arresters
(*iii*) Isolators and circuit breakers
 (*iv*) Different types of insulators
  (*v*) Cables
 (*vi*) Capacitors
(*vii*) Line hardware and accessories
(*viii*) Other equipment like reactors, etc.

Different tests conducted on the above equipment are:
  (*i*) Power frequency withstand tests – wet and dry
 (*ii*) Impulse tests
(*iii*) d.c. withstand tests
 (*iv*) Switching surge tests
  (*v*) Tests under polluted atmospheric conditions
 (*vi*) Partial discharge and RIV measurements

In addition, high current tests at power frequency and impulse current tests on transformers, line conductors, and lightning arresters are necessary. Details of some of these tests have already been discussed in Chapter 10.

Apart from the above facilities which are needed for routine testing, the laboratories are expected to have facilities for studying dielectric properties of insulation and insulating materials.

11.3 ACTIVITIES AND STUDIES IN HIGH VOLTAGE LABORATORIES

High voltage laboratories, in addition to conducting tests on equipment, are used for research and development work on the equipment. This includes determination of the safety factor for dielectrics and reliability studies under different atmospheric conditions such as rain, fog, industrial pollution, etc., at voltage higher than the test voltage required. Sometimes, it is required to study problems associated with test lines and other equipment under natural atmospheric or pollution conditions, which cannot be done indoors.

Research activities usually include the following:
  (*i*) Breakdown phenomenon in insulating media such as gases, liquids, solids, or composite systems,
 (*ii*) withstand voltage on long gaps, surface flashover studies on equipment with special reference to the equipment and materials used in power systems,
(*iii*) electrical interference studies due to discharges from equipment operating at high voltages,
 (*iv*) studies on insulation co-ordination on h.v. power systems, and
  (*v*) high current phenomenon such as electric arcs and plasma physics.

Usually, high voltage laboratories involve tremendous cost. Hence, planning and layout have to be carefully done so that with the testing equipment chosen, the investment is not high and the maximum utility of the laboratory is made.

11.4 CLASSIFICATION OF HIGH VOLTAGE LABORATORIES

High voltage laboratories, depending on the purpose for which they are intended and the resources (finances) available can be classified into three types.
   (i) Small laboratories
   (ii) Medium size laboratories
   (iii) Large general — laboratories
Some salient features of these various types of laboratories are discussed below.

(i) Small Laboratories

A small laboratory is one that contains d.c. or power frequency test-equipment of less than 10 kW/10 kVA rating and impulse equipment of energy rating of about 10 KJ or less. Voltage ratings can be about 300 kV for a.c., single unit or 500 to 600 kV a.c. for cascade units, ±200 to 400 kV d.c. and less than 100 kV impulse voltage. Normally the equipment is meant for housing in a room or hall of size 15 m × 10 m × 8 m. Sometimes the equipment ratings are limited such that they can be accommodated in a room of height 5 m to 6 m only. Such laboratories are meant for Engineering Colleges and Universities who decide to build such a facility with small resources for doing high voltage tests or research or for imparting training. In such a case, it is preferable that the Engineering college or University associate with a local industry or R & D organization. It is important to decide and define the responsibilities of the parties concerned as to how the test facilities and time can be shared. Another idea is to have the university to decide to own the laboratory fully but throw open the facilities of regular technical training and high voltage testing for the clients. Here it may be mentioned that many high voltage problems can be solved by tests at moderate voltage levels. Such laboratories can be built with an investment of 2 to 10 million rupees (at 1991 prices).

(ii) Medium Size Laboratory — An Industrial Laboratory

In case of medium size laboratories, their main function will be for doing routine tests. The demand on future tests and test resources will be known to the same extent as that of the future production targets. Careful planning of such laboratories should include (i) ground transport, (ii) handling equipment like cranes etc., (iii) rationalization of test procedures by making instruments easily accessible, and (iv) Providing room for the possibility of increasing the maximum voltage ratings etc. Such a laboratory may initially contain a power frequency testing facility in the range of 200 to 600 kV depending on the ratings and the size of the equipment being manufactured and proposed to be tested, such as cables, transformers etc., but its kVA rating will be much higher (100 to 1000 kVA). The impulse voltage generator required would have a rating of 20 to 100 kJ. or more. Other test equipments like the impulse current generator for testing surge diverters and d.c. test facilities for testing cables and capacitors can also be made available. In industrial laboratories not much emphasis is generally given for undertaking research work and little flexibility may be available for incorporating new equipments.

(iii) Large Size Laboratories

This type of laboratories are meant to carryout testing and undertake research work as envisaged in Sections 11.2 and 11.3 and will contain almost all high voltage and high current test equipments and facilities. The basic facilities available will be

- (*i*) One or more h.v. test halls,
- (*ii*) Corona and pollution test chambers,
- (*iii*) Outdoor test area for tests on large sized equipment, transmission lines and towers etc.,
- (*iv*) Controlled atmospheric test rooms/chambers,
- (*v*) Computer facilities, conference halls, library etc. with good office facilities, and
- (*vi*) Provision for overnight tests and stay.

The size and ratings of the test equipment will be quite large and are dealt with in the next Section 11.5. The building and equipment include the workshop, material handling equipment like cranes, ladders, air cushion platforms etc. and large control and electric supply facilities (up to few KVA or MVA). The personnel connected with such a laboratory will include a director or manager, few group leaders, and section heads separately for research, testing, measurements, electronics and computer facilities etc. In addition, there will be supporting staff comprising of test engineers, technicians, librarians, office staff and skilled and semi-skilled workmen. The cost of such laboratories will be several millions of rupees.

11.5 SIZE AND RATINGS OF LARGE SIZE HIGH VOLTAGE LABORATORIES

As stated earlier, and large size laboratory contains equipment of very high ratings with enough flexibility incorporated. In the following sections, details of the ratings and size of the equipment and their layout are briefly indicated.

11.5.1 Withstand Voltages, Test Voltages and the Rating of Equipment in High Voltage Laboratories

The ratings and size of test equipment chosen in the h.v. laboratories depends on the test facilities to be provided. Normally, the design of the laboratories for 230 kV system voltage and below does not pose any problems, but laboratories intended for system voltages of 400 kV and above require special attention.

In Tables 11.1 and 11.2 various test voltages for different transmission system voltages are given.

For research and development work, the voltage levels needed are usually about 1.3 times the maximum test voltage needed. Hence, the laboratories intended for different system voltages should have the test voltages as available in Tables 11.3 and 11.4.

Table 11.1 Test Voltages for Equipment (a.c. Systems)

System nominal voltage	Line to ground voltage	Power frequency withstand voltage	Impulse withstand voltage	Switching surge with- stand vol- tage	Pollution test voltage
kV (rms)	kV (peak)	kV (rms)	kV (Peak)	kV (peak)	kV (rms)
400	335	530	1425	875	280
525	430	670	1800	1100	330
765	625	960	2300	1350	500
1100	900	1416	2800	1800	700
1500	1220	1920	3500	2200	950

Table 11.2 Test Voltages for Equipment (d.c. Systems)

Nominal voltage	d.c. with- stand vol- tage	Reverse polarity test voltage	Impulse withstand voltage	Switching surge with- stand voltage	Pollution test voltage
kV	kV	kV	kV (peak)	kV (peak)	kV
± 400	800	± 600	1350	1000	440
± 600	1200	± 900	1900	1500	660
± 800	1600	± 1200	2300	2000	880

From the values given in Tables 11.3 and 11.4, one can conclude that laboratories intended for testing and development of equipment for 100 kV a.c systems require test transformers of 1.5 of 2.0 MV, impulse generator rated for 5 to 6 MV, and h.v.d.c. rectifiers of 1.2 to 1.5 MV.

Table 11.3 Test Voltages Required for Different System Voltages (a.c. Systems)

Nominal voltage	Power frequency voltage	Pollution test voltage	Impulse (standard/ voltage)	Switching surge voltage
kV (rms)	kV (rms)	kV (rms)	kV (peak)	kV (peak)
400	800	300	2400	1150
765	1000	500	3000	1750
1100	1400	700	3700	2300
1500	1900	1000	4600	2800

Table 11.4 Test Voltages Required for Different System Voltages (d.c. Systems)

Nominal voltage	d.c. voltage	Pollution test voltage	Impulse (standard) voltage	Switching surge voltage
kV	kV	kV	kV (peak)	kV (peak)
± 400	800	500	1750	1300
± 600	1200	700	2500	2000
± 800	1600	900	3000	2600

High voltage laboratories intended for system voltages of 400 kV or less need not go for such super high voltage rated equipment. The insulation levels for 400 kV system equipment are given below.

(a) Impulse withstand voltages : Line to Earth   = 1425 kV (peak)
    (standard impulse voltage) : Phase to phase   = 1640 kV (peak)

(b) Power frequency withstand   One minute dry = 680 kV (rms)
    voltages

Momentary dry = 800 kV (rms)

30 sec. wet = 630 kV (rms)

Visible corona level   = 320 kV (rms)

For the above data it may be concluded that a factor of more than 3 for impulse voltages and a factor of 2 to 2.5 for power frequency voltages (highest line to ground peak voltages) are adopted for system voltages of 400 kV and less. Hence, the rating of the equipment should be at least about 900 kV (rms) for power frequency and 2000 kV (peak) for impulse voltages. The rating of the equipment will be still less, if the laboratories are intended for system voltages of 132 kV or 230 kV and may be arrived at by considering the test voltages required.

11.5.2 Voltage and Power Ratings of Test Equipment

(a) D.C. Testing Equipment

High Voltage d.c. tests are performed using cascaded rectifiers. Careful consideration is necessary when tests on polluted insulation are to be performed which require currents of 50 of 200 mA, but strong predischarge streamers of 0.5 to 1.0 A of milliseconds duration may occur. Hence, the generator must have adequate internal reactance in order to maintain the test voltage without too high a voltage drop. The voltage ratings are given in Table 11.4, and the power rating may vary from a few kW to a few hundred kW.

(b) Power Frequency Testing Equipment

It is known that the flashover voltage of an insulator in air or oil or in some fluid depends on the capacitance of the supply system, due to the fact that a voltage drop may not maintain the predischarges before breakdown. Hence, a minimum of about 1000 pF or more in parallel with the energized insulator is needed to determine the real flashover or puncture voltage, and the generator has to supply at least 1 A in the case of clean and 5 A in the case of polluted insulator at test voltage on short circuit. Approximate values of the self-capacitances of different equipments are given below:

Insulators	less than 100 pF
Bushings	100 to 400 pF
Current transformers	200 to 600 pF
Power transformers (1 MVA and above)	1000 to 8000 pF
Cables per 10 m length	1000 to 3000 pF

The output of testing transformer will be given by

$$P = (2\pi fC)\ V^2 \times 10^{-9}\ \text{kVA}$$

where,
$f =$ supply frequency,
$C =$ capacitance in pF, and
$V =$ test voltage at the transformer terminals in kV (rms).

The transformer self-capacitance and the capacitances of various high voltage leads (bushings), etc. should also be included in determining the load capacitance. From the above figures it is implied that the minimum power rating of a 1 MV testing transformer will be about 300 kVA. Usually, the power rating of a testing transformer in kVA (single unit) is approximately taken to be equal to the voltage rating in kV.

(c) Impulse Generators

The maximum charging voltage of an impulse voltage generator is given by the stage voltage multiplied by the number of stages. The peak value of the impulse voltage V_s for a standard 1.2/50 μs wave is

$$V_s = nV_{d.c.} \left[0.95 - \frac{C_L}{(C_L + C_g)} \right]$$

where,
$V_{d.c.} =$ Charging voltage,
$n =$ number of stages in the generator,
$C_L =$ load capacitance, and
$C_g =$ generator capacitance.

For $C_g/C_L \geq 5$, the peak value of the impulse generator output voltage will be approximately $V_s \approx 0.7\ nV_{d.c.}$. In other words, the generator rating has to be at least 1.3 times more than the desired output voltage. The energy rating of the impulse generator at its maximum voltage rating is given by

$$W = \frac{1}{2} C_g V^2 \times 10^{-9}\ \text{kJ}$$

where,
$W =$ stored energy,
$C_g =$ capacitance of the generator in pF, and
$V =$ total charging voltage in kV.

In order to test transformers which have large capacitance, a minimum of 30,000 to 40,000 pF of generator capacitance is needed. A simple calculation will show that a minimum of 135 kJ is required for a 3 MV impulse generator, if the IEC specification for impulse waveshape is to be maintained. The minimum energy rating of a 6 MV impulse generator will be about 600 kJ. From this it may be concluded that the energy rating in kilojoules may be approximated to be equal to 0.1 times the voltage rating in kV.

There is no problem to pile up a large size capacitance in the form of a number of capacitors and to charge them in parallel and discharge them in series to give the required peak of the standard impulse wave. But many difficulties exist in reducing the internal inductance of the circuit to a minimum to obtain a steep front and to avoid oscillations. As an example, a 4 MV impulse generator test circuit has a length equal to the height of the generator plus twice the distance between the test object and the generator. The overall inductance of such a circuit including the internal inductance of the generator will easily be more than 140 μH. With such a generator, it is impossible to test an object of capacitance of 5000 pF with a front time of 1.2 μs and with less than 5% overshoot. Hence, a very careful design and a very careful consideration of the test circuit only can give the optimum test conditions which are not far from theoretical specifications.

The necessity of rapid change of the test circuit from standard impulse to switching surges requires careful studies for placing of series and parallel resistors when producing switching surges like 100/1000 μs or 200/2000 μs waveform, in which the efficiency of the generator is very much reduced. Also, the front time and the tail time resistors have to be carefully rated, as they have to dissipate larger amounts of energy than in the case of standard impulses.

(d) Other High Voltage Testing Equipment

Usually, the other testing equipment that will be available is, (i) impulse current generators for testing lightning arresters, (ii) test facilities for measuring RIV and partial discharges, (iii) sphere gaps for measurement and calibration purposes, and (iv) high voltage Schering bridge for dielectric testing. Usually, the impulse current generators are rated between 100 to 250 kA with an energy rating of 50 to 100 kJ. This is more than adequate for testing with lightning stroke currents. Partial discharge and RIV measurements require testing transformers free from internal discharges. The detection equipment should be capable of detecting 0.01 pico coloumb of charge in a test object capacitance of 100 pF and 2 to 3 pico coloumbs at 1 μF test capacitance. Therefore, the test transformers should have internal discharges of the same order or less at the specified voltage value. Now-a-days, it is possible to design a.c testing transformers with necessary shielding, etc. with internal discharges less than 5 pC at 500 kV.

Where sphere gaps are used, it is important to bestow thought regarding the proper size and space requirements. Proper attention must be given to (i) type and magnitude of the voltage to be measured, (ii) range of operation keeping in view that the sparking distance is less than 0.5 times the diameter of the spheres, and (iii) space requirements as specified in *IS: 1876-1961* and other specifications.

11.5.3 Size and Dimensions of the Equipment in High Voltage Laboratories

High voltage laboratories may be either (a) indoor type or (b) outdoor type. The indoor type has the advantage of protection of testing equipment against variable weather conditions, simplicity in design and control of the test equipment, and provision of observation facilities during testing. But outdoor laboratories have the advantage of less cost due to the absence of building cost and the planned facility layout cost. But outdoor test areas have limitations such as (i) absence of lifting and supporting facilities, (ii) climatic conditions which may restrict or impede testing, (iii) reproducibility of results not being guaranteed due to uncontrolled atmospheric conditions, and (iv) artificial and wet test studies which are difficult due to wind variation, etc.

When high voltage laboratories are planned as indoor laboratories, the following figures fix the dimensions of the laboratories:

(i) Size of the test equipment for a.c., d.c., or impulse generators
(ii) Distances or clearances between the test object and ground during test conditions and also between all the high voltage terminals and earthed or grounded surroundings such as walls, roofs of buildings, and other test equipment not energized.

Table 11.5 Approximate Dimensions of Testing Apparatus and Test Objects

Nominal system voltage for the equipment kV (rms)	a.c. test transformer height m	Impulse generator height m	Test object dimensions (maximum)		
			length m	breadth m	height m
400	10	6	7	2	11
765	15	8	11	2	17
1100	18	12	17	2	24
1500	21	15	28	2	38

In Table 11.5 are given the approximate size and dimensions of the test transformers and impulse generators for different system voltages. The table also gives minimum room to be provided for the equipment.

Regarding clearances, that is, the minimum distance between the high voltage surfaces and the ground points; they are of utmost importance in high voltage testing. The approximate working clearances recommended are as follows:

a.c. power frequency voltages: 200 kV (rms)/m
d.c. voltages : 275 kV/m
Impulse voltages: 500 kV/m

For switching surges, the clearance is worked out from the following approximate formula

$$d = (2V)^2$$

where d is in m, and V in MV.

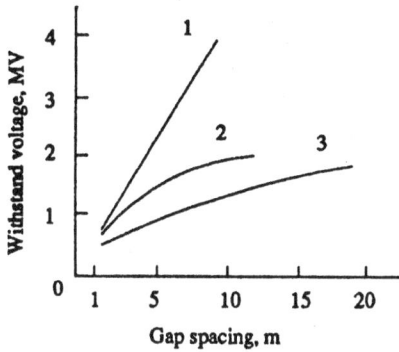

1. Impulse 1.2/50 µs 50% withstand voltage, positive polarity
2. 50 Hz a.c. (peak) 50% withstand voltage
3. Switching surge 120/4000 µs 50% withstand voltage, positive polarity

Fig. 11.1 Withstand voltage of rod-plane configuration (CESI-MILANO)

The above clearances are safe, as long as the test voltages do not exceed 1.5 MV for a.c. and d.c. voltages and 2.5 MV for impulse and switching surge voltages. For higher voltages the clearances have to be worked out by considering the withstand voltages for rod-plane configuration. The characteristic is given in Fig. 11.1. The necessary distances to the surroundings for switching surges of a long duration will be about 12 m for 760 kV and 30 m for 1500 kV system voltage equipment. Hence, from the above data and from Fig. 11.1 it is evident that the h.v. laboratories rated for 400 kV and above are practically conditioned by the necessary clearances for switching surge tests.

11.5.4 Layout of High Voltage Laboratories

The layout of a h.v. laboratory is an important aspect for providing an efficient testing facility. Laboratory arrangements differ very much from a single equipment to multi d.c., a.c., and impulse arrangements in different testing programmes. Each laboratory has to be designed individually considering the type of equipment to be tested, the available space, other accessories needed for the tests, the storage space required, etc. Earthing, control gear, and the safety precautions require most careful consideration.

Laboratory Building

The building construction is not critical except where ionization tests are conducted. To minimize the floor loading problems and to simplify earthing arrangement, ground level location is preferred. The floor should withstand the loading imposed by the equipment and test objects. Arrangements should be made to ensure that the laboratory is free from dust, draught, and excessive humidity. Laboratory windows may require blackout arrangement for visual corona tests, etc. The control room should be located in such a way as to include good overall view of the laboratory and test area. The main access door to the test area must accommodate the test equipment and the test object and have adequate interlocking arrangements and warning system to ensure safety to the personnel. A typical layout of a high voltage laboratory accommodating a 1.0 MV a.c. testing transformer and a 3 MV impulse generator is shown in Fig. 11.2. The dotted circles indicate the clearances necessary.

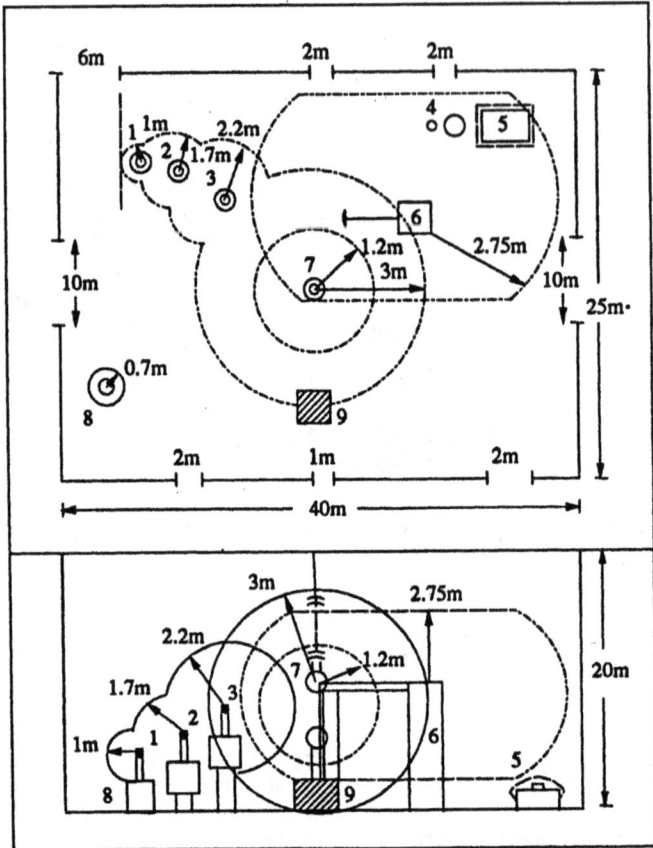

1,2,3 — Cascade transformer set	: 1 MV
4 — d.c. charging unit	: 200 kV
5 — Impulse current generator	: 200 kA
6 — Impulse voltage generator	: 3 MV
7 — Sphere gap	: 2 metres diameter
8 — d.c. test set	: 300 kV
9 — Control room Faraday cage	: 5 × 3.5 × 3 m^3

Clearances for various equipment are indicated by chain circles

Fig. 11.2 Layout of a typical high voltage laboratory with a 1 MV cascade transformer and 3 MV impulse generator

11.5.5 High Voltage Laboratories in India and Abroad

High Voltage test facilities and large size high voltage laboratories are available at only a few places in the country, as each of them costs several crores of rupees. It has

Table 11.6 High Voltage Laboratories Abroad

Sl. No.	Location	Size			Power frequency test facilities		Impulse voltage test facility		Switching surge voltage facility (MV)
		Length (m)	Breadth (m)	Height (m)	Voltage (MV)	Current (A)	Voltage (MV)	Energy (KJ)	
1.	Les Renardiers Electricite De France, France	65	65	45	2.25	1	7.2	450	6.0
2.	Hydro-Quebec, Montreal, Canada	82	67	57	3.30	2	6.4	400	6.0
3.	CESI, Milan, Italy	45	40	35	2.25	—	4.8	200	3.0
4.	U.S.S.R.	115	80	60	3.00	—	7.2	—	—
5.	Hemsdorf, GDR	—Outdoor—			2.25	2	7.2	—	—
6.	Australia	—"—			1.50	—	8.0	—	—
7.	Hitachi, Japan	60	40	31	1.65	—	4.0	600	—
8.	ASEA, Sweden	47	25	25	1.50	—	3.2	140	—
9.	CERL, U.K.	41	28	22	1.20	—	4.0	100	—
10.	CEPEL, Brazil	44	30	27	—	—	—	—	—

Table 11.7 High Voltage Laboratories in India

Sl. No.	Location	Size			Power frequency test facilities		Impulse voltage test facility		Switching surge voltage facility (MV)
		Length (m)	Breadth (m)	Height (m)	Voltage (MV)	Current (A)	Voltage (MV)	Energy (KJ)	
1.	Bharat Heavy Electricals Ltd., Bhopal	67	35	35	1.5	2	4.0	400	2.0
2.	Central Power Research Institute, Bangalore	50	40	35	1.8 (outdoor test facility also available)	2	2.4	30	1.5
3.	Indian Institute of Science, Bangalore	37.5	30	20	1.05	1	3.0	50	1.6
4.	Indian Institute of Technology, Madras	28.0	10	9.7	0.80	—	1.5	37.5	—
5.	Government Engineering College, Jabalpur (MP)	36.0	26	30	0.50	—	1.6	26.4	—
6.	Anna University, Madras	25	15	15	0.30	—	1.2	—	—
7.	Jadavpur University, Calcutta	25	15	20	0.25	—	1.4	16.0	—
8.	Engineering College, Jawaharlal Nehru Technological University, Kakinada (A.P)	20	12	8	0.50	—	1.4	16.0	—
9.	Central Power Research Institute's UHV Lab, Hyderabad	outdoor facility			.1.60	6.0	5.4	750	3.7

been stated (ref. 6) that a fully screened high voltage laboratory with all test and research facilities will cost Rs. 100 crores or even more. In Table 11.6 details of a few large size high voltage laboratories in the world are listed along with the ratings of the equipment available. Table 11.7 gives the details of a few high voltage laboratories in India. Some of the laboratories like those at the Indian Institute of Science, Bangalore, Central Power Research Institute, Bangalore, and Indian Institute of Technology, Madras have adequate research facilities available apart from the normal test facilities.

Apart from the above laboratories, medium sized industrial type laboratories are available with organizations like the National Test House, Alipore, Calcutta, and Bharat Heavy Electricals Limited, R & D Unit, Hyderabad etc., which carry out normal testing and also provide facilities for conducting research.

11.6 GROUNDING OF IMPULSE TESTING LABORATORIES

An earth or ground system means an established stable reference potential normally taken to be zero potential. There are three types of grounds (i) the ideal ground, (ii) single point ground (Fig. 11.3a), and (iii) the bus ground (Fig. 11.3b). Of all these, the best ground is the ideal ground which cannot be realized in practice. The next preferred ground is the single point ground, and the bus ground is least satisfactory. Ideal ground can be approximated by an equipotential plane realized by a finite conducting material. The laboratory is covered by a sheet of copper metal welded into a single unit. But this is very costly and is used rarely. A single point ground is commonly used. In this (see Fig. 11.3a) an earthing grid is installed within the laboratory floor, and connection from the grid is given by a large sized copper conductor to a point identified as a common ground point. The ground connections of various equipments and other components of the high voltage test circuit are made to the common ground. High voltage impulse tests give rise to high currents of several kiloamperes, and the rate at which the currents may change ranges between 10^7 to 10^9 A/s. If proper care is not taken, flashover or damage to control gear and risk of life to persons can occur. In order to avoid these difficulties, copper strips are used instead of round conductors to minimize the inductance in the ground circuit. Secondly, metal grid embedded in a concrete floor gives rise to less resistance and inductance in the ground circuit. The ground is effective only when large size strips are used with close spacing. The ground system should ensure the following conditions:

(i)   imperfections of grounding system are to be avoided, as they cause excessive voltage difference between points and cause flashovers, damage, or danger to human life,

(ii) the imperfections will cause excessive loop currents along the sheaths of measuring cables, which will introduce errors in measurements,

(iii) the grounding system should be such that the voltage drop along the ground system, the voltage at a loop, and the circulating currents in the loops are avoided or minimized,

(iv) metal conduits should be used for the measuring and control cables to avoid neutral inductance effect between the ground grid and the cables.

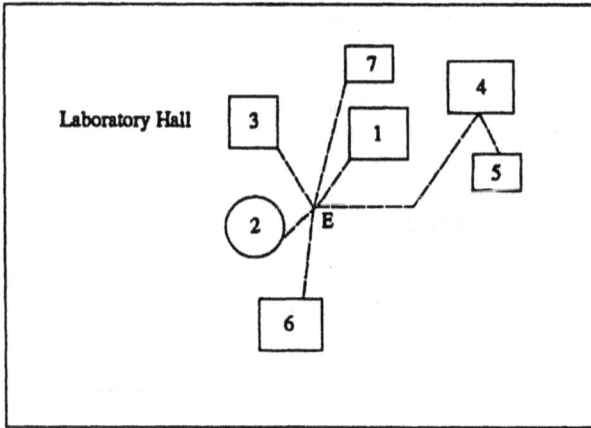

(a)   Single point grounding system

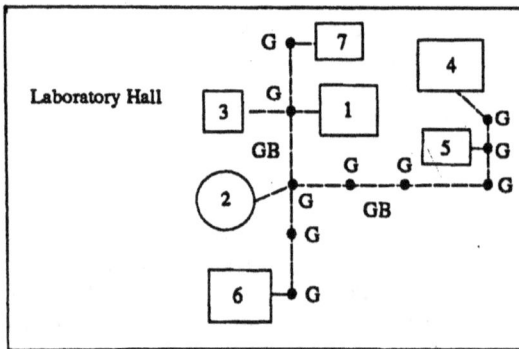

(b)   Bus grounding system

Fig. 11.3   1 — Impulse voltage generator; 2 — Sphere gap; 3 — Position of test object for impulse testing; 4 — Impulse current generator; 5 — Position of test object for impulse current generator; 6 — Control room and oscilloscope position; 7 — Charging rectifier set; E — Single point earth position; G — Ground connection position; GB — Ground bus

A typical good earthing system consists of a copper network with meshes of 1 m width laid down below the ground level around the impulse test area and well connected. This network is extended over the entire area comprising all equipment such as testing transformers, charging h.v. rectifier set, test bay, etc. The grid should be electrically connected to all the metallic frames and reinforcing iron in the concrete walls and pillars of the building at their bottom points. Impulse test area must be provided with a spread, stretched, or expanded copper grid on the floor of thickness of about 2 mm with ground rods driven into the earth to a depth equal to the height of the impulse generator. The rods are welded to the inside copper grid as well as surface copper grid. Earth connection facility is to be provided for every 16 sq. m area so that shortest lead can be used from any position inside the laboratory.

Where ionization measurements are to be made, the earthing system should keep the RIV level from external sources to the 'lowest' value. In addition, the high frequency energy produced during impulse tests should not cause any trouble around the test area. If this is to be met the entire laboratory should be built into a Faraday cage.

The general layout of the laboratory with its conduit pipes for control and measuring cables is shown in Fig. 7.55 (see Chapter 7). Such a layout will avoid all interferences.

11.6.1 Electromagnetic Shielding and Earth Return in High Voltage Laboratories

A high voltage laboratory, small, medium or large in size should have some type of screening against electrostatic and electromagnetic field interference. The screening is essential if partial discharge measurements are to be made in the laboratory. An attenuation of less than 40 db is needed for attenuation of electrical signals in the frequency range of 1MHz, while a still lower attenuation is needed for electromagnetic signals. In larger test laboratories attenuation levels due to interferences are higher and arise mostly due to imperfect screening. One way to check the screening is to tune a portable pocket-radio and walk around the laboratory tuning the radio to different frequencies between 500 kHz to 10 MHz. The signal should not be heard, However, it is often found that the signal level increases significantly when a cable or an electrical outlet is crossed. If it is possible, the same check may be carried out with the automatic volume control (gain control) disconnected. The sources of disturbance inside the laboratories are (*i*) switching transients due to switching on or switching-off of loads like lifts or cranes, transformers etc., (*ii*) rectifier circuits, and (*iii*) shielded cables acting as antennas for outside signals. Care should be taken to see that the above are avoided. The best screening is obtained if the roof, the walls and the flooring area are screened with an expanded metal wire mesh and joined together. Further, all electric conductors are fully screened in metal conduits which are run below the floor metal network.

For the purpose of measurement, one point in the test circuit such as the base of a test object or an impulse voltage generator should be made a reference ground or a zero potential point. This is usually disturbed in measurements for fault indication in transformers during impulse testing, tests with chopped waves etc. as spikes are introduced into the test circuits due to low-voltage conductors that run into the control room from the impulse test area which carry the operating signals of large magnitude to control impulse generator, sphere gaps etc. As such a potentially difference is created momentarily during the transient period between the base of impulse voltage generator and that of the base of the measuring voltage divider and test object. These differences are carried on to the measuring device which will give erroneous results. These voltage differences can be reduced by reducing the impedance of the ground side of the test circuit. The most effective method for reducing the voltage differences is to have the return conductor in the form of a metal sheet placed on the top of the floor. Some laboratories use coarse copper nets or aluminium nets or aluminium sheets, but they are not very effective.

The high voltage laboratory must be earthed to (i) protect the equipment against the lightning strokes, and (ii) to protect the equipment from short circuits inside the laboratory from the power supply source. If not properly earthed, these will give rise to potentials which are different at different points in the laboratory thus causing unnecessary danger to human life and damage to the equipment.

The acoustical attenuation of the building is also important. It is necessary in large laboratories to have comfortable and clear communication between persons at different locations inside the laboratory. Reverberation inside the laboratory should be avoided. To get the desired effect, the laboratory should have perforated holes and fiberglass or some such material fixed to the walls. The above aspects need careful consideration in the design of a high voltage laboratory.

QUESTIONS

Q. 11.1 List out the common test facilities available in high voltage laboratories.
Q. 11.2 What are the criteria used in selecting the ratings of the testing equipment for h.v. laboratories?
Q. 11.3 Why is grounding very important in a h.v.laboratory? Describe a typical grounding system used.
Q. 11.4 Estimate the clearances required and the approximate dimensions of the test room for a high voltage laboratory with the following equipment.
A.C. testing transformer   : 25 kVA, 250 kV
                          Size : 1.2 m dia × 3m (including bushing height)
Impulse voltage generator : 800 kV, 24 kJ
                          Size : 1.5m × 1.5m × 3m
Charging unit requires a space of 1m × 1m × 1m.
Accessories include a 75 cm sphere gap, 900 kV capacitance potential divider and a 200 kV gas filled standard capacitor.
Q.11.5 What are the extra precautions that are to be taken while grounding an impulse current generator. Give a typical grounding arrangement for a 160 kJ, 200 kA impulse current generator.

REFERENCES

1. August F. Metraux. "Some problems and actual limits of test techniques at extra high voltages", *Haefely Publication, EIS 14* (1969).
2. August F. Metraux, "Earthing of impulse stations", *Haefely Publication, 508040' IE* (1962).
3. Govinda Raju, G.R., "Planning of H.V. laboratories", *Lecture Notes on H.V Laboratory Testing Techniques*, Summer School, IISc, Bangalore (1977).

4. Gopalakrishna, H.V., "Design of H.V. laboratories", *Lecture Notes on H.V. Testing Techniques*, Summer School, IISc, Bangalore (1975).
5. N.R. Hylten Cavallius and T.N. Giao, "Floor net used as ground return in high voltage test areas", *Tr. IEEE, PAS, PAS-88*, 996, July (1969).
6. Hylten Cavallius, N., *High Voltage Laboratory Planning*, Emael Haefely & Co. Ltd., Basel, Switzerland, 1988.

Subject Index

Author Index